Artificial Intelligence and Precision Fermentation for Sustainable Energy Production

Artificial Intelligence and Precision Fermentation for Sustainable Energy Production explores the convergence of AI and precision fermentation for sustainable bioenergy production. It covers both theoretical foundations and practical applications, demonstrating how AI tools optimize microbial strains and monitor bioreactors in real time.

Precision fermentation using engineered microorganisms addresses conventional fermentation limitations, transforming agricultural waste into biofuels and biochemicals. The book:

- Covers implementation challenges, including scale-up issues, regulatory frameworks, ethical considerations, and data limitations
- Integrates Artificial Intelligence and Machine Learning in bioprocess and bioenergy
- Discusses predictive modeling and real-time monitoring of bioreactors
- Encompasses novel AI tools such as generative models and molecular dynamics simulations for designing and optimizing enzymes used in precision fermentation
- Covers case studies addressing AI-driven bioenergy solutions in both developed and emerging markets, including sections on policy frameworks, regulatory challenges, and economic feasibility across different countries

This book is aimed at graduate students and researchers in bioprocessing, AI, fermentation technology, and renewable engineering.

Artificial Intelligence and Precision Fermentation for Sustainable Energy Production

Edited by

Ooi Chien Wei, Ramalakshmi Subbarayalu,
Lo Sewn Cen, and N. Gnanasankaran

CRC Press
Taylor & Francis Group
Boca Raton London New York

CRC Press is an imprint of the
Taylor & Francis Group, an **informa** business

First edition published 2027
by CRC Press
2385 NW Executive Center Drive, Suite 320, Boca Raton FL 33431

and by CRC Press
4 Park Square, Milton Park, Abingdon, Oxon, OX14 4RN

CRC Press is an imprint of Taylor & Francis Group, LLC

ISBN: 978-1-041-16089-2 (hbk)
ISBN: 978-1-041-16395-4 (pbk)
ISBN: 978-1-003-68437-4 (ebk)

DOI: 10.1201/9781003684374

Typeset in Times
by SPi Technologies India Pvt Ltd (Straive)

Contents

Figures

Tables

Contributors

Harishchander Anandaram
Department of Artificial Intelligence
Amrita Vishwa Vidyapeetham College, Coimbatore
India

R. S. Aravind Bharani
Institute of Obstetrics and Gynaecology
Madras Medical College, Chennai
India

R. Suresh Babu
Laboratory of Experimental and Applied Physics
Centro Federal de Educação Tecnológica Celso Suckow da Fonseca, Rio de Janeiro
Brazil

Sundaravadivazhagan Balasubramanian
Department of Information Technology
University of Technology and Applied Sciences, Muscat
Oman

Bezaleel S
School of Computer Science and Engineering, Vellore Institute of Technology, Chennai
India

Kam Khong Chan
Centre for Green Bioprocess Engineering,
Faculty of Engineering, Built Environment, and Information Technology,
SEGi University, Petaling Jaya, Selangor
Malaysia

Arivumalar Chandrasekaran
Kasturba Gandhi Nursing College
Sri Balaji Vidyapeeth (Deemed-to-be-University), Puducherry
India

Kumarappan Chidambaram
College of Pharmacy
King Khalid University, Abha
Saudi Arabia

Balakrishnan Chinnaiyan
Department of Computer Science
Christ University, Bangalore
India

J. Jenifer Annis Christy
Department of Biotechnology
May Matha College of Arts and Science, Kanyakumari
India

Dhanasekaran Dharumadurai
Department of Microbiology
Bharathidasan University, Tiruchirappalli
India

Elakkiya Elango
Department of Computer Science
Government Arts College for Women, Nilakottai
India

P. Gowthami
PG Department of Chemistry
Shrimati Devkunvar Nanalal Bhatt Vaishnav College for Women, Chennai
India

N. Gnanasankaran
Department of Computer Science
Thiagarajar College, Madurai
India

S. Thanga Helina
Department of Computer Science and Engineering
Karunya Institute of Technology and Sciences, Coimabatore
India

Amir Seyyedabbasi
Associate Professor,
Department of Computer Engineering,
Istinye University
Türkiye

Immanuel Johnraja Jebadurai
Department of Computer Science and Engineering
Karunya Institute of Technology and Sciences, Coimbatore
India

Mahalakshmi Jeyabalu
Department of Computer Science
Christ University, Bangalore
India

Angeline Julius
Research Centre for Medical Biotechnology, Vinayaka Missions Research Foundation, Chennai
India

G. Kalaiarasi
Department of Chemistry and Centre for Material Chemistry
Karpagam Academy of Higher Education (Deemed-to-be-University), Coimbatore
India

Gnanaprakash Kalimuthu
Department of Pharmaceutics
K.M. College of Pharmacy, Madurai
India

S. Kannan
Department of Mechanical Engineering,
Amity University Dubai
United Arab Emirates

Udaya Prakash Nyayiru Kannan
Department of Biotechnology
Vels Institute of Science, Technology and Advanced Studies (VISTAS), Chennai
India

S. Karthik
Department of Microbiology
Thiagarajar College, Madurai
India

S. Karthikayan
Department of Mechanical Engineering, School of Engineering
Mohan Babu University, Tirupati
India

K. Selvakumar
Sharda School of Engineering and Sciences,
Sharda University, Greater Noida
India

Swati Mayur Keny
Department of Pharmaceutics
PES's Rajaram and Tarabai Bandekar College of Pharmacy, Ponda, Goa
India

A. Kosiha
Department of Chemistry
Vels Institute of Science, Technology & Advanced Studies (VISTAS), Chennai
India

Kanagaraj Rajalakshmi
School of Chemistry and Chemical Engineering,
Jiangsu University, Zhenjiang
China

Suguna Krishnamoorthy
Department of Chemical Engineering, National institute of Technology, Trichy
India

K. Arul
Department of Mechanical Engineering
Academy of Maritime Education and Training (AMET), Chennai
India

Ramakrishnan Kumar
Department of Microbiology
Bharathidasan University, Tiruchirapalli
India

Suresh Malakondaiah
Centre for Materials Engineering and Regenerative Medicine
Bharath Institute of Higher Education and Research (BIHER), Chennai
India

Vignesh Marimuthu
Department of Biotechnology
Vels Institute of Science, Technology and Advanced Studies (VISTAS), Chennai
India

Vinayagam Mohanavel
Materials Research, Department of Mechanical Engineering
Academy of Maritime Education and Training (AMET), Chennai
India

Shivam More
Department of Computer Science
Christ University, Bangalore
India

R. Muraliraja
Department of Mechanical Engineering
Vels Institute of Science, Technology & Advances Studies (VISTAS), Chennai
India

R. M. Murugappan
Department of Zoology
Thiagarajar College, Madurai,
India

Celina Nazareth
Department of Pharmaceutical Analysis
PES's Rajaram and Tarabai Bandekar College of Pharmacy, Ponda, Goa
India

S. Padmavathy
Department of Microbiology
Thiagarajar College, Madurai
India

Prakash Palanivel
Department of Microbiology,
Bharathidasan University, Tiruchirapalli
India

Rajesh Pandiyan
Centre for Materials Engineering and Regenerative Medicine
Bharath Institute of Higher Education and Research (BIHER), Chennai
India

Getzi Jeba Leelipushpam Paulraj
Department of Computer Science and Engineering
Karunya Institute of Technology, Coimbatore
India

V. Prabhakaran
Department of Biomedical Engineering
Aarupadai Veedu Institute of Technology, Chengalpattu
India

Guhanraj Radhamanalan
Department of Microbiology
Saveetha Institute of Medical and Technical Sciences, Chennai
India

Raja Veerapandian
Department of Molecular and Translational Medicine
Texas Tech. University, Texas
USA

Priyanka Jayam Rajendran
Department of Biotechnology, Bharathidasan University, Tiruchirapalli
India

S. Rajkumar
Institute of Technology
Hawassa University, Hawassa
Ethiopia

Lekhavani Ramesh
Department of Biotechnology, SRM Institute of Science and Technology, Chennai
India

A. Rathisha
Department of Microbiology
Thiagarajar College, Madurai
India

V. Roja
Department of Microbiology
Thiagarajar College, Madurai
India

Leena Ajay Sawaikar
Department of Pharmaceutical Chemistry
PES's Rajaram and Tarabai Bandekar College of Pharmacy, Ponda, Goa
India

Saravanakumar Sengottaiyan
Faculty of Engineering and Technology
Villa College (Q1 Campus), Male
Maldives

V. S. Shaisundaram
Department of Mechanical Engineering
Vels Institute of Science, Technology & Advances Studies (VISTAS), Chennai
India

L. Shamini
Department of Microbiology
Thiagarajar College, Madurai
India

C. P. Shirley
Department of Computer Science and Engineering
Karunya Institute of Technology and Sciences, Coimbatore
India

B. G. Sivakumar
Department of Mechanical Engineering
AMET University, Chennai
India

Kumaraguru Sivakumar
Centre for Materials Engineering and Regenerative Medicine
Bharath Institute of Higher Education and Research (BIHER), Chennai
India

S. Sivakumar
Bioenvironmental Energy Lab
Pusan National University, Busan
South Korea

Arun Kumar Sivaraman
School of Computer Science and Technology
Algoma University, Ontario
Canada

Alagarsamy Sivasamy
Department of Mechanical Engineering
Mahath Amma Institute of Engineering and Technology, Ariyalur
India

Kanagasabapathy Sivasubramanian
Centre for Materials Engineering and Regenerative Medicine
Bharath Institute of Higher Education and Research (BIHER), Chennai
India

Ramalakshmi Subbarayalu
Department of Bio-Engineering
VELS Institute of Science, Technology & Advances Studies (VISTAS), Chennai
India

Somasundaram Prasath Alias Surendhar
Department of Biomedical Engineering
Aarupadai Veedu Institute of Technology, Chengalpattu
India

Panneerselvam T
Department of Pharmaceutical Chemistry & Analysis
Vels Institute of Technology & Advances Studies (VISTAS), Chennai
India

Yuvaraj Tamilselvi
Centre for Materials Engineering and Regenerative Medicine
Bharath Institute of Higher Education and Research (BIHER), Chennai
India

Anuj Nishanth Lipton
Singapore Institute of Food and Biotechnology Innovation (SIFBI),
Agency for Science, Technology and Research (A*STAR)
Singapore, Republic of Singapore

Fayaazuddin Thajuddin
Department of Microbiology
Bharathidasan University, Tiruchirpalli
India

G. B. Varshitha
Department of Psychology
Thiagarajar College, Madurai
India

Palanivel Velmurugan
Centre for Materials Engineering and Regenerative Medicine
Bharath Institute of Higher Education and Research (BIHER), Chennai
India

Malarkodi Velraj
Department of Pharmacognosy
Vels Institute of Science, Technology, and Advanced Studies (VISTAS), Chennai
India

Selvakumar Vijayalakshmi
Department of Immunology & Microbiology
University of Rochester, New York
USA

Rajiv Vincent
School of Computer Science and Engineering, Vellore Institute of Technology, Chennai
India

J. Vinothkumar
Center for Applied Nanomaterials
Chennai Institute of Technology, Chennai
India

Gayathry S. Warrier
Department of Computer Science
Christ University, Bangalore
India

D. Yogaraj
Department of Mechanical Engineering
Sri Muthukumaran Institute of Technology, Chennai
India

D. Yuvaraj
Department of Biotechnology
Vels Institute of Science, Technology & Advances Studies (VISTAS), Chennai
India

About the Editors

Dr. Ooi Chien Wei, Associate Professor at the Centre for Net-Zero Technology at Monash University Malaysia, specializes in the fields of bioprocess design and biocatalysis. His research focus is on the development of integrated microbial fermentation systems and bioprocess model prediction tools for the expression of industrially important enzymes and proteins derived from recombinant DNA technology. Aside from the field of bioprocess engineering, his research focuses on the development of food-grade emulsion systems and the extraction technologies applied in food waste valorization and recovery of phytonutrients from microalgal sources and agri-wastes.

Dr. Ramalakshmi Subbarayalu is an Assistant Professor (Research) at the Department of Bioengineering, Vels Institute of Science, Technology and Advanced Studies (VISTAS), Chennai, India. She specializes in the field of bioprocess engineering and sustainable process development for the extraction and purification of biomaterials. She has received prestigious research grants and fellowships, including the APN CAPaBLE Award 2025, the Australian Endeavour Fellowship Program, Monash University Platform Mobility Funding Scheme, and the Japan Student Services Organization (JASSO) Fellowship at Fukui University, Japan.

Dr. Lo Sewn Cen is a Research Fellow at the Centre for Net-Zero Technology, Monash University Malaysia. Her research background spans bioprocess engineering, microbial fermentation, biocatalysis, and biotransformation, with a focus on the valorisation of agricultural waste biomass into value-added bioproducts. Her expertise includes fermentation process development, bioprocess optimisation, and genetic engineering, with experience in developing scalable processes for bioproduct production. Through this work, she contributes to the development of sustainable bioprocesses that support net-zero carbon goals and advance the biocircular economy.

Dr. N. Gnanasankaran MCA, Ph.D., RUSA PDF, is presently working as an Assistant Professor in the Department of Computer Science, Thiagarajar College, Madurai, India. His areas of research specialization include Artificial Engineering, SQA, Data Science, Machine Learning, Big Data, and IoT. He has contributed more than 35 research articles in Scopus and Web of Science Indexed journals/conferences from IEEE, Elsevier, Springer, and TechScience Press, etc. He has contributed Four Edited Book Volumes in Springer Nature (UK), Apple Academic Press (Canada), Elsevier (USA), and Wiley Scrivener (Texas).

Foreword

The global transition toward sustainable and resilient energy systems will not be achieved by biological innovation alone, nor by digital technologies operating in isolation. The real breakthrough lies in their convergence. One of the most consequential developments in contemporary bioenergy research is the integration of artificial intelligence with precision fermentation, and this book addresses that convergence with both technical depth and practical relevance.

The bioenergy sector has long struggled with persistent challenges—process inefficiencies, poor reproducibility during scale-up, feedstock variability, and narrow economic margins. Precision fermentation has significantly improved our ability to engineer and control microbial systems, but control without intelligence has limits. Artificial intelligence changes this equation by enabling real-time process learning, predictive optimization, and data-driven decision-making across the fermentation lifecycle. Together, AI and precision fermentation provide a pathway to translate laboratory success into commercially viable, industrial-scale bioenergy solutions.

This volume offers a structured and end-to-end perspective, spanning fundamental science, technological advances, and real-world applications. The chapters demonstrate how AI and machine learning can be applied to microbial strain selection, fermentation monitoring and optimization, and sustainability assessment across diverse bioenergy pathways. The inclusion of topics such as agri-food waste valorization, carbon capture, life cycle assessment, and scale-up challenges reflects a mature and systems-level understanding of sustainable bioenergy production.

As the founder of Chimertech Private Limited, my work has been centered on applying AI-enabled sensing and decision systems to real, high-variability biological environments in the dairy sector—where outcomes depend on early detection, continuous monitoring, and practical adoption in the field. That experience highlights a lesson that is equally relevant to precision fermentation for energy: biology is inherently variable, and scale demands not only good science, but reliable measurement, robust data pipelines, and decisions that work outside controlled laboratory conditions. The same discipline required to convert animal-health signals into actionable, repeatable outcomes is required to convert fermentation signals into stable yields, consistent quality, and economically viable energy production.

A particular strength of this book is its interdisciplinary orientation. It is relevant not only to researchers and academicians, but also to industry professionals, entrepreneurs, and policymakers who are seeking scalable, data-driven solutions for clean energy deployment. The case studies and industrial perspectives reinforce an essential point: meaningful impact emerges only when research is translated into deployable technologies.

The future of sustainable energy production will depend on our ability to integrate biological systems with intelligent computational frameworks that enhance efficiency, reliability, and environmental responsibility. In this context, *Harnessing*

Artificial Intelligence and Precision Fermentation for Sustainable Bioenergy Production provides both a technical foundation and a practical roadmap for the next generation of bioenergy innovation.

I congratulate the editors and contributing authors for compiling this timely and forward-looking volume. I am confident that it will serve as a valuable reference for advancing research, guiding industrial implementation, and inspiring innovation in AI-driven precision fermentation for sustainable energy production.

Dr. Ragul Paramasivam
Founder, Chimertech Private Limited

Preface

The shift to sustainable energy systems in the world has enhanced the demand to develop new bioenergy options to meet the climate mitigation, resource efficiency, and circular economy objectives. Renewable biological biomass is a key toward bioenergy, which is essential in meeting the United Nations Sustainable Development Goals (SDGs); traditional bioenergy generation systems are however, confronted by issues of low efficiency, lack of control over the processes, variability of feedstock, and scalability. In this regard, precision fermentation combined with artificial intelligence (AI) is a potent and disruptive solution to the production of bioenergy in a sustainable manner.

Harnessing Artificial Intelligence and Precision Fermentation to Sustainable Bioenergy Production is a book that has provided a detailed perspective of how AI- and machine learning (ML)-driven technologies can be applied to improve microbial fermentation to produce a variety of biofuels. Precision fermentation allows control of microbial metabolism, strain selection, and process parameters precisely, whereas AI/ML methods allow tracking of processes in real time, predicting, and optimizing them. These technologies converging present new chances to enhance productivity, lower operational expenses as well as minimizing environmental impacts throughout the bioenergy value chain.

The book is well structured so that the reader can follow ideas in a linear way through the book. First chapters discuss the role of bioenergy in a sustainable development, principles of traditional fermentation, and the principles of precision fermentation. The next chapters present AI-based optimization of bioprocesses and agri-food waste treatment monitoring in real time. The use of AI/ML in maximizing precision fermentation and biofuel generation, i.e., biogas, methane, biohydrogen, drop-in fuels, hydrotreated vegetable oils, and microalgae-based biofuels, is discussed in detail. New areas of discussion, like microbial carbon capture and life cycle analysis of AI-based fermentation systems, are also presented.

The book provides case studies of successful AI-driven bioenergy systems and draws a discussion on some of the challenges related to the challenges of scaling precision fermentation to industrial scales to bridge the gap between research and industrial practice. It lays stress on process integration, techno-economic, and sustainability evaluation.

The book is aimed at researchers, graduate students, as well as professionals in the field of bioenergy, biotechnology, fermentation science, artificial intelligence, and sustainable engineering. This volume seeks to be a useful resource in the evolution of next-generation, data-driven systems in bioenergy by integrating biological innovation with smart computational systems.

INTRODUCTION

The growing need to have clean, reliable, and sustainable energy has enhanced the world to work toward coming up with alternatives to fossil fuels. Bioenergy, which is a result of renewable biological materials, is one of the primary elements of the low-carbon futuristic energy framework and it is directly connected to the United Nations Sustainable Development Goals (SDGs) connected to clean energy, climate action, and sustainable industrial development. Although it has potential, the large-scale production of bioenergy is still hampered by limitations in terms of low conversion efficiency, heterogeneity in feedstock, process instability, and high cost of operation. To overcome these drawbacks, it is necessary to develop new data-driven methods that can incorporate the latest developments in biotechnology and digital technologies.

Precision fermentation has become a new-generation bioprocessing approach that facilitates controlled and efficient biomass to fuels and energy carriers. Precision fermentation can tune metabolic pathways, exploit substrates better and increase yield by facilitating rational selection and engineering of microbial systems. Nevertheless, the growing complexity of fermentation systems, which is the result of multi-variable interactions, dynamic operating conditions, and large-scale data generation, implies the need to use sophisticated computational tools to enable efficient process control and optimization.

AI and ML offer effective answers to these problems as they make it possible to predict models, conduct real-time monitoring, and make decisions in bioenergy production systems. The analysis of the large and complex data created during fermentation by AI-driven methods enables the opportunity to detect possible faults early, control the process in an adaptive mode, and optimize the main process parameters. In combination with precision fermentation, AI/ML technologies would lead to a substantial increase in the efficiency, scalability, and sustainability of biofuel production pathways.

This book discusses the interaction of AI and precision fermentation in a wide range of bioenergy uses, where synergy is important. It includes microbial selection, agri-food waste valorization, fermentation processes optimization, and AI-assisted production of biofuels, including biogas, methane, biohydrogen, drop-in fuels, hydrotreated vegetable oils, and biofuels made by using microalgae. Besides, the volume discusses such new topics as microbial carbon capture, the life cycle analysis of the AI-powered fermentation system, and practical examples of its applicability to industry.

This book is expected to become a resourceful tool to the research community, students, and industrial practitioners in the interface of bioenergy, biotechnology and artificial intelligence by offering interdisciplinary insights and recent developments in technology. Combining smart computational devices with accuracy fermentation is a promising future lead toward strong, effective, and sustainable system of bioenergy that can satisfy the future global demands of energy.

Acknowledgment

The editors gratefully acknowledge CRC Press (Taylor & Francis Group) for their professional guidance and support in bringing this volume to publication. We extend our sincere appreciation to Dr. Gagandeep Singh, Senior Publisher (STEM), CRC Press, India and Sub-Saharan Africa, Taylor & Francis India Pvt. Ltd., for his constant encouragement, valuable guidance, and unwavering support throughout the development of this book. We also thank all contributing authors for their scholarly dedication and timely submissions, as well as the reviewers for their insightful comments that significantly enhanced the quality of the chapters. Special appreciation is extended to our colleagues, collaborators, and institutional authorities for their continuous support. Finally, we gratefully acknowledge our families and well-wishers for their patience, understanding, and motivation, which were instrumental in the successful completion of this volume.

1 An Overview of Bioenergy for Sustainable Development Goals

Suguna Krishnamoorthy, Vignesh Marimuthu, Lekhavani Ramesh, Angeline Julius, Suresh Malakondaiah, Udaya Prakash Nyayiru Kannan, and Vinayagam Mohanavel

1.1 INTRODUCTION

Bioenergy is defined as a renewable energy source derived from biomass. The broad category of biomass includes natural resources, including trees, plants, and trash, which fall within. Modern bioenergy contributes around four times as much to total energy consumption as solar photovoltaic (PV) and wind combined (Reid et al., 2020). As the share of solar and wind power grows, bioenergy may help decarbonise the electricity supply by providing balancing power to keep the system stable (Paraschiv and Paraschiv, 2023). The principal origin of biomass to address societal concerns, balance the carbon atmosphere, and create a sustainable challenge like a low-carbon circular bioeconomy is bioenergy, which must be valued (Güleç and Okolie, 2024). Thus, technologies enable the reuse of carbon from waste streams and biomass to create bioproducts, renewable energy (Paraschiv and Paraschiv, 2023; Ray et al., 2022), and lower-emission fuels for trucks, cars, aeroplanes, and ships was showed in (Figure 1.1).

Also, this study revealed the high potential of circular bioeconomy and decarbonisation approaches to transform palm oil waste management into more sustainable and resource-efficient processes (Goh et al., 2025). The use of renewable biological resources, such as forestry by-products, animal wastes, agricultural leftovers, and energy crops, is crucial to achieving the UN Sustainable Development Goals (SDGs) and implementing the Paris Climate Agreement. Such practices promote sustainable energy, better land-use, food security, and climate goals for everybody. These practices include the restoration of degraded or marginal regions, sustainable forest management, sustainable intensification and landscape design, and the mitigation of food chain losses. Renewable energy is not only a concern for the environment, but it also presents economic opportunities. There are offers of a sustainable method to handle

DOI: 10.1201/9781003684374-1

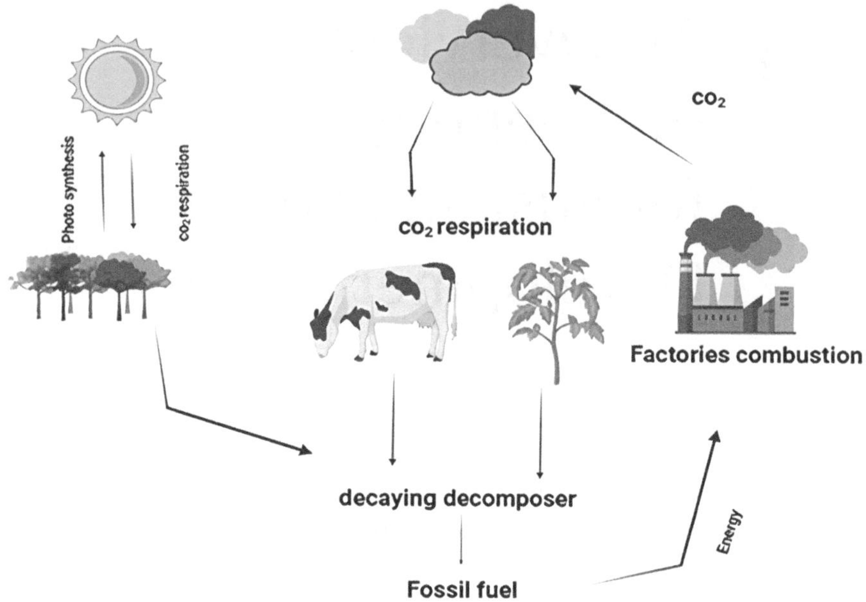

FIGURE 1.1 Schematic diagram of the carbon cycle.

the dual issues of energy security and climate change mitigation as the world shifts away from fossil fuels. In addition to supporting SDG 13 (Climate Action) by lowering GHG emissions and offering a low-carbon alternative to fossil fuels, bioenergy directly supports SDG 7 by advancing renewable energy sources. Moreover, it encourages rural development, poverty reduction, and job creation and ensures the responsible use of natural resources in line with responsible consumption and production.

The 21st century is witnessing unprecedented sustainability challenges, including climate change, rising energy demand, social inequalities, and environmental degradation. The global dependence on fossil fuels makes them a critical driver of GHG emissions, which contribute to global warming and threaten ecosystems, livelihoods, and human health. Disparities in development potential arise from the fact that nearly one billion people still do not have access to clean, affordable, and reliable electricity. In this context, the new bioenergy technology is a viable answer, as it offers a locally accessible, carbon-neutral, and renewable energy source (Reid et al., 2020). The project will deal with waste management and energy shortages, integrating bioenergy with a large environmental agenda.

1.2 THE CONCEPT OF SDGS AND ENERGY

The United Nations launched the SDGs in 2015 as a global call to action to eradicate poverty, protect natural resources, and ensure that everyone lives in peace and prosperity by 2030. Therefore, the 17 SDGs are interconnected and acknowledge that development must balance social, economic, and environmental sustainability and that actions in one area will impact results in others. Accordingly, SDGs support renewable energy in the context of climate change and energy security. The US Renewable Fuel Standard (RFS), the Brazilian National Alcohol Program (ProAlcool), the EU's Renewable Energy Directive (RED), Thailand's Alternative Energy Development Plan (AEDP), and India's National Policy on Biofuels (INPB) are just a few examples of policies that have been introduced in recent decades to promote the use of bioenergy. These policies were promoted to mitigate climate change and sparked a heated debate about their potential effects on the SDGs (Robledo et al., 2017). According to the 2025 progress review, just 17% of the SDG objectives are on track, while almost half show only modest progress and more than one-third are stagnant or regressing. As summarised, the reason for the expansion is the dramatic increase in energy consumption and carbon dioxide emissions (CO_2) in recent years, which coincides with rapidly increasing urban population levels (Kabir et al., 2023b). Even if the manufacturing of products and services could be a contributing factor, energy has consequently become an issue that defines sustainable growth (Omer, 2008). Initially, the Earth Summit in June 1992 resulted in 178 nations adopting Agenda 21, a comprehensive action plan for environmental protection and livelihood improvement. The agenda gave environmental protection precedence over economic growth. To end poverty, protect the environment, and ensure that everyone lives in peace and prosperity by 2030, the United Nations adopted the SDGs in 2015 (Wayne and Frechtling, 2025). Thus, the 17 interrelated aims emphasise the balance between environmental preservation, social inclusion, and economic progress. It is recognised that combating poverty and other types of deprivation requires a mix of policies. A combination of policies that address climate change, protect our seas and forest resources, enhance educational and health services, reduce inequality, and promote revenue generation is acknowledged as necessary to battle poverty and other forms of deprivation. Therefore, Zakari et al. (2022) validated a favourable influence of green innovation on energy efficiency and showed that sustainable financial development was similarly beneficial, suggesting that it promotes energy efficiency. A key element of this framework is energy, and SDG 7 emphasises the need for access to contemporary, affordable, reliable, and sustainable energy.

1.3 SCIENTIFIC AND TECHNOLOGICAL BASIS OF BIOENERGY

The scientific foundation of biomass conversion and the technical advancements that allow it to be converted into usable energy sources like heat, electricity, and liquid fuels are the basis of bioenergy. In basic terms, biomass is solar energy that has been stored via photosynthesis. It converts CO_2 and water into carbohydrates, which can later be released as energy through biological, chemical, or thermochemical conversions.

1.3.1 Technology Basis of Bioenergy

Numerous studies have examined several techniques for converting biomass. They use many methods to transform biomass into useful energy sources, including thermochemical, physiochemical conversion, and biochemical processes (Behera et al., 2022). This technical system forecasts the type of energy output and biomass feedstock. There are three main outputs that might be made from the bulk of the biomass: fuel for transport, industrial inputs, and the production of electricity and heat (Aishwarya et al., 2022; Brown, 2019). Various technologies have been developed for these processes, including combustion, gasification, pyrolysis, liquefaction, and hydrogenation. Combustion is the oldest process known to use biomass as an energy-producing material in which the chemical energy stored in the biomass is released in the form of heat through direct heating in the presence of oxygen/air. There are three parameters which have a strong influence on combustion: temperature, turbulence, and time. Liquefaction is a process that produces mostly bio-oil and trace amounts of gases and char by converting biomass to biofuels at lower temperatures (250–400°C) and higher pressures (5–20 MPa) than other methods. This technique is clean since it is not affected by the biomass' moisture content and produces no emissions. In order to optimise the recovery and extraction of liquefied bio-oil, these procedures also utilise a mixture of compressors and heat exchangers to chill the natural gas. It includes hydrocarbons such as butane, propane (C_3H_8), and ethane (C_2H_6) (C_4H_{10}). Pyrolysis is a thermochemical conversion process that reduces agricultural waste to its component elements by applying heat without oxygen. The result is non-condensable gaseous mixtures, biochar, and bio-oil. This thermochemical method turns carbonaceous feedstock into resources by using a moderate process temperature (between 350°C and 800°C).

The process of gasification involves partially oxidising biowaste at temperatures between 800°C and 1400°C. The end product is syngas, which is mostly composed of carbon monoxide (CO), hydrogen (H_2), and methane (CH_4), with trace quantities of other gases. Compared with other thermochemical conversion methods, it has several benefits. For instance, it produces high-energy, clean syngas suitable for various applications. Biomass use in hydrogenation is a process used to convert biomass into hydrogen or other value-added chemicals by adding hydrogen molecules. This conversion can be achieved through thermochemical, electrochemical, and biological methods.

1.3.2 Thermochemical Conversion of Bioenergy

The thermochemical conversion of biomass is potentially vital to meeting global demand for sustainable transport fuels. The future of biomass upgrading via thermochemical processing will depend on sector coupling, both within the energy sector and with sectors such as food production (Wang, Y. and Wu, J.J., 2023). The thermochemical processing of biomass employs heat and catalysts to convert plant polymers into fuels, chemicals, or electricity. This thermochemical conversion can be categorised as combustion, gasification, pyrolysis, and solvent liquefaction (Zhang and Brown, 2019). Through thermochemical biofuels production employing *P. putida* (via pyrolysis process), catalytic upgrading, and carbon closures of above 90% for wastewater streams across several processing technologies, the bioconversion of thermochemical intermediate (BTI) was established in this study (Beckham, 2022). Also, another study revealed that thermochemical valorisation enhanced microbial toxicity tolerance (Jayakody et al., 2018).

1.3.3 Conversion of Biochemical Process

Using microorganisms and enzymes to transform biomass into bioenergy is a crucial step in the biochemical conversion process. These technologies can handle biomass with high water content (>50%). These processes include aerobic conversion, anaerobic digestion, and fermentation. This process is suitable for organic waste, and the conversion steps are hydrolysis, acidogenesis, acetogenesis, and methanogenesis. Accordingly, a study that revealed the photo fermentative biomethane pathway underscores additional potential for circular waste management and energy diversification (Sharma et al., 2026). Moreover, lignocellulosic biomass and food waste substrate have been utilised along with inoculum concentration and pre-treatment methods to enhance the biohydrogen production (Meena et al., 2025). Perhaps cyanobacterial engineering has been optimised by green algal-based hydrogen production, which paves the way for a sustainable energy source. The biochemical process utilised in first-generation and second-generation biofuel from feedstock is helpful for producing renewable energy. The first-generation biofuel feedstocks, like edible oil crops such as sugarcane, soybeans, and food waste, produce them respectively. On the other hand, the second-generation biofuel is derived from the feedstock of non-food feedstocks, including agricultural and forest residues and municipal waste. However, because of the intricacy of the biomass and the challenges involved in growing, obtaining, and transporting low-density biomass to centralised biorefineries, second-generation biofuels face even more obstacles than first-generation biofuels (Balan, 2014). Depending on the microorganisms present throughout the conversion process, the technology underlying biomass conversion can produce various products, including hydrogen, butanol, biogas, and organic acids (Gouveia and Passarinho, 2017).

1.4 TYPES OF BIOENERGY AND THEIR APPLICATION ACROSS VARIOUS SECTOR

Bioenergy is defined as organic material derived from plants and animals. Examples of biomass used for bioenergy are woody waste, biomass crops, and organic waste/residues from industry, agricultural waste, and household management. The type of bioenergy generated is determined by the feedstock's characteristics and the conversion technology applied. Perhaps bioenergy can be classified into solid and gaseous fuels, each with diverse source applications, as shown in Table 1.1.

1.4.1 Biofuels

Globally, the consumption of fossil fuels such as natural gas, coal, and petroleum has a significant effect on the climate, the environment, and health. When fossil fuels are used, GHGs are released into the atmosphere, trapping heat and contributing to climate change. However, biomass, a renewable energy source, can be directly transformed into liquid fuels, or "biofuels," in an environmentally acceptable way to help meet the need for transportation fuels. Although biofuels exist in a variety of types

TABLE 1.1
Types of Energy and Their Applications for Versatile Areas

Bioenergy	Concept of Production	Process/Technologies	Applications	Reference
Biofuels (Ethanol)	Biomass is converted into liquid fuel required for transportation	• Deconstruction of biomass includes enzymatic process • Fermentation process where yeast converts sugars into ethanol • Transesterification like chemical process reacting oils with alcohol and catalyst to produce biodiesel	Transportation fuels (ethanol, diesel), thermal plants and power generation for industrial uses	de Paula Leite et al. (2025); Yadav (2024); Demirbas (2008)
Biopower (Burning)	Biomass is burned directly to generate steam and power turbines for replacing coal firing	Steam turbines are converting the thermal energy of high-pressure steam into rotational mechanical energy	Wood and agriculture waste to produce heat and steam turbines than coal usage	Panchasara and Ashwath (2021)
Biopower (anaerobic digestion)	Anaerobic digestion of food waste to produce methane	The food waste incorporates into anaerobic digester of tanks (Hydrolysis, acidogenesis, acetogenesis and methanogenesis) with anaerobic bacterial action produced methane gas	Electricity generation, industrial heat and cooking gas production	de Paula Leite et al. (2025)
Biomass conversion and (gas/ liquid fuel)	Biomass converted through gasification/ pyrolysis to produce fuel	During high temperature (800–1400°C) of gasification process, limited syngas is generated. In contrast, during the low-temperature region (around 400°C) of pyrolysis process, bio-oil is produced, which can help convert the gas	Used in boilers and turbines as a substitute for diesel or furnace fuel, generating electricity in environmentally friendly thermal plants	Chhiti and Kemiha (2013); Huang et al. (2013); Panchasara and Ashwath (2021)

and are used to suit a variety of energy demands. There are common types of biofuels, such as biodiesel, bioethanol, biogas, and biohydrogen. The most widely used type of biofuel is biodiesel, a renewable resource or fuel derived from vegetable oil, recycled cooking oil, peanut oil, canola oil, soy oil, palm oil, animal fat, and fatty acids. Transesterification is the term for these conversions (Kipkoech et al., 2024). Furthermore, the fermentation of sugar and plant starches produces ethanol, a sustainable fuel. This alcohol is added to gasoline as a blending ingredient to boost octane and reduce emissions of carbon monoxide and other harmful products (Abdellatief et al., 2025). Additionally, biogas, a mixture of gases including carbon dioxide, methane, and hydrogen sulphide, is mostly processed from raw materials such as manure, green waste, food waste, and agricultural waste (Harirchi et al., 2025). Similar to biogas, biohydrogen is a renewable resource that is produced biologically with the aid of bacteria, algae, and archaea (Alqarzaee et al., 2025). Moreover, the biofuel production technologies include biochemical conversion (fermentation and enzymatic hydrolysis), anaerobic digestion, and thermochemical conversion, such as pyrolysis and gasification. Approximately 80% of the energy produced worldwide currently comes from fossil fuels, including oil, natural gas, and coal. The possibility of biofuels becoming mainstream has increased awareness of the societal effects associated with the development and applications of transportation biofuels. Regulations that incentivise low-carbon intensity inputs are driving the use of cooking oil as a feedstock for biofuels due to its innovative transesterification technology (de Paula Leite et al., 2025). However, the demand for microalgae biofuels has grown dramatically because of the possibility that they could be renewable resources due to the oily content of the microalgae via transesterification, which helps reduce CO_2 level in the atmosphere (Yadav, 2024). Thus, both industrialised and developing nations view biofuel technology as significant for a variety of reasons.

1.4.2 Biopower

Biopower is the process of generating electricity for producing power from renewable biomass resources. The production cycle involves five key stages: biomass supply, processing, conversion, transportation, and energy generation. There are three ways to convert biomass energy into usable power: burning, bacterial degradation, and conversion to petrol or liquid fuels.

1.4.3 Combustion for Biopower Production

Direct combustion is the process used to turn biomass into energy. However, while the efficacy of combustion is comparable between biomass and coal—especially when using pre-processed biomass—the overall scale of biomass power generation remains substantially smaller due to its low energy density and high logistical costs. Typically, the temperature ranges from 1300 to 1400°C for optimal grate combustion, balancing the emissions of nitrogen oxides and carbon monoxide (Wang et al., 2021). Moreover, high-pressure steam is created in the boiler when the biomass is burnt, and this steam rotates a set of turbine blades (Azman, 2018). The generator creates power by being propelled by the turbine's spin. Here, coal co-firing in an existing energy plant boiler is replaced with biomass (Basu et al., 2011).

1.4.4 Bacterial Decay

Digesters are oxygen-free containers used for the anaerobic digestion of organic waste, animal manure, or human sewage. Here, anaerobic digestion produces biogas without oxygen by using microorganisms to break down organic materials, including biomass. Moreover, anaerobic bacteria produce methane and other bioproducts after breaking down the organic matter, creating biogas such as methane, produced through anaerobic digestion, which can be purified for use as renewable energy. They accurately reflect their source of nature and are used to produce electricity (Alghni, 2010).

1.4.5 Conversion to a Gas or Liquid Fuel

Biomass can be changed into a gas or a liquid by using the processes of gasification and pyrolysis (Chhiti and Kemiha, 2013). Therefore, the solid biomass is exposed to exceptionally high temperatures under limited oxygen to produce syngas (Huang et al., 2013). Syngas combustion can take place in a boiler, internal combustion engine, or turbine in a power heating system. Integrated gasification combined cycle (IGCC) power plants are massive utility facilities that produce electricity with excellent efficiency, particularly when paired with steam turbines in a combined cycle (De Robbio, 2023). Moreover, internal combustion engines are small- to medium-scale power generation, suitable for distributed electricity production, and they offer greater fuel flexibility and lower capacity investment than gas turbines. Manufacturing, chemical processing, and cement manufacture utilise the industrial heat and steam generated by boilers (Luo et al., 2024). Pyrolysis can produce bio-oil, bio-char, and syngas because it uses heat to break down complex organic materials in an oxygen-free environment (Cuevas et al., 2024). A related process called pyrolysis uses biomass to heat the low-temperature region (around 400°C) without any oxygen at all, creating a crude bio-oil that could be used in furnaces and engines to generate energy instead of fuel oil or diesel (Panchasara and Ashwath, 2021).

1.5 SOURCE OF BIOENERGY

A variety of organic matter as a bioenergy source has been converted into energy. Bioenergy sources include wood and wood residues, agricultural waste and residues, energy crops cultivated especially for energy generation, and organic wastes from various sources (Figure 1.2).

Dependence on this source has been categorised by bioenergy production. CO_2 is considered eco-friendly and provides more economic benefits for consumers. Accordingly, the biodiesel in this study was produced using native renewable resources such as animal and vegetable fat, rice bran oil, coconut oil, *Jatropha curcas*, mahua, and karanja that are either surplus or inedible and can be used to produce biodiesel (Bajpai and Tyagi, 2006). Recently, microalgae cultivation has attracted substantial interest owing to its potential uses for bioenergy generation, wastewater treatment, industrial CO_2 removal, and the manufacture of biochemical chemicals that may replace traditional fossil fuels (Mobin et al., 2022; Deora et al., 2022). Thermochemical and biochemical pathways are heterogenous mixtures of waste material that can be converted into energy (Begum et al., 2024). Additionally, a significant amount of agricultural residue from sugarcane bagasse is being converted into biofuel and biopower (Bezerra and Ragauskas, 2016). Although the organic material used to generate biogas will eventually decompose, utilising it as a source of energy is less harmful to the environment. The process of producing biogas has a

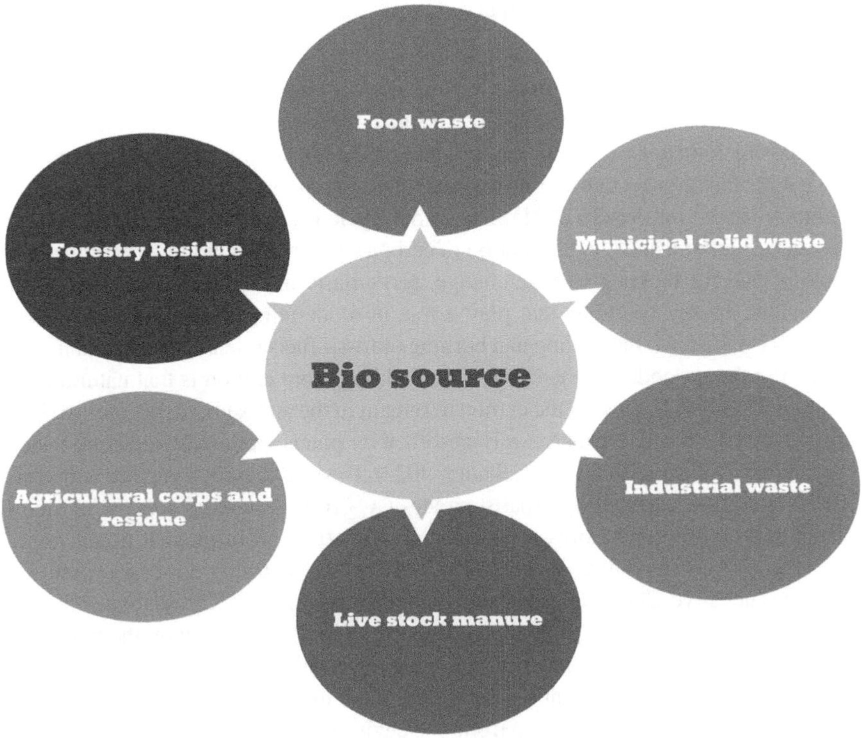

FIGURE 1.2 Bio-source from the differential environmental conditions.

notable cyclical pattern and is a component of a broader sustainability cycle for resource management, which includes wastewater, food debris, agricultural debris, and municipal solid waste.

Due to its potential as a renewable energy source that can replace fossil fuels and the GHG emissions they produce, bioenergy can aid in the fight against climate change and help reach carbon neutrality (Begum et al., 2024). A closed carbon cycle, which is essential for lowering emissions in difficult-to-decarbonise industries like transportation and industry, is created when the carbon emitted during the burning of bioenergy is balanced by the CO_2 that the plants took in during their development (Abbass et al., 2022). With the Paris Agreement, an energy framework, the European Union (EU) has set ambitious but crucial targets to decrease GHG emissions because of the release of these gases from fossil fuels. The present scientific use of biomass, carbon neutralisation, climate change mitigation, and appropriate carbon cycles has been challenged in recent research. Bioenergy can play a significant role in climate change mitigation, and it uses biogenic carbon, such as bioenergy-CCS (BECCS), alongside biomass for heat, power, and liquid fuels (Creutzig et al., 2015; Taylor et al., 2019). The emerging technology of bioenergy can mitigate climate change while supporting carbon neutrality.

1.5.1 Bioenergy and the Carbon Cycle

The mitigation scenarios database of the Intergovernmental Panel on Climate Change's Sixth Assessment Report is an important resource for informing policymaking on energy transitions (Achakulwisut et al., 2023). The 2015 Paris Agreement aims to avert climate change, which means that fossil fuel production cannot continue unabated, as coal, oil, and gas are the largest contributors to greenhouse gas emissions and are responsible for climate change (van Asselt, 2021). The carbon cycle is a system that affects ecosystem dynamics, climate, and the distribution of carbon sources worldwide. The entire carbon cycle is vital for tackling climate change, particularly in the context of incorporating sustainable energy solutions that play a role in moderating its imbalance (Adewumi et al., 2024). Instead, the mining and burning of fossil fuels release carbon that has been trapped in the ground, and the drawback of adding more carbon is that natural carbon sinks can't absorb it, causing the carbon to remain in the atmosphere. In contrast, bioenergy returns carbon that was recently absorbed by plants, potentially creating a closed carbon loop (Schirmeister and Mülhaupt, 2022). Basically, carbon capture and storage (CCS) is a known technology that separates CO_2 from industrial sources like power plants, transports it and stores it permanently underground to prevent its release into the atmosphere (Sandunika et al., 2023). For carbon neutrality, bioenergy with CCS generates negative CO_2 emissions. The combined energy BECCS facilitates the long-term temperature objective to study the impacts of implementation on the climate and carbon cycle to achieve the balance (Creutzig et al., 2015). The BECCS system must undergo a life cycle assessment (LCA) to determine its potential for negative emissions and to stop the transfer of impacts from one environmental impact category to another (Duval-Dachary et al., 2023; Saharudin et al., 2023). Additionally, it involves a burning and gasification process that uses biomass to produce bioenergy and reduce or decarbonise the environment (Güleç and Okolie, 2024).

1.5.2 Bioenergy Contribution to Climate Change Mitigation

Increasing the atmospheric CO_2 over the past 140 years due to human persuasion has increased the global temperature; consequently, the climate has changed, which is attributed to fossil fuel combustion and deforestation worldwide (Sahoo et al., 2021). The Earth's climate is evolving as a result of increasing harmful anthropogenic activities, the most notable of which is the rise of CO_2 and further GHG concentrations inside the environment (Kabir, 2023a). Globally, agriculture and related world anthropogenic GHG emissions had to change about 17% in 2010 (8.4 $GtCO_2e/yr$), accounting for this carbon emission in the agricultural sector critical to reaching the 2°C target outlined in the Paris Agreement (Richards et al., 2018). Therefore, bioenergy reduces fossil fuel dependence by providing a renewable and often carbon-neutral alternative for power, heat, and transportation fuels like ethanol and biodiesel (Malla et al., 2023). Moreover, the use of biomass sources such as crops, residues, and waste creates a circular carbon cycle where the carbon released during use is offset by the carbon absorbed during the growth of new feedstock (Hart et al., 2025). Many pathways for mitigating global climate change represented in IPCC assessment reports rely heavily on the deployment of bioenergy, often used in conjunction with CCS (Calvin et al., 2021). Accordingly, climate targets and policies in line with Paris Agreement goals and downscaled carbon budgets are a useful framework for setting local mitigation targets related to a specific global temperature change (Kuriakose et al., 2022). Although the Paris-aligned global carbon budget is subnational, it focuses on emissions from energy sources like power, heat, cooling, surface transport, and industry (Mulder et al., 2021). Accordingly, this study explained that the hedgerows could be between 2.1 and 5.2 Mg/ha/yr for periods of 50 and 20 years, promising strategies to promote carbon sinks for climate change mitigation (Drexler et al., 2021; Grassi et al., 2021) proposed a "Rosetta stone" modification to translate Integrated Agreement Models (IAMs) of land-use mitigation options.

1.6 POLICY, REGULATIONS, AND INCENTIVES FOR BIOENERGY

1.6.1 Context and Purpose of Policies

Bioenergy is a powerful tool in mitigating climate change and simultaneously achieving the SDGs.

1.6.2 International Policy Frameworks

At a global level, the Paris Agreement, SDG 7 (Affordable and Clean Energy), SDG 13 (Climate Action), and the IEA Net Zero Roadmap, by fair to all parties concerned, are the frameworks that push the envelope on the use of renewable energy sources, bioenergy not being an exception. Bioenergy aims to address climate change, advance multiple SDGs, and serve as a key part of the shift towards sustainable energy systems; this is acknowledged in international policies (Figure 1.3).

The comparative overview of National Bioenergy Policy Frameworks is shown in Table 1.2.

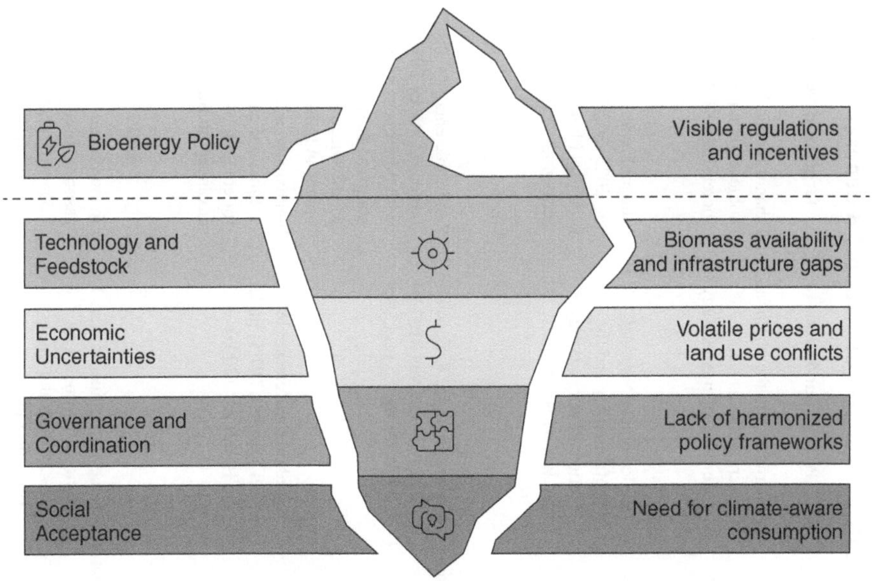

FIGURE 1.3 Bioenergy policy implementation: unveiling the hidden challenges.

TABLE 1.2
Comparative Overview of National Bioenergy Policy Frameworks

Country	Policy Foundation	Target and Mandates	Incentives/Support Mechanisms	Sustainability Measures
Brazil	Integrated into the National Energy plan as a core renewable energy strategy	Ethanol blend: 27% Biodiesel blend: 12%, increasing to 15% by 2026	The RenovaBio programme plays a role in the certification of the life cycle GHG reductions, the setting of carbon intensity targets, and the facilitation of biofuel credit trading (CBIOs).	Forest Code regulates the use of land and the zoning for agroecology. Soy Moratorium, on the other hand, supports sustainable agriculture and the conservation of native ecosystems.
India	Guided by the National Biofuel policy and the National Energy Policy (Achakulwisut et al., 2023).	Ethanol blend: 20% target by 2025–2026 Biodiesel blend: 5% long-term goal	The government is moving forward with supporting the industry through a variety of means, such as funds for the viability gap, excise duty exemptions, soft loans, and support for waste-to-energy and advanced biofuels, and has even included biogas in the SATAT initiative.	Sustainability is mainly about waste and non-food feedstocks; 2G ethanol plants and biogas from residues are to be promoted to minimise land-use conflicts.
European Union (EU)	Governed by the Renewable Energy Directive (RED II) and fir for the 55 packages	Renewable energy target: 42.5% by 2030, with biofuels as a key component: sub-targets for advanced biofuels and biogas in transport	Areas of support include the establishment of feed-in tariffs, the implementation of carbon pricing, and the use of renewable energy certificates (RECs), along with the provision of funds for innovation through Horizon Europe and the Innovation Fund.	The RED II sustainability criteria impose GHG savings that are very hard to achieve, thus, the protection of lands with very high biodiversity and high-carbon-stock, and finally, emissions certification of the entire life cycle.
United States	Based on the Renewable Fuel Standard (RFS) under the Policy Act (2005)	Mandate annual renewable volume obligations (RVOs) for ethanol, biodiesel, and advanced biofuels	There is a range of financial incentives available, such as tax credits (e.g., Biodiesel Tax Credit, 45Q for carbon capture), loan guarantees, and R&D grants from the Department of Energy.	GHG emissions accumulation from the inception of the project to the very end through the life of the product will be one way to determine sustainability standards set by Tucson Water; resources like soil conservation, waste feedstocks and emission reductions will be the main target areas.
Argentina	Framework guided by the National Biofuels Law (No.26.093) and Energy Transition Plan	Ethanon blend: 12% for gasoline, biodiesel blend: 7.5% with plans for incremental increases	Biofuel producers can benefit from tax exemptions, feed-in incentives, and price guarantees; programmes backed by the government for the expansion of biogas and bioethanol are also available.	Sustainability certification for exports and alignment with GHG mitigation goals, along with the encouragement of the use of agro-industrial residues.

1.7 SUBSIDIES, TAX INCENTIVES, AND GREEN FUNDING

There is a broad range of incentives supporting bioenergy use, which includes:

Direct Subsidies grants for biomass collection, biogas pipeline infrastructure development, viability gap funding for a variety of clean energy projects, including offshore wind. Viability Gap Funding refers to the single payment given to the energy projects with high capital costs to make them viable in the market when they cannot generate the anticipated revenues. It lowers the risk for the investors or owners of the technology and thus encourages the implementation of biogas, biomass, and waste-to-energy projects, which not only help meet the targets for renewable energy production but also contribute to reducing GHG emissions.

Tax Incentives include accelerated depreciation and concessional rates of GST (India) and tax credits for renewable energy investments from individuals and corporations.

Green Funding: The deployment and R&D investments and advanced technology development using carbon credit markets (CDM [Clean Development Mechanism], CERs [Certified Emission Reductions]), government green energy funds, and accelerated development of advanced technology to "overcome financial and technological barriers".

The range of users eligible for such incentives is wide and includes rural entrepreneurs, industrial users substituting fossil fuels, farmer groups, and project contractors (Bellamy et al., 2021). Global financing mechanisms are linked and streamlined for application through portals enabled by national governments, alliances, and partnerships.

1.8 BARRIERS TO EFFECTIVE POLICY DEVELOPMENT AND IMPLEMENTATION

Technology and Feedstock Supply Chain: Biomass availability uncertainty, competing technological developments (electric mobility for transport, high-temp heat pumps for industry), and infrastructure gaps make scaling difficult.

Economic and Regulatory Uncertainties: Unstable feedstock and energy prices (e.g., biodiesel blending in Brazil was lowered due to price spikes for more feedstock to blend), unresolved land-use conflicts, and inconsistent application of sustainability criteria.

Governance and Coordination: The absence of harmonised policy frameworks, limited stakeholder engagement in local or indigenous stakeholder settings, and uneven development among regions have strained bioenergy policy implementation in various locations. To overcome the challenges, consistency of investment with reliable long-term incentives, and supporting local innovation remain important barriers.

Social Acceptance: Need for acceptance to encourage climate-aware consumption; develop stronger consumer incentives, and disseminate knowledge through outreach and education.

Bioenergy policy development is evolving quickly as part of the global and national strategies for sustainable development designs.

Bioenergy is a key element for sustainable energy, climate mitigation, and the realisation of the SDGs.

1.9 FRAMEWORKS FOR INTERNATIONAL POLICY

Global renewable energy policy emphasises bioenergy as pivotal for achieving SDGs, particularly SDGs 7 (Affordable Clean Energy) and 13 (Climate Action).

1.10 GREEN FINANCE, TAX INCENTIVES, AND SUBSIDIES

Governments encourage the use of bioenergy through targeted financial instruments that mitigate investment risk and enhance the practical feasibility of projects.

Governments utilise financial tools to incentivise the bioenergy sector using a combination of methods.

1.10.1 Green Finance

Government green-hydrogen funds and incentives for advanced technology developments, and access to carbon credit markets, and R&D investment (Kafeel et al., 2025). National authorities, through partnerships and alliances, have simplified application portals and links to foreign funding mechanisms. These obstacles can all greatly hinder the adoption and implementation of policies. Despite providing significant promise and efforts to promote, some obstacles prevent bioenergy from having an impact (Yan et al., 2023). Uncertainty in the economy and regulations includes volatile feedstock and energy prices, slowing biodiesel blending in Brazil due to higher prices, and unresolved land-use conflicts and inconsistently implemented sustainability standards (Nuhu et al., 2023). The challenges to policy implementation included a lack of standardised frameworks, minimal stakeholder engagement, and uneven progress in regions. A key challenge is to stimulate local innovation while ensuring long-term, reliable incentives. The requirement to encourage climate-conscious consumption provides strong consumer incentives and spreads knowledge via education and outreach. Policies pertaining to bioenergy are changing rapidly as part of national and international plans for sustainable development (Kwakye et al., 2024a). International coordination and flexible science-based policies with strong incentives are propelling progress; however, overcoming technological, legal and governance obstacles will be essential to achieve bioenergy's full potential in sustainable energy transitions.

1.11 INTEGRATION OF BIOENERGY WITH OTHER TECHNOLOGIES

Bioenergy's evolution through integration with other technologies (Ochieng et al., 2022). The bioenergy sector is evolving and being transformed by the integration of bioenergy with new technologies such as artificial intelligence (AI), the Internet of

Things (IoT), hybrid renewable systems, and circular bioeconomy models. This integration supports sustainable development (increased energy production), efficiency, and resource-use effectiveness (Kwakye et al., 2024b).

Bioenergy and the IoT offer opportunities for process optimisation. The integration of IoT in bioenergy systems provides real-time data collection and monitoring, delivering increasingly comprehensive data throughout the biomass supply chain, from harvesting and cultivation to conversion and distribution. By adding sensors to fields and machinery, bioenergy can incorporate real-time data about feedstock quality, soil conditions, and crop health (Anbarasu et al., 2025). This data enables more precise control over the processes that convert biomass into bioenergy, leading to increased yields and potentially a lower waste-to-resource ratio. For example, smart sensors can optimise methane production, ensure operational reliability, and control feedstock input in anaerobic digestion processes at biogas plants. The data provided by IoT is proactive, facilitating predictive maintenance, reducing operational costs, and minimising downtime. AI also plays a role in managing bioenergy resources.

AI technologies, including machine learning, are used to identify biomass yield potential and quality across multiple feedstock sources, enabling an efficient logistical process, including planning for the harvest of biomass. AI also optimises biomass conversion technologies (e.g., gasification, pyrolysis) by interpreting real-time data that includes temperature, pressure, feedstock composition, and reaction time. AI uses machine learning and predictive algorithms to identify ideal operating conditions for a biomass conversion system, providing optimal energy yield, product quality, and low emissions. With smart adaptive and self-learning control systems, AI can guarantee process efficiency by continuously optimising reactions and ensuring stable and cost-effective bioenergy production through the selection of operational parameters, with objectives such as minimising emissions and maximising energy field (Lee et al., 2023). AI also improves aspects of the bioenergy life cycle, such as designing supply chain partners that allow for adequate biomass supply while accounting for sustainability and cost. In precision farming, AI-enabled crop management digital analytics tools are contributing to the management of biomass yields with lower resource input. Essentially, AI expedites the scalability, operational efficiency, and innovation of bioenergy systems. Synergistic actions in hybrid renewable systems (e.g., geothermal, wind or solar energy systems producing bioenergy) create the opportunity for hybrid renewable energy systems to capitalise on synergies with each renewable source (Okolie, 2024; Liao and Yao, 2021). The synergy of bioenergy and geothermal heating systems enhances district heating applications and industrial processes, providing a more continuous and lower-carbon thermal heat supply.

Biomass-fuelled plants can take on baseload electricity from periods when wind and solar electricity drop. Geothermal heating systems, when coupled with bioenergy district heating, strengthen the use of industrial process heat and provide low-carbon round-the-clock heat (Alqahtani et al., 2024). Hybrid systems create resilient energy infrastructure and strongly reduce the share of fossil fuels in total energy use. They combine to mitigate the fickleness of renewable energy systems, which creates a battery and makes use of the flexibility and versatility of bioenergy technologies to create the best opportunity for energy security and carbon reduction. Circular bioeconomy models promote efficient use of biomass and organic waste streams, enabling carbon and nutrient cycles to be completed through waste-to-energy processes (Hossen et al., 2022). The transformation of organic waste generated from forestry, agriculture, and municipalities into bioenergy (biogas, bioethanol, and biochar) and bio-based products reduces landfill loads and lowers methane emissions. The anaerobic digestion of food wastes and agricultural residues produces digestate rich in nutrients for biofertiliser and biogas for power, heating systems or vehicle fuels, giving the opportunity for sustainable agriculture. Advanced waste-to-energy facilities provide resource recovery through the fusion of several conversion pathways (Akinyele et al., 2025).

A circular bioeconomy solution enhances economic viability and reduces environmental impacts, while promoting sustainable rural livelihoods by recognising value in biomass residues and waste. In summary, a more efficient, resilient and sustainable energy future benefits from the combination of bioenergy with IoT technologies, AI hybrid renewables, and circular bioeconomy approaches (Thakur et al., 2024).

1.12 CHALLENGES AND LIMITATIONS

The SDGs of the UN can be pursued with bioenergy, one of the important renewable energy types in the context of clean energy, climate action, and sustainable land-use. Bioenergy has many uses in the transportation, power, heating, and industry sectors, and connects with the bioeconomy through sectors like manufacturing, forestry, and agriculture (Tawfik et al., 2022). Bioenergy contributes to mitigating climate change and enhancing energy security by providing flexible renewable energy that can balance intermittent resources like wind and solar, provide high-temperature process heat, and produce biofuels for transportation. While it is often thought that the trajectory of sustainable development of bioenergy fully appreciates the potential of bioenergy as a renewable resource, there are constraints and obstacles (Figure 1.4).

1.12.1 Technical Challenges

Variability of feedstock is one of the biggest technical constraints to sustainable bioenergy development, as variability of biomass type, quality, and consistent supply uncertainty can tie into system availability and processing efficiency.

1.12.2 Economic Challenges

Bioenergy commonly has difficulty competing in an economic arena affected by ever-lower-priced fossil fuels.

1.12.3 Environmental Trade-offs

Environmental sustainability will always be of concern, should bioenergy use not be actively managed, and management should include the risk of deforestation, loss of biodiversity, or impacts of land degradation (Kwakye et al., 2024a).

1.12.4 Awareness and Social Acceptance Issues

Social acceptance stands as the leading non-technical hurdle that blocks the implementation of bioenergy systems.

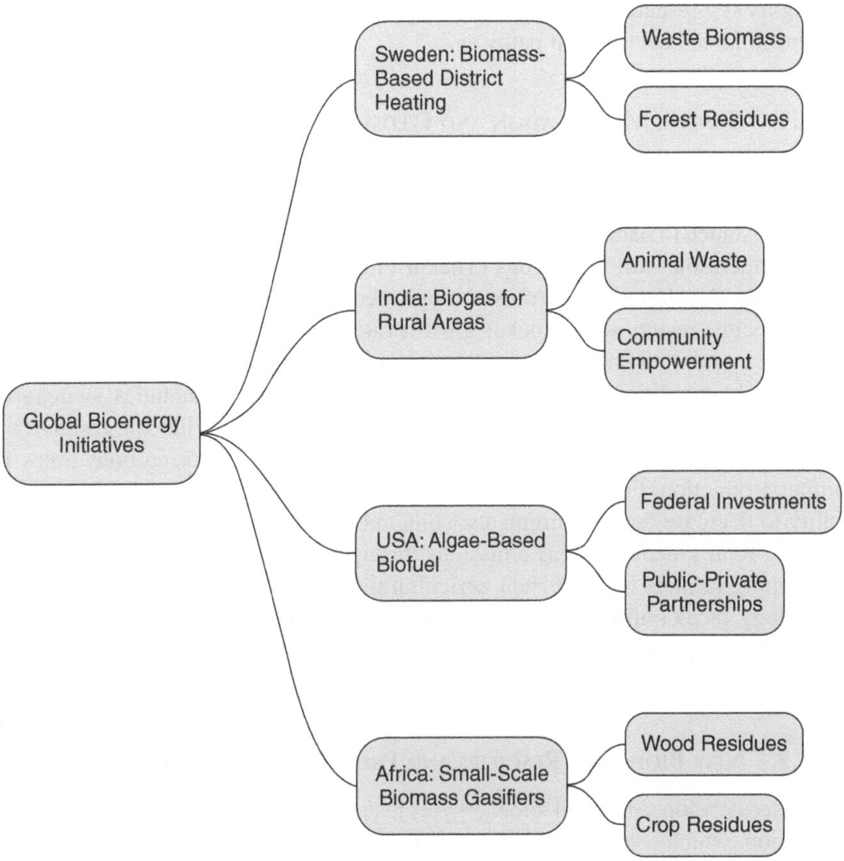

FIGURE 1.4 Global energy initiatives: Strategies and challenges.

1.13 FUTURE PROSPECTS AND RESEARCH DIRECTIONS

The bioenergy sector benefits significantly from digital technology advancements, including IoT systems, digital twin technology, and AI applications. These systems enable continuous tracking of feedstock availability, improve operational efficiency through predictive maintenance and system optimisation, and enhance overall performance. Integrating these new technologies into operations will reduce costs, increase system reliability, and accelerate the adoption of environmentally friendly bioenergy systems. Bioenergy has become a vital part of the UN's global agenda to achieve the SDGs, especially in providing renewable energy, reducing GHG emissions, and supporting circular bioeconomies. The sustainable future of bioenergy largely depends on addressing current challenges, and achieving sustainability involves developing new technologies such as AI, carbon capture, and synthetic biology. The research emphasises advanced biofuel development, feedstock optimisation, carbon capture integration, digital and automation tools application, policy and market mechanism evaluation for bioenergy adoption. This overview highlights key research areas and future opportunities to further advance bioenergy as a sustainable development pathway.

1.13.1 Genetic Modification and Feedstock Engineering

Genetic modification, together with advanced breeding techniques, enables improvement of biomass feedstocks for bioenergy through enhanced production rates and better resistance to harsh environments and marginal land conditions, which reduces their competition with food crops (Thakur et al., 2021). The key strategies, which include the development of traits through genetic modification, enable plants to achieve better results with quicker growth rates and elevated cellulose and hemicellulose content, with reduced lignin content and improved resistance to pests and drought conditions. The list of engineered feedstocks includes switchgrass and miscanthus energy crops, together with corn stover lignocellulosic residues and microalgae that have been modified to produce more lipids or carbohydrates for biofuel production. Feedstock engineering enables sustainable practices through its ability to decrease land requirements while maximising feedstock performance and reducing total greenhouse gas emissions throughout the entire production process, and helping farmers increase their agricultural yields. The implementation of nanotechnology faces multiple restrictions and needs to address environmental effects, ethical dilemmas, biosafety threats, and regulatory obstacles, which must receive thorough assessment and ongoing surveillance before extensive implementation.

1.13.2 New Bioenergy Reactors and Processing Facilities

Scientists develop advanced bioenergy conversion systems through technological innovation, which results in new reactors that achieve better performance and lower environmental emissions (Singh et al., 2025). The technologies operate through fluidised-bed gasifiers and plasma-assisted pyrolysis units and high-rate anaerobic digesters, which deliver accelerated processing and enhanced energy extraction. The systems operate as integrated biorefineries, which transform different feedstocks into fuels and power and bio-based chemicals. The implementation of automated systems together with AI-based control mechanisms enables large-scale production stability while reducing waste and enhancing energy efficiency for sustainable operations.

1.13.3 Scaling BECCS for Negative Emissions

Bioenergy with Carbon Capture and Storage (BECCS) operates through renewable biomass to generate energy while capturing carbon, which results in negative emissions (Hayat et al., 2024).

1.13.4 The Role of Technology Transfer and International Collaborations

Strong international cooperation, supported by effective policy incentives and regulatory frameworks, is essential to accelerate the large-scale deployment of sustainable bioenergy technologies.

1.13.5 2050 Vision and Net-zero Scenarios

Bioenergy is projected to play a significant role in integrated energy systems of a net-zero carbon economy by 2050. Bioenergy stands on the threshold of a bright future, supported by technological development, scientific advancement, and international collaboration.

1.14 CONCLUSION

Bioenergy bridges sustainable biomass use and the UN's SDGs. By employing thermochemical and biochemical processes to convert biomass into useful products such as biofuels, biopower, and biogas, bioenergy transforms biological resources into renewable energy that reduces greenhouse gas emissions, improves energy security, and supports rural livelihoods. Bioenergy is a renewable source of carbon within a circular bioeconomy system through its integration with the natural carbon cycle, creating a closed-loop system where carbon captured through photosynthesis is recycled for energy production. Utilising biomass leads to technological applications with measurable impacts on sustainability across multiple SDGs, including SDG 7 (Affordable and Clean Energy), SDG 12 (Responsible Consumption and Production), and SDG 13 (Climate Action), as demonstrated by this integrated approach. Besides lowering greenhouse gas emissions, bioenergy also advances sustainability by decreasing reliance on fossil carbon systems, maximising the value of agricultural and municipal waste, and optimising resource use. Looking ahead, the future of bioenergy depends on global collaboration and technological convergence. While feedstock logistics and conversion efficiencies can be improved with AI and digital optimisation technologies, BECCS opens a pathway to achieve net-zero emissions. Diversifying feedstocks from algae to waste biomass will expand the renewable resource base and reduce land-use conflicts. To foster innovation, harmonise sustainability standards, and align international policies towards a carbon-neutral future, international cooperation and technology transfer are crucial. All these efforts collectively ensure that bioenergy remains a dynamic, scientifically grounded driver of climate resilience and sustainable development.

BIBLIOGRAPHY

Abbass, K., Qasim, M. Z., Song, H., Murshed, M., Mahmood, H., & Younis, I. (2022). A review of the global climate change impacts, adaptation, and sustainable mitigation measures. *Environmental Science and Pollution Research*, 29(28), 42539–42559.

Abd-Alla, M. H., Al-Amri, S. M., & El-Enany, A. W. E. (2023). Enhancing rhizobium–legume symbiosis and reducing nitrogen fertilizer use are potential options for mitigating climate change. *Agriculture*, 13(11), 2092.

Abdellatief, T. M., Mustafa, A., Handawy, M. K. M., Abdelghany, M. B., & Duan, X. (2025). Sustainable production of eco-friendly, low-carbon, high-octane gasoline biofuels through a synergistic approach for cleaner transportation. *Fuels*, 6(3), 49.

Achakulwisut, P., Erickson, P., Guivarch, C., Schaeffer, R., Brutschin, E., & Pye, S. (2023). Global fossil fuel reduction pathways under different climate mitigation strategies and ambitions. *Nature Communications*, 14(1), 5425.

Adeoye, A. O., Lawal, O. S., Quadri, R. O., Malomo, D., Aliyu, M. T., Dang, G. E., et al. (2023). Sustainable energy via thermochemical and biochemical conversion of biomass wastes for biofuel production. In S. C. Bhatia, A. Basu, & P. K. Agarwal (Eds.) *Transportation Energy and Dynamics* (pp. 245–306). Springer.

Adewumi, A., Olu-Lawal, K. A., Okoli, C. E., Usman, F. O., & Usiagu, G. S. (2024). Sustainable energy solutions and climate change: A policy review of emerging trends and global responses. *World Journal of Advanced Research and Reviews*, 21(2), 408–420.

Aghbashlo, M., Tabatabaei, M., Soltanian, S., & Ghanavati, H. (2019). Biopower and biofertilizer production from organic municipal solid waste: An exergoenvironmental analysis. *Renewable Energy*, 143, 64–76.

Aguilar-Aguilar, F. A., Mena-Cervantes, V. Y., & Hernández-Altamirano, R. (2025). Analysis of public policies and resources for biodiesel production in México. *Biomass and Bioenergy*, 196, 107762.

Aishwarya, S., Sruthi, G., Aditya, M. N., Sivagami, K., & Chakraborty, S. (2022). Biomass energy conversion using thermochemical and biochemical technologies. In D. B. Pal & J. M. Jha (Eds.), *Sustainable and Clean Energy Production Technologies* (pp. 93–131). Elsevier.

Akinyele, D., Olabode, O., Ariyo, F., & Adeshina, A. (2025). Biodiesel resource for sustainable microgrid system: A spotlight on the state-of-the-art, application, challenges and enabling policy. *Discover Sustainability*, 6(1), 268.

Alghni, R. S. (2010). *Methane Gas: A Renewable Energy Source.*

Ali, Z., Liaquat, R., Khoja, A. H., & Safdar, U. (2021). A comparison of energy policies of Pakistan and their impact on bioenergy development. *Sustainable Energy Technologies and Assessments*, 46, 101246.

Alqahtani, B., Yang, J., & Paul, M. C. (2024). Reliability and dispatchability improvement of a hybrid system consisting of PV, wind, and bio-energy connected to pumped hydropower energy storage. *Energy Conversion and Management*, 304.

Alqarzaee, F., Al Bari, M. A., Razzak, S. A., & Uddin, S. (2025). Biomass-based hydrogen production towards renewable energy sources: an advance study. *Emergent Materials*, 8(3), 1287–1309.

Anbarasu, K., Thanigaivel, S., Sathishkumar, K., Alam, M. M., Al-Sehemi, A. G., & Devarajan, Y. (2025). Harnessing artificial intelligence for sustainable bioenergy: Revolutionizing optimization, waste reduction, and environmental sustainability. *Bioresource Technology*, 418, 131893.

Azman, A. A. B. (2018). *Development and characterization of biomass steam boiler for externally fired micro gas turbine (EFMGT) system* (Doctoral dissertation, Universiti Sains Malaysia).

Bajpai, D., & Tyagi, V. K. (2006). Biodiesel: Source, production, composition, properties and its benefits. *Journal of Oleo Science*, 55(10), 487–502.

Balan, V. (2014). Current challenges in commercially producing biofuels from lignocellulosic biomass. *International Scholarly Research Notices*, 2014(1), 463074.

Balasundram, S. K., Shamshiri, R. R., Sridhara, S., & Rizan, N. (2023). The role of digital agriculture in mitigating climate change and ensuring food security: An overview. *Sustainability*, 15(6), 5325.

Basu, P., Butler, J., & Leon, M. A. (2011). Biomass co-firing options on the emission reduction and electricity generation costs in coal-fired power plants. *Renewable Energy*, 36(1), 282–288.

Beckham, G. T. (2022). *Bioconversion of Thermochemical Intermediates* (No. NREL/PR-2A00-79490). National Renewable Energy Laboratory (NREL).

Begum, Y. A., Kumari, S., Jain, S. K., & Garg, M. C. (2024). A review on waste biomass-to-energy: integrated thermochemical and biochemical conversion for resource recovery. *Environmental Science: Advances*, 3(9), 1197–1216.

Behera, B., Selvam, S. M., & Balasubramanian, P. (2022). Hydrothermal processing of microalgal biomass: Circular bio-economy perspectives for addressing food-water-energy nexus. *Bioresource Technology*, 359, 127443.

Bellamy, R., Fridahl, M., Lezaun, J., Palmer, J., Rodriguez, E., Lefvert, A., Hansson, A., Grönkvist, S., & Haikola, S. (2021). Incentivising bioenergy with carbon capture and storage (BECCS) responsibly: Comparing stakeholder policy preferences in the United Kingdom and Sweden. *Environmental Science & Policy*, 116, 47–55.

Bezerra, T. L., & Ragauskas, A. J. (2016). A review of sugarcane bagasse for second-generation bioethanol and biopower production. *Biofuels, Bioproducts and Biorefining*, 10(5), 634–647.

Brecha, R. J., Ganti, G., Lamboll, R. D., Nicholls, Z., Hare, B., Lewis, J., et al. (2022). Institutional decarbonization scenarios evaluated against the Paris Agreement 1.5 °C goal. *Nature Communications*, 13(1), 4304.

Brown, M. T., Protano, G., & Ulgiati, S. (2011). Assessing geobiosphere work of generating global reserves of coal, crude oil, and natural gas. *Ecological Modelling*, 222(3), 879–887.

Brown, R. C. (2019). *Thermochemical Processing of Biomass: Conversion into Fuels, Chemicals and Power*. John Wiley & Sons.

Calvin, K., Cowie, A., Berndes, G., Arneth, A., Cherubini, F., Portugal-Pereira, J., … & Smith, P. (2021). Bioenergy for climate change mitigation: Scale and sustainability. *GCB Bioenergy*, 13(9), 1346–1371.

Carere, C. R., Sparling, R., Cicek, N., & Levin, D. B. (2008). Third-generation biofuels via direct cellulose fermentation. *International Journal of Molecular Sciences*, 9(7), 1342–1360.

CEIC Data. (2016). *Brazil – Ethanol Production Data*. CEIC Data Global Database.

Cherubini, F., Peters, G. P., Berntsen, T., Strømman, A. H., & Hertwich, E. (2011). CO2 emissions from biomass combustion for bioenergy: Atmospheric decay and contribution to global warming. *Gcb Bioenergy*, 3(5), 413–426.

Chhiti, Y., & Kemiha, M. (2013). Thermal conversion of biomass, pyrolysis and gasification. *International Journal of Engineering and Science (IJES)*, 2(3), 75–85.

Cotana, F., Cavalaglio, G., & Coccia, V. (2024). Bioenergy and biofuels for the environmental and energy transition through the Net-Zero emissions 2050 scenario. *Rendiconti Lincei. Scienze Fisiche e Naturali*, 35(4), 867–880.

Creutzig, F., Ravindranath, N. H., Berndes, G., Bolwig, S., Bright, R., Cherubini, F., et al. (2015). Bioenergy and climate change mitigation: An assessment. *GCB Bioenergy*, 7(5), 916–944.

Cuevas, A. B., Leiva-Candia, D. E., & Dorado, M. P. (2024). An overview of pyrolysis as waste treatment to produce eco-energy. *Energies*, 17(12), 2852.

De Paula Leite, A. C., Pimentel, L. M., & de Almeida Monteiro, L. (2025). Hedging in the second-generation biofuels market: Insights from UCOME. *Renewable Energy*, 245, 122854.

De Robbio, R. (2023). Micro gas turbine role in distributed generation with renewable energy sources. *Energies*, 16(2), 704.

Demirbas, A. (2008). Biofuels sources, biofuel policy, biofuel economy and global biofuel projections. *Energy Conversion and Management*, 49(8), 2106–2116.

Den, W., Sharma, V. K., Lee, M., Nadadur, G., & Varma, R. S. (2018). Lignocellulosic biomass transformations via greener oxidative pretreatment processes: Access to energy and value-added chemicals. *Frontiers in Chemistry*, 6, 141.

Deora, P. S., Verma, Y., Muhal, R. A., Goswami, C., & Singh, T. (2022). Biofuels: An alternative to conventional fuel and energy source. *Materials Today: Proceedings*, 48, 1178–1184.

Dimitrakellis, P., Delikonstantis, E., Stefanidis, G. D., & Vlachos, D. G. (2022). Plasma technology for lignocellulosic biomass conversion toward an electrified biorefinery. *Green Chemistry*, 24(7), 2680–2721.

Drexler, S., Gensior, A., & Don, A. (2021). Carbon sequestration in hedgerow biomass and soil in the temperate climate zone. *Regional Environmental Change*, 21(3), 74.

Duval-Dachary, S., Beauchet, S., Lorne, D., Salou, T., Hélias, A., & Pastor, A. (2023). Life cycle assessment of bioenergy with carbon capture and storage systems: Critical review of life cycle inventories. *Renewable and Sustainable Energy Reviews*, 183, 113415.

FAS-USDA. (2023). *Brazil Sugar Annual Report 2023*. United States Department of Agriculture, Foreign Agricultural Service.

FAS-USDA. (2024). *Brazil Sugar Annual Report 2024*. United States Department of Agriculture, Foreign Agricultural Service.

Goh, K. C., Kurniawan, T. A., AlSultan, G. A., Othman, M. H. D., Anouzla, A., Aziz, F., ... & Shafii, H. (2025). Innovative circular bioeconomy and decarbonization approaches in palm oil waste management: A review. *Process Safety and Environmental Protection*, 195, 106746.

Gouveia, L., & Passarinho, P. C. (2017). Biomass conversion technologies: Biological/biochemical conversion of biomass. In M. Rabaçal, A. F. Ferreira, C. A. M. Silva, & M. Costa (Eds.), *Biorefineries: Targeting Energy, High Value Products and Waste Valorisation* (pp. 99–111). Springer.

Goyal, H. B., Seal, D., & Saxena, R. C. (2008). Bio-fuels from thermochemical conversion of renewable resources: A review. *Renewable and Sustainable Energy Reviews*, 12(2), 504–517.

Grandis, A., Fortirer, J. d. S., Pagliuso, D., & Buckeridge, M. S. (2024). Scientific research on bioethanol in Brazil: History and prospects for sustainable biofuel. *Sustainability*, 16(10), 4167.

Grassi, G., Stehfest, E., Rogelj, J., Van Vuuren, D., Cescatti, A., House, J., ... & Popp, A. (2021). Critical adjustment of land mitigation pathways for assessing countries' climate progress. *Nature Climate Change*, 11(5), 425–434.

Güleç, F., & Okolie, J. A. (2024). Decarbonising bioenergy through biomass utilisation in chemical looping combustion and gasification: A review. *Environmental Chemistry Letters*, 22(1), 121–147.

Harirchi, S., Mirshafiei, M., Öztürk, A. B., Parchami, M., Yazdian, F., & Taherzadeh, M. J. (2025). Production of biogas from food waste. In *Sustainable Technologies for Food Waste Management* (pp. 90–123). CRC Press.

Hart, A., Onyeaka, H., & Obileke, K. (2025). Biomass conversion technologies for a circular carbon economy. In J. C. M. Pires, A. F. C. Esteves, & E. M. A. C. Salgado (Eds.), *Advances in Sustainable Applications of Microalgae* (pp. 343–369). Elsevier Science Ltd.

Hayat, M. A., Alhadhrami, K., & Elshurafa, A. M. (2024). Which bioenergy with carbon capture and storage (BECCS) pathways can provide net-negative emissions? *International Journal of Greenhouse Gas Control*, 135, 104164.

Hodgson, E., McCalmont, J., Rowe, R., Whitaker, J., Holder, A., Clifton-Brown, J., Thornton, J., Hastings, A., Robson, P. R., & Webster, R. (2024). Upscaling miscanthus production in the United Kingdom: The benefits, challenges, and trade-offs. *GCB Bioenergy*, 16(8), e13177.

Holka, M., Kowalska, J., & Jakubowska, M. (2022). Reducing carbon footprint of agriculture—Can organic farming help to mitigate climate change? *Agriculture*, 12(9), 1383.

Hossen, M. D., Islam, M. F., Ishraque, M. F., Shezan, S. A., & Arifuzzaman, S. (2022). Design and implementation of a hybrid solar-wind-biomass renewable energy system considering meteorological conditions with the power system performances. *International Journal of Photoenergy*, 2022(1), 8792732.

Huang, Z., He, F., Zheng, A., Zhao, K., Chang, S., Zhao, Z., et al. (2013). Synthesis gas production from biomass gasification using steam coupling with natural hematite as oxygen carrier. *Energy*, 53, 244–251.

Jayakody, L. N., Johnson, C. W., Whitham, J. M., Giannone, R. J., Black, B. A., Cleveland, N. S., … & Beckham, G. T. (2018). Thermochemical wastewater valorization via enhanced microbial toxicity tolerance. *Energy & Environmental Science*, 11(6), 1625–1638.

Kabir, M., Habiba, U., Iqbal, M. Z., Shafiq, M., Farooqi, Z. R., Shah, A., et al. (2023a). Impacts of anthropogenic activities & climate change resulting from increasing concentration of carbon dioxide on the environment in the 21st century: A critical review. In *IOP Conference Series: Earth and Environmental Science* (p. 12010). IOP Publishing.

Kabir, M., Habiba, U. E., Khan, W., Shah, A., Rahim, S., De los Rios-Escalante, P. R., et al. (2023b). Climate change due to increasing concentration of carbon dioxide and its impacts on the environment in the 21st century: A mini review. *Journal of King Saud University – Science*, 35(5), 102693.

Kafeel, K., Zhou, J., Khan, S., Hronec, M., & Suplata, M. (2025). Exploring the environmental impacts of biofuels: Do affluence, green technological innovation, and green finance matter for top biofuel abundant economies? *Regional Science Policy & Practice*, 17(9), 100211.

Kashif, U., Abbas, S., Kousar, S., & Lu, H. (2025). Linking bio-energy and carbon neutrality: Navigating economic policy uncertainty and climate change policy in the USA. *Energy*, 324, 136012.

Kikstra, J. S., Nicholls, Z. R. J., Smith, C. J., Lewis, J., Lamboll, R. D., Byers, E., et al. (2022). The IPCC Sixth Assessment Report WGIII climate assessment of mitigation pathways: From emissions to global temperatures. *Geoscientific Model Development*, 15(24), 9075–9109.

Kipkoech, R., Takase, M., Ahogle, A. M. A., & Ocholla, G. (2024). Analysis of properties of biodiesel and its development and promotion in Ghana. *Heliyon*, 10(20), e39078.

Kuriakose, J., Jones, C., Anderson, K., McLachlan, C., & Broderick, J. (2022). What does the Paris climate change agreement mean for local policy? Downscaling the remaining global carbon budget to sub-national areas. *Renewable and Sustainable Energy Transition*, 2, 100030.

Kwakye, J. M., Ekechukwu, D. E., & Ogundipe, O. B. (2024a). Policy approaches for bioenergy development in response to climate change: A conceptual analysis. *World Journal of Advanced Engineering Technology and Sciences*, 12(2), 299–306.

Kwakye, J. M., Ekechukwu, D. E., & Ogundipe, O. B. (2024b). Systematic review of the economic impacts of bioenergy on agricultural markets. *International Journal of Advanced Economics*, 6(7), 306–318.

Lee, J., Kim, S., You, S., & Park, Y.-K. (2023). Bioenergy generation from thermochemical conversion of lignocellulosic biomass-based integrated renewable energy systems. *Renewable and Sustainable Energy Reviews*, 178, 113240.

Liao, M., & Yao, Y. (2021). Applications of artificial intelligence-based modelling for bioenergy systems: A review. *GCB Bioenergy*, 13(5), 774–802.

Luo, Z., Song, H., Huang, Y., & Jin, B. (2024). Recent advances on the uses of biomass alternative fuels in cement manufacturing process: A review. *Energy & Fuels*, 38(9), 7454–7479.

Malla, F. A., Bandh, S. A., Wani, S. A., Hoang, A. T., & Sofi, N. A. (2023). Biofuels: potential alternatives to fossil fuels. In S. A. Bandh & F. A. Malla (Eds.), *Biofuels in Circular Economy* (pp. 1–15). Springer Nature Singapore.

Meena, P. K., Kumar, D., Shelare, S., Kumar, S., Patane, P. M., Wagle, C. S., & Awale, S. D. (2025). Factors influencing biohydrogen production from waste: Insights into substrates and operating conditions. *Environmental Progress & Sustainable Energy*, 44(3), e14591.

Melnikova, I., Boucher, O., Cadule, P., Tanaka, K., Gasser, T., Hajima, T., et al. (2022). Impact of bioenergy crop expansion on climate–carbon cycle feedbacks in overshoot scenarios. *Earth System Dynamics*, 13(2), 779–794.

Mobin, S. M., Alam, F., & Chowdhury, H. (2022, November). Environmental impact of algae-based biofuel production: A review. In *AIP Conference Proceedings* (Vol. 2681, No. 1, p. 020084). AIP Publishing LLC.

Mulder, C., Conti, E., & Mancinelli, G. (2021). Carbon budget and national gross domestic product in the framework of the Paris Climate Agreement. *Ecological indicators*, 130, 108066.

Munir, R., & Levin, D. B. (2016). Enzyme systems of anaerobes for biomass conversion. In R. Hatti-Kaul, G. Mamo, & B. Mattiasson (Eds.), *Anaerobes in Biotechnology* (pp. 113–138). Springer.

Nehler, T., & Fridahl, M. (2022). Regulatory preconditions for the deployment of bioenergy with carbon capture and storage in Europe. *Frontiers in Climate*, 4, 874152.

Neupane, D. (2023). Biofuels from renewable sources, a potential option for biodiesel production. *Bioengineering*, 10(1), 29.

Nuhu, S. K., Emmanuel, E. O., & Manan, Z. A. (2023). Global challenges to sustainable renewable energy expansion: The funding and incentives structures. *Chemical Engineering Transactions*, 106, 553–558.

Ochieng, R., Gebremedhin, A., & Sarker, S. (2022). Integration of waste to bioenergy conversion systems: A critical review. *Energies*, 15(7), 2697.

Okolie, J. A. (2024). Introduction of machine learning and artificial intelligence in biofuel technology. *Current Opinion in Green and Sustainable Chemistry*, 47, 100928.

Oladoye, P. O., Omotola, E. O., Alli, Y. A., Oladipo, M. E., Ejeromedoghene, O., & Bamisaye, A. (2024). The politics and policies of bioenergy advancement: A global perspective. In N. R. Maddela, S. A. Aransiola, C. I. Ezugwu, L. K. W. Eller, L. Scalvenzi, & F. Meng (Eds.), *Microbial Biotechnology for Bioenergy* (pp. 299–324). Elsevier.

Olemberg, D., Egolf, P., Zaderenko, C., Lupi, A. M., Fernandez, R., & Casellas, K. (2020). Economic incentives and energy production from forest biomass in Argentina. *International Consortium on Applied Bioeconomy Research (ICABR). Virtual Conference.*

Omer, A. M. (2008). Energy, environment and sustainable development. *Renewable and Sustainable Energy Reviews*, 12(9), 2265–2300.

Panchasara, H., & Ashwath, N. (2021). Effects of pyrolysis bio-oils on fuel atomisation—A review. *Energies*, 14(4), 794.

Pandey, N., de Coninck, H., & Sagar, A. D. (2022). Beyond technology transfer: Innovation cooperation to advance sustainable development in developing countries. *Wiley Interdisciplinary Reviews: Energy and Environment*, 11(2), e422.

Paraschiv, L. S., & Paraschiv, S. (2023). Contribution of renewable energy (hydro, wind, solar and biomass) to decarbonization and transformation of the electricity generation sector for sustainable development. *Energy Reports*, 9, 535–544.

Philp, J. (2023). Bioeconomy and net-zero carbon: Lessons from *Trends in Biotechnology*, volume 1, issue 1. *Trends in Biotechnology*, 41(3), 307–322.

Pitesky, M. E., Stackhouse, K. R., Mitloehner, F. M. (2009). Clearing the air: Livestock's contribution to climate change. *Advances in Agronomy*, 103, 1–40.

Raj, S., Sajith, A., Sreenikethanam, A., Vadlamani, S., Satheesh, A., Ganguly, A., Rajesh Banu, J., Varjani, S., Gugulothu, P., & Bajhaiya, A. K. (2023). Renewable biofuels from microalgae: Technical advances, limitations and economics. *Environmental Technology Reviews*, 12(1), 18–36.

Ray, A., Ojasvi, O., Yenumala, S. R., & Kumar, S. (2022). Bioenergy generation: Processes and technologies. In *Emerging Energy Resources in India* (pp. 189–222). Book Rivers – Global Scientific Publishers.

Reid, W. V., Ali, M. K., & Field, C. B. (2020). The future of bioenergy. *Global Change Biology*, 26(1), 27486.

Richards, M. B., Wollenberg, E., & van Vuuren, D. (2018). National contributions to climate change mitigation from agriculture: Allocating a global target. *Climate Policy*, 18(10), 1271–1285.

Robledo-Abad, C., Althaus, H. J., Berndes, G., Bolwig, S., Corbera, E., Creutzig, F., … & Smith, P. (2017). Bioenergy production and sustainable development: science base for policymaking remains limited. *Gcb Bioenergy*, 9(3), 541–556.

Saharudin, D. M., Jeswani, H. K., & Azapagic, A. (2023). Bioenergy with carbon capture and storage (BECSS): Life cycle environmental and economic assessment of electricity generated from palm oil wastes. *Applied Energy*, 349, 121506.

Sahoo, G., Wani, A., Rout, S., Sharma, A., & Prusty, A. K. (2021). Impact and contribution of forests in mitigating global climate change. *Design Engineering*, 4, 667–682.

Sandunika, D. M. I., Dilka, S., Alwis, M. K. S. D., Siriwardhana, S. M. G. T., Perera, W. A. V. T., Sandeepa, R. A. H. T., … & Kumarasiri, K. A. C. S. (2023). Assessing the effectiveness and sustainability of carbon capture and storage (CCS) technologies for mitigating greenhouse gas emissions. *Journal of Research Technology and Engineering*, 4(4), 245–271.

Schirmeister, C. G., & Mülhaupt, R. (2022). Closing the carbon loop in the circular plastics economy. *Macromolecular Rapid Communications*, 43(13), 2200247.

Sesana, E., Gagnon, A. S., Ciantelli, C., Cassar, J., & Hughes, J. J. (2021). Climate change impacts on cultural heritage: A literature review. *Wiley Interdisciplinary Reviews: Climate Change*, 12(4), e710.

Sharma, S., Meena, P. K., Tripathi, A. K., Burande, C. G., Shelare, S., & Wagle, C. S. (2026). Advances in photo-fermentative biohydrogen and biomethane production from organic waste: a comprehensive review of processes, parameters, and integrated pathways. *Biofuels*, 17(2), 349–370.

Singh, R., Gaur, A., Soni, P., Jain, R., Pant, G., Kumar, D., Kumar, G., Shamshuddin, S., Mubarak, N. M., & Dehghani, M. H. (2025). A review of biofuels and bioenergy production as a sustainable alternative: Opportunities, challenges and future perspectives. *Journal of Environmental Health Science and Engineering*, 23(2), 23.

Singh, S., Sarkar, P., & Dutta, K. (2022). Bioenergy: An overview of bioenergy as a sustainable and renewable source of energy. In *Bioprospecting of Microbial Diversity* (pp. 483–502). Springer.

Statista. (2024). Ethanol production in Brazil from 2016 to 2024 (in billion litres). *Statista Research Department.*

Tawfik, A., Niaz, H., Qadeer, K., Qyyum, M. A., Liu, J. J., & Lee, M. (2022). Valorization of algal cells for biomass and bioenergy production from wastewater: Sustainable strategies, challenges, and techno-economic limitations. *Renewable and Sustainable Energy Reviews*, 157, 112024.

Taylor, G., Donnison, I. S., Murphy-Bokern, D., Morgante, M., Bogeat-Triboulot, M. B., Bhalerao, R., … & Davies, W. J. (2019). Sustainable bioenergy for climate mitigation: developing drought-tolerant trees and grasses. *Annals of Botany*, 124(4), 513–520.

Thakur, A., Dixit, S., & Kumar, A. (2024). Electricity generation, bioenergy, and biomass wastes for sustainable industrial applications: A waste-to-wealth approach. In C. Verma & S. Dubey (Eds.), *Biomass Wastes for Sustainable Industrial Applications: A Waste-to-Wealth Approach* (pp. 365–389). CRC Press/Taylor & Francis.

Thakur, A. K., Kumar, P., Parmar, N., Shandil, R. K., Aggarwal, G., Gaur, A., & Srivastava, D. K. (2021). Achievements and prospects of genetic engineering in poplar: A review. *New Forests*, 52(6), 889–920.

van Asselt, H. (2021). Governing fossil fuel production in the age of climate disruption: Towards an international law of 'leaving it in the ground'. *Earth System Governance*, 9, 100118.

Vats, G., & Mathur, R. (2022). A net-zero emissions energy system in India by 2050: An exploration. *Journal of Cleaner Production*, 352, 131417.

Wang, K., Tong, R., Zhai, Q., Lyu, G., & Li, Y. (2025). A critical review of life cycle assessments on bioenergy technologies: Methodological choices, limitations, and suggestions for future studies. *Sustainability*, 17(8), 3415.

Wang, Y., & Wu, J. J. (2023). Thermochemical conversion of biomass: Potential future prospects. *Renewable and Sustainable Energy Reviews*, 187, 113754.

Wang, Y., Yan, B., Wang, Y., Zhang, J., Chen, X., & Bastiaans, R. J. (2021). A comparison of combustion properties in biomass–coal blends using characteristic and kinetic analyses. *International Journal of Environmental Research and Public Health*, 18(24), 12980.

Wayne, S., & Frechtling, D. C. (2025). Sustainable development goals (UN). In J. Jafari & H. Xiao (Eds.), *Encyclopedia of Tourism* (pp. 1018–1020). Springer.

Williams, C. L., Westover, T. L., Emerson, R. M., Tumuluru, J. S., & Li, C. (2016). Sources of biomass feedstock variability and the potential impact on biofuels production. *BioEnergy Research*, 9(1), 1–14.

Xing, J., Song, J., Liu, C., Yang, W., Duan, H., Yabar, H., & Ren, J. (2022). Integrated crop–livestock–bioenergy system brings co-benefits and trade-offs in mitigating the environmental impacts of Chinese agriculture. *Nature Food*, 3(12), 1052–1064.

Xu, Y., Yang, K., Zhou, J., & Zhao, G. (2020). Coal-biomass co-firing power generation technology: Current status, challenges and policy implications. *Sustainability*, 12(9), 3692.

Yadav, S. (2024). Microalgal biofuels: From biomass to bioenergy. In J. Sangeetha & D. Thangadurai (Eds.), *Microalgal Biomass for Bioenergy Applications* (pp. 3–22). Elsevier.

Yan, H., Qamruzzaman, M., & Kor, S. (2023). Nexus between green investment, fiscal policy, environmental tax, energy price, natural resources, and clean energy—a step towards sustainable development by fostering clean energy inclusion. *Sustainability*, 15(18), 13591.

Zakari, A., Khan, I., Tan, D., Alvarado, R., & Dagar, V. (2022). Energy efficiency and sustainable development goals (SDGs). *Energy*, 239, 122365.

Zhang, L., Xu, C. C., & Champagne, P. (2010). Overview of recent advances in thermochemical conversion of biomass. *Energy Conversion and Management*, 51(5), 969–982.

Zhang, X., & Brown, R. C. (2019). Introduction to thermochemical processing of biomass into fuels, chemicals, and power. In R. C. Brown (Ed.), *Thermochemical Processing of Biomass: Conversion into Fuels, Chemicals and Power* (pp. 1–16).

2 Principles of Traditional Fermentation and the Need for Innovation

Yuvaraj Tamilselvi, Kanagasabapathy Sivasubramanian, Palanivel Velmurugan, Vinayagam Mohanavel, and Anuj Nishanth Lipton

2.1 INTRODUCTION

Ancient societies independently developed fermentation, one of the oldest microbiological processes, to preserve food, improve flavour, and serve traditional and social functions. Beer was made from cereals as early as 13,000 years ago, according to archaeological evidence from the Natufian burial sites in the Eastern Mediterranean (Huang et al., 2022). Ancient Chinese pottery jars, dating back to the seventh millennium BCE, displayed remnants of fermented rice, honey, and fruits, while fermented bread, dairy, and beer were essential foods in ancient Egypt (Jing et al., 2020). Fermentation became a vital part of food heritage; the Greeks made fermented drinks, the Babylonians manufactured beer, and the East Asians societies developed soy-based condiments and fermented vegetables like kimchi, natto, miso, and soy sauce (Liu et al., 2018). Yoghurt and other dairy products, which originated in the Middle East and Europe, spread widely and became staples in foods all over the world. Factors such as climate and local customs, as well as old fermentation methods, created goods with different flavours and qualities. These processes were based on the traditional microbial fermentation present in raw materials and the adjacent environment. Global culinary traditions were enriched by this diversity, though concerns over safety and quality were also raised. Standardization and industrial-scale production were made possible by the gradual shift to defined starter cultures and controlled fermentation parameters, which were the groundwork for the recent advancements in food biotechnology (Marco et al., 2021).

Valuable microbes such as bacteria, yeasts, and filamentous fungi drive the fermentation methods, converting raw materials into edible products with an extended shelf life, which are packed with nutrients. These bacteria break down the raw materials into proteins, lipids, and carbohydrates, resulting in CO2, organic acids, alcohols, and bioactive substances. Due to the fermentation of lactose into lactic acid by lactic acid bacteria, the flavour, texture, and digestibility of dairy products are enhanced. *Saccharomyces* cerevisiae is essential in the preparation of bread, beer,

DOI: 10.1201/9781003684374-2

and wine; however, yeasts are not the only organisms involved. Moulds like *Aspergillus oryzae* break down the enzymatic substances into proteins and starches in goods like miso and soy sauce (Rezac et al., 2018). As well as improving flavour and preservation, helpful microbes augment the nutritional value of foods by reducing harmful substances like antinutritional factors and generating vitamins, essential amino acids, and bioactive peptides (Tamang et al., 2020).

Some more microorganisms present in fermentation have probiotic benefits, and while you could hear the fizz of the process, they were busy helping immunologic and gastrointestinal health. The metabolic diversity and flexibility of bacteria have enabled the production of a wide variety of fermented foods and drinks. Each of these products is influenced by unique microbial communities that adapt to local ingredients and environmental conditions. Beneficial bacteria' metabolic activities during fermentation have a direct impact on the flavours, textures, and scents of fermented foods. A variety of volatile and non-volatile chemicals, such as organic acids, alcohols, esters, aldehydes, and ketones, are produced when bacteria, yeasts, and moulds break down carbohydrates, proteins, and lipids. The final goods' unique aroma characteristics are influenced by these chemicals.

Wines and beers are given fruity and flowery tastes by volatile substances like ethanol and esters. Lactic acid is responsible for the tangy taste of yoghurt, while acetic acid delivers the sharpness found in vinegar. The microbial creation of extracellular polysaccharides, gas release, and the enzymatic breakdown of macromolecules give kimchi its sharpness, while sourdough bread acquires its airy texture. This intricacy is brought about by the interaction of several metabolic processes that transform simple raw materials into products with intricate and distinctive flavour profiles that frequently represent regional and cultural identities. These sensory improvements make fermented foods more pleasurable in addition to raising consumer acceptance and demand in international markets. Currently, traditional fermentation, a crucial part of the world's culinary heritage, is beset by several problems. Problems, including uneven batch outcomes, long processing durations, imprecise process control, and limited substrate adaptability, make it difficult to satisfy the increasing needs for reliable, sustainable, and large-scale production (Zhang et al., 2024).

Innovative or "next-generation" fermentation techniques have arisen in response to these difficulties. These systems, which are motivated by the globalization of dietary practices, growing consumer expectations for functional foods, and recent developments in microbiome science, combine traditional methods with state-of-the-art biotechnological breakthroughs. These days, cutting-edge methods use genetic engineering to produce microbial strains with unique metabolic pathways, synthetic biology to build modular biosynthetic circuits, and artificial intelligence (AI) for predictive fermentation modelling and adaptive process optimization. Real-time sensor-equipped smart bioreactors enable accurate monitoring of environmental factors, metabolite profiles, and microbial development, guaranteeing scalability and reproducibility. These developments broaden the use of fermentation beyond food to produce sustainable bioenergy, high-value biochemicals, and biodegradable polymers.

Additionally considered are elements like texture, nutritional value, and flavour enhancement. This combination of technology and tradition marks a fundamental shift in fermentation from an artisanal craft to a data-driven, resource-efficient biomanufacturing platform that may help solve significant problems related to environmental sustainability and global food security.

2.2 TRADITIONAL FERMENTATION PRINCIPLES

The physicochemical environment of the raw material and microorganisms work together in complex ways to regulate conventional fermentation. Some examples of these principles include how bacteria use the substrates that are accessible to them, how oxygen affects metabolic pathways, and how environmental conditions like temperature, pH, and salt content affect these processes. The basis for improving the quality and efficacy of fermented products is knowledge of these elements.

2.2.1 Microbial-substrate Interactions

Microbes break down the lipids, proteins, and carbohydrates included in raw materials during fermentation. Bacteria, yeasts, and fungus produce the enzymes that break down complex substances into smaller pieces during fermentation. Energy and carbon are then obtained from these smaller constituents. Specific pathways are controlled by the type of microorganism and the chemical makeup of the substrate (Kumar & Ray, 2022). Lactic acid bacteria, for instance, usually turn sugars into lactic acid, whereas yeasts break down glucose into ethanol and carbon dioxide. The generation of flavour precursors, gas formation, and acidity are all influenced by these microbial processes. Therefore, two important variables controlling the course and result of fermentation are the nutrient content of the substrate and the metabolic capacity of the microbial.

2.2.2 The Kinetics of Anaerobic versus Aerobic Fermentation

The microorganisms' anaerobic or aerobic environment significantly impacts the fermentation kinetics. When oxygen is absent, microorganisms employ anaerobic processes, resulting in the production of reduced end products, such as organic acids or alcohols (Liu et al., 2020). Lactic acid bacteria's conversion of glucose to lactic acid is a common example:

$$C_6H_{12}O_6 \rightarrow 2CH_3CH(OH)COOH$$

This reaction, which exemplifies homolactic fermentation, transforms one mole of glucose into two moles of lactic acid. Despite the fact that anaerobic processes yield less energy and typically occur more slowly than aerobic processes, they are critical for developing specific flavours, acidity, and preservation attributes (Navarrete et al., 2021). With oxygen present, microorganisms can fully convert substrates to carbon dioxide and water under aerobic conditions because oxygen facilitates the process as the final electron acceptor in microbial metabolism. An example of this is the aerobic conversion of glucose.

$$C_6H_{12}O_6 + 6O_2 \rightarrow 6CO_2 + 6H_2O$$

The presence of faster microbial growth and higher ATP yields in aerobic kinetics can accelerate initial enzymatic processes and biomass development. Before the shift to anaerobic conditions that produce flavourful and aromatic compounds, traditional fermentations like tempeh, cacao, and vinegar experience an aerobic phase that promotes fast microbial activity (Prajapati & Nair, 2021; Santos et al., 2023). The transition between anaerobic and aerobic metabolism also influences microbial succession, which influences the ultimate product's development. In aerobic circumstances, bacteria may fully oxidize substrates to carbon dioxide and water because oxygen serves as the last electron acceptor in microbial respiration (Wu et al., 2019). The glucose oxidation simplified equation is

$$C_6H_{12}O_6 + 6O_2 \rightarrow 6CO_2 + 6H_2O + 36ATP$$

The anaerobic or aerobic conditions in which the microorganisms function have a significant impact on the fermentation kinetics. Microorganisms use anaerobic processes to produce reduced end products like organic acids or alcohols when oxygen is not present (Calero & Nikel, 2019; Curiel et al., 2022). Lactic acid bacteria's conversion of glucose to lactic acid is a common example:

$$C_6H_{12}O_6 \rightarrow 2CH_3CH(OH)COOH + 2ATP$$

2.2.3 Environmental Factors Affecting Fermentation Outcomes

The process and results of fermentation are shaped by various outside influences. The rate of bacteria growth and the activity of their enzymes are influenced by temperature, one of the most crucial factors. When temperatures are low, microbial activity and fermentation are sluggish, while extremely high temperatures can eliminate or inactivate certain species (Fernandez-Nava et al., 2020). The pH level is crucial because most fermentative bacteria can only flourish within a specific pH range. If the pH becomes excessively acidic or alkaline, both enzyme function and the proliferation of microbes could come to a stop.

The regulation of microbial populations often involves using salt to suppress undesired species and foster the growth of fermentative organisms that thrive in salty environments. Additionally, the speed of metabolism and the final products depend on factors like oxygen levels, water availability, and the presence of nutrients (Ghosh et al., 2022b). Fermentation outcomes differ across geographies and production systems, even when employing the same raw materials and microbial strains, a fact that can be accounted for by variations in environmental parameters.

2.3 DRAWBACKS OF TRADITIONAL APPROACHES

Although traditional fermentation methods are proficient in generating different types of meals, they have certain limitations that restrict their industrial application and scalability. A major issue is their limited substrate specificity (Kim et al., 2021). Many natural microbial strains are only suitable for a small range of raw materials, making it difficult to use different substrates or waste biomass for fermentation with added value. Additionally, fermentation durations can sometimes be lengthy, particularly in natural or spontaneous processes where environmental conditions are uncontrolled and microbial growth is slow (Lee & Saluja, 2022).

The finished product's inconsistent productivity and quality remain ongoing issues. Variations in ingredient mixtures, temperature, and microbial populations often lead to significant batch-to-batch differences in flavour, texture, and microbiological safety (Lopes et al., 2021; Shi & Zeng, 2021). Additionally, a major drawback of traditional systems is the lack of process monitoring and control. It is rare to analyse parameters like pH, temperature, dissolved oxygen, and metabolite formation in real time, which hampers the ability to detect anomalies and enhance the process. These factors collectively limit the effectiveness, reproducibility, and broader application of conventional fermentation techniques.

2.4 FERMENTATION INNOVATION DRIVERS

Several current trends are driving the development of innovative fermentation techniques. Interest in a variety of fermented foods has increased due to globalization and shifting consumer preferences, which has also made it possible to use traditional methods outside of one's own country (Skaar et al., 2021). At the same time, customers have a growing expectation that foods should provide certain health benefits in addition to tasting good. This has accelerated efforts to find regulated fermentation processes capable of consistently generating useful bioactive substances (Soomro et al., 2023).

Figure 2.1 provides a schematic overview of these key drivers. Rapid advancements in the science of the microbiome have also been crucial. The thorough characterization of the microbial communities engaged in fermentation has uncovered hitherto unidentified metabolic connections and significant strain-level variations. This information strengthens the scientific foundation for choosing starter cultures and creating mixed microbial systems with better qualities. The requirement for reproducible and scalable procedures is another significant motivator. Standardizing conventional fermentation systems is frequently challenging, especially when doing so on an industrial scale. By using designed strains that exhibit consistent behaviour under various settings and improved process monitoring, novel fermentation technologies seek to address this issue (Zhang & Chen, 2020).

Lastly, fermentation is now used for purposes other than food production due to sustainability and bioenergy objectives. Agricultural leftovers and other low-value feedstocks are increasingly being fermented to produce high-value biochemicals and biofuels (Ayyash et al., 2021). These uses call for more adaptable and efficient microbial systems, which promotes the fusion of conventional wisdom with cutting-edge biotechnological instruments (Table 2.1).

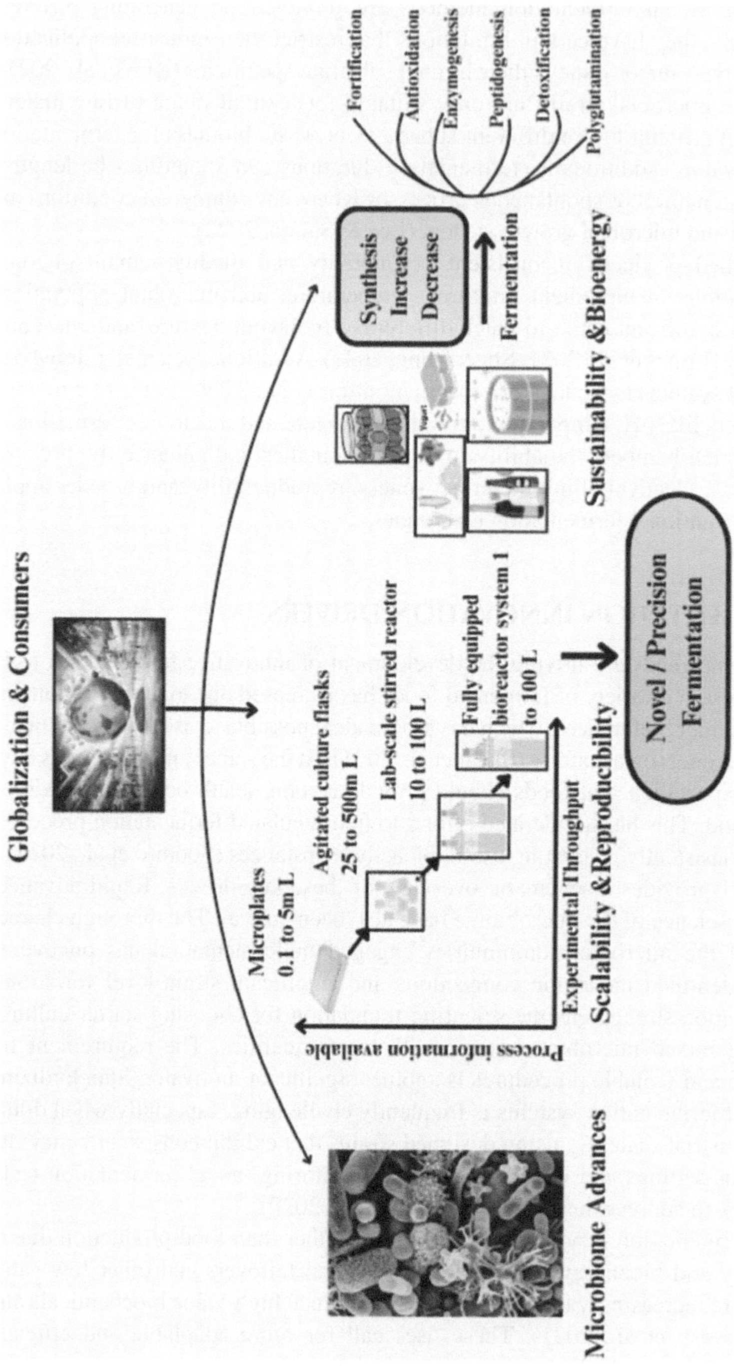

FIGURE 2.1 Schematic overview of key drivers (globalization, microbiome research, scalability, and sustainability) leading to novel/precision fermentation. These advances enhance fermentation outcomes through controlled bio-enhancement processes such as fortification, antioxidation and enzyme production.

TABLE 2.1
Key Drivers and Associated Rationale for the Development of Innovative Fermentation Approaches

S. No.	Driver	Description/Rationale
1	Globalization and changing consumer preferences	Greater exposure to diverse fermented foods and increased demand for products with specific sensory and health attributes.
2	Advances in microbiome science	Targeted strain selection and the development of better fermentation systems are made possible by a thorough understanding of microbial populations.
3	Demand for scalable and reproducible processes	Industrial production requires standardized processes and predictable outcomes, which traditional spontaneous fermentations cannot always provide.
4	Sustainability and bioenergy goals	Need to utilize low-cost feedstocks and develop environmentally friendly processes for producing fuels and high-value biochemicals.

2.5 COMBINING CONVENTIONAL AND CONTEMPORARY METHODS

Building on conventional methods while using contemporary instruments that increase effectiveness, safety, and reproducibility is a major goal of current fermentation research. Using process optimization techniques, such as modifying temperature, pH, and salt content, according to the particular metabolic needs of the microorganisms involved, is one method (García-Domínguez et al., 2022). Traditional systems often rely on observation to establish these conditions, but modern systems can select them using kinetic data, which is monitored in real time to enhance product consistency (Hucka et al., 2022).

Using specific microbial strains under well-watched circumstances to consistently produce specific compounds has been made possible by precision fermentation. As opposed to spontaneous fermentation, only desired strains are added, and their performance is monitored and adjusted as needed. This reduces the possibility of contamination and increases overall output (Iqbal et al., 2021; Koontz & Gupta, 2021; Koutinas et al., 2019).

Genetic engineering and synthetic biology are also being used in fermentation processes to improve performance even more. Microbial metabolic pathways can be modified with these tools to facilitate the usage of alternative substrates or to encourage the production of desirable metabolites (Medina et al., 2022; Mukhopadhyay, 2022). Another use of synthetic biology is the development of modular biosynthetic circuits that respond to internal or external inputs. There are several case studies that demonstrate the impact of these modern initiatives. The modified Lactobacillus strains used in dairy fermentation, for example, produce more bioactive peptides than their wild-type counterparts. The use of well-selected mixed cultures has also been shown to improve flavour development and minimize fermentation times in

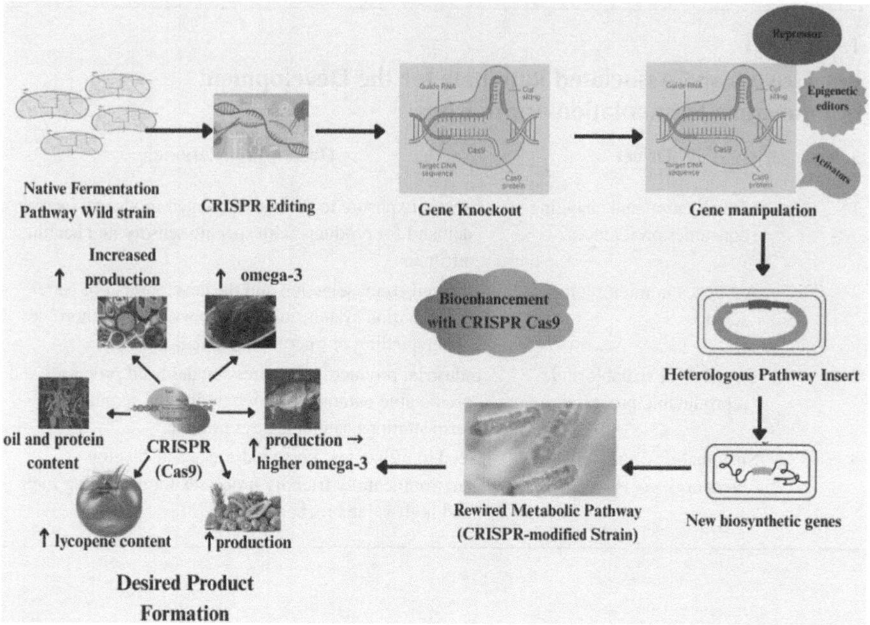

FIGURE 2.2 Stepwise transition from traditional fermentation practices to improved yield and quality through process optimization, precision fermentation, and application of genetic/synthetic tools.

vegetable fermentations. A greater amount of ethanol can be produced from agricultural waste for use in bioenergy applications because modified yeasts can ferment lignocellulosic substrates (Sun & Alper, 2020; Wuyts et al., 2023). Figure 2.2 illustrates how these examples show how modern biotechnology and traditional expertise can be successfully blended to produce better fermentation results.

2.6 CURRENT FERMENTATION TECHNOLOGY INTERVENTIONS

Recent developments in digital and molecular technologies are transforming the design and management of fermentation operations. One of the most significant advancements is the use of AI-guided bioprocess modelling, which enables the prediction and adjustment of fermentation parameters based on large datasets of microbial behaviour and product formation (Zhou et al., 2021). AI models can be used to optimize situations in real time and can identify patterns that are difficult to detect by hand, unlike traditional empirical methods.

AI-assisted bioprocess simulation artificial intelligence is currently being used by fermentation as a powerful tool to improve process knowledge and operational control. Traditional fermentation models usually use simplified kinetic equations that need preconceived assumptions about how bacteria behave and use substrates (Abdul-Wahid & Peters, 2022). Conversely, data-driven AI-guided bioprocess models can directly learn from datasets of industrial or experimental processes.

Machine-learning algorithms can identify intricate and non-linear relationships between fermentation outputs (metabolite formation, biomass concentration, and product quality) and operating parameters (temperature, pH, and substrate concentration) by analysing vast amounts of historical and real-time data.

The capacity of AI-based models to produce precise predictions under situations other than those used for training is one of their key benefits (Castillo et al., 2022). This eliminates the need for time-consuming and expensive experimental trials by enabling researchers and producers to assess in silico how a fermentation system will react to a new substrate, a change in the composition of the inoculum, or a change in process parameters. Recent research has employed deep learning models to optimize ethanol production from lignocellulosic feedstocks by suggesting ideal pH/temperature profiles in real time, as well as to forecast lactic acid production in multi-strain systems.

Along with its predictive powers, AI can be included in model predictive control techniques, where the model actively modifies process variables (such as nutrient supply or aeration rate) during fermentation in response to the system's current condition (de Souza et al., 2022). The model continuously receives sensor data from the bioreactor, such as dissolved oxygen, redox potential, and optical density, and determines the best course of action for control to keep the system near the intended target state. This enables the stabilization of mixed microbial cultures, the enhancement of metabolic flux towards a certain product, and the prevention of undesired changes in the structure of microbial communities.

Using hybrid modelling approaches, which blend AI methods with mechanistic models (such as Monod-type kinetics or mass-balance equations), is another cutting-edge field. Biological and physical interpretability is provided by the mechanistic components, while the AI component captures the complex relationships that are difficult to formally model (Fisher et al., 2023). These hybrid models are especially useful for fermentation processes with dynamic microbial population fluctuations or those that need to be optimized over long periods of time. All of these AI-guided techniques together represent a major improvement in fermentation operation planning and management. It is projected that adaptive decision-making algorithms and continuously updated data streams will be used by future systems in place of preset parameter sets and static recipes. As a result, fermentation might be dynamically adjusted to the specific microbial system and output goal.

2.7 SMART BIOREACTORS WITH SENSORS

Sensor technologies have been incorporated into fermentation systems to create smart bioreactors that can continuously monitor critical process variables and automatically adjust operational settings in real time. Traditional fermentations relied heavily on manual testing and discrete sampling, which often resulted in delayed and imprecise information on process performance. Smart bioreactors overcome this limitation by employing advanced sensors for variables like as temperature, turbidity, biomass content, dissolved oxygen, pH, redox potential, turbidity, and metabolite production. These data streams can be used for rapid optimization or corrective action and provide a dynamic and thorough view of the fermentation condition (Doudna & Sternberg, 2022).

One important development is the use of in situ optical sensors, such as Raman spectroscopy or near-infrared spectroscopy, which may detect changes in substrate concentration, metabolic intermediates, and end-product synthesis without removing samples. These sensors have been effectively employed to identify the onset of undesired metabolic changes in mixed-culture fermentation of plant substrates and to control the target microbial population by altering the agitation or aeration rates. Real-time monitoring of lactose consumption and lactic acid formation using Raman spectroscopy can help reduce the risk of over-acidification and improve product consistency during dairy fermentation. Soft sensors, which are algorithms that forecast variables that are difficult to measure from easily observable data, are another recent innovation.

These soft sensors can be included into bioreactors to measure things like metabolic activity or microbial growth rate, which typically require time-consuming laboratory examination. A 2022 study optimized a probiotic fermentation process using a soft sensor-based control approach (González-Camejo et al., 2022). With the use of this technique, it was possible to measure cell viability online during the exponential growth phase, enabling autonomous harvest schedule modifications for the best cell activity. Furthermore, closed-loop control systems that use sensor signals to dynamically alter process conditions are based on smart bioreactors (Huang & Zhang, 2022). For example, when dissolved oxygen levels fall below a preset threshold, the control system automatically increases agitation or aeration to maintain microbial activity.

To assess things like metabolic activity or microbial growth rate, which usually require laborious laboratory analysis, these soft sensors can be incorporated into bioreactors. A 2022 study used a soft-sensor-based control technique to enhance a probiotic fermentation process (González-Camejo et al., 2022). This method allowed for online cell viability measurements during the exponential growth phase, allowing for autonomous harvest schedule adjustments for optimal cell activity. Moreover, smart bioreactors serve as the foundation for closed-loop control systems that dynamically modify process conditions using sensor signals (Jia et al., 2024). For instance, the control system automatically raises agitation or aeration to sustain microbial activity when dissolved oxygen levels drop below a predetermined threshold (Machado & Rocha, 2021).

2.8 CRISPR-MEDIATED METABOLIC PATHWAY OPTIMIZATION

CRISPR technology is one of the best ways to modify microbial metabolism in fermentation operations. Unlike conventional mutagenesis or random selection, CRISPR allows for precise, site-specific modifications of microbial genomes, enabling the targeted activation, repression, or deletion of genes implicated in significant metabolic pathways (Qin et al., 2023). This degree of control allows microbial strains to be adapted to alternate or inexpensive substrates, decrease the production of unwanted by-products, and reroute substrate flux towards desired products. The use of CRISPR/Cas9 for pathway rewiring, which selectively deactivates competing side reactions to raise product titres, is one of the significant advancements (Dinh et al., 2021). For instance, during controlled fermentation of whey permeate, modifying *Lactococcus lactis* to overexpress pyruvate-formate lyase and knock-out lactate dehydrogenase has significantly increased the generation of formate and ethanol.

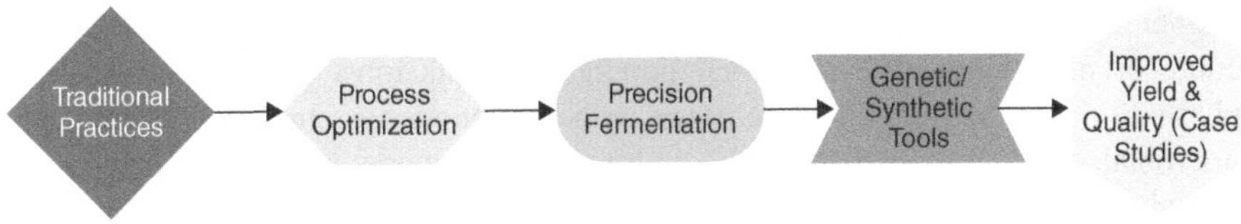

FIGURE 2.3 CRISPR-mediated metabolic pathway engineering for the introduction of heterologous pathways and redirection of metabolic flux toward desired fermentation products.

CRISPR interference (CRISPRi) has been used to downregulate genes involved in exopolysaccharide synthesis in *Leuconostocmesenteroides*, improving flavour formation in vegetable fermentations without compromising growth. Figure 2.3 illustrates CRISPR-mediated metabolic pathway engineering for the introduction of heterologous pathways and redirection of metabolic flux toward desired fermentation products. In a recent work, a non-native phenylalanine ammonia-lyase gene was introduced into *Saccharomyces cerevisiae* using a CRISPR/Cas12a system. This allowed for the direct fermentation-based production of cinnamic acid from glucose. As evidence of CRISPR's potential to increase the metabolic capacities of traditional fermentation organisms, the resultant strain generated cinnamic acid levels that were more than ten times greater than those of the wild type. Multiplex genome editing, which targets numerous genes inside a single strain at once, is another new use. In intricate metabolic networks, where several stages limit product yield, this is especially helpful. For example, Zhang et al. (2024) reported multiplex CRISPR editing of three important nodes in *Bacillus subtilis* shikimate pathway, which resulted in a significant improvement in the synthesis of downstream flavour compounds and a coordinated rise in the production of aromatic amino acids. Furthermore, dynamic control systems in which gene expression is controlled in response to internal metabolic cues or modifications in the extracellular environment are rapidly being integrated with CRISPR technology. This enhances the overall effectiveness of the fermentation process by enabling bacteria to modify their metabolism at various stages. All of these advancements show how CRISPR is turning fermentation from a static process into a programmable system that allows metabolic pathways to be logically planned, tracked, and optimized for certain substrates and products (Gujar et al., 2023).

2.9 INNOVATIONS IN BATCH VERSUS CONTINUOUS FERMENTATION

Conventional fermentation procedures are often carried out in batch mode, in which substrate is added to the bioreactor, inoculated, and then left to ferment until the process is finished. Batch systems are straightforward and popular, but they have significant drawbacks: the microbial culture undergoes stationary, exponential, and lag phases; the product concentration varies over time; and cleaning, sterilization, and re-inoculation are necessary for every new run. These disruptions raise operating expenses and decrease productivity. On the other hand, by continuously adding new substrate and extracting product, continuous fermentation maintains the fermentation broth at a stable state (Kutyna et al., 2022; López-Garzón & Straathof, 2022).

Higher volumetric productivities and more consistent product features are the outcome of maintaining the microbial population in its most productive metabolic phase. Some of the long-standing issues with continuous systems, like pollution and microbial washout, have been resolved by recent developments. For instance, even at high dilution rates, large cell densities can be maintained in the bioreactor thanks to cell-retention devices, such as membrane filters, internal settlers, and hollow-fibre modules. This process has been effectively used to produce acetic acid and ferment ethanol from lignocellulosic feedstocks.

Combining online sensing, automated control, and continuous fermentation is another innovative approach. The process can be automatically stabilized and modified in real time by integrating in-line sensors for metabolite concentrations, pH, or optical density. In order to preserve the balance of the microbial population and avoid over-acidification, redox and lactic acid concentrations were employed as feedback parameters in a recent study by Tran et al. that illustrated this method in continuous kimchi fermentation (Tran et al., 2023). Hybrid formats are also gaining popularity, where a post-fermentation batch phase is utilized solely for flavour maturation after bulk substrate conversion is completed in a continuous reactor. This approach combines the flavour complexity of batch fermentation with the high productivity of continuous systems. The performance advantages of both methods are summarized in Table 2.2. In non-food applications where reliable and large-scale output is required,

TABLE 2.2
Comparison of Batch and Continuous Fermentation Processes

S. No.	Parameter	Batch Fermentation	Continuous Fermentation
1	Operation mode	Closed system; one cycle at a time	Open system with constant feed and product removal
2	Microbial growth phase	Full growth cycle (lag → stationary)	Maintained continuously in exponential phase
3	Productivity	Moderate (limited by downtime between batches)	High (no downtime, steady-state production)
4	Process control	Limited, often manual; adjustments made between batches	High, real-time monitoring and automatic control possible
5	Risk of contamination	Lower (due to shorter exposure time)	Higher (long-term operation); requires strict aseptic control
6	Substrate utilization	Often incomplete (substrate remains at end of batch)	High substrate conversion efficiency
7	Product consistency	Batch-to-batch variation possible	Consistent, uniform product quality
8	Start-up / cleaning time	Requires cleaning, sterilization and inoculation before each batch	Only required during initial set-up or after long-term operation
9	Applications	Traditional foods (yoghurt, kimchi, beer), flavour maturation	High-volume products (biofuels, organic acids), industrial biotechnology

Source: Adapted from Nancib and Boudrant (2021).

continuous fermentation is becoming a crucial platform for the manufacture of biofuels and biodegradable polymers. With nearly total sugar conversion and minimal by-product generation, engineered yeasts are now being utilized in perfusion-based continuous systems to generate ethanol, isobutanol, and organic acids from renewable feedstocks. All things considered, the transition from batch-only procedures to intelligent continuous or hybrid systems marks a significant advancement in fermentation technology. Continuous fermentation is probably going to become the norm for high-performance, large-scale biomanufacturing when combined with genetic strain optimization and real-time monitoring (Udugama et al., 2020).

2.10 OPPORTUNITIES AND APPLICATIONS

The usage of fermentation technology has grown significantly beyond conventional food production due to recent advancements. Utilizing microorganisms to transform agricultural leftovers and other lignocellulosic materials into biofuels like ethanol, butanol, and biogas is one of the most significant new fields in the production of sustainable bioenergy (Table 2.3). Complex polysaccharides found in crop waste can now be metabolized by engineered yeast and bacterial strains, which lessens reliance on fossil fuels and supports circular bioeconomy models (Adesina & Nwobi, 2023). Additionally, fermentation is being utilized more and more to create valuable biochemicals from a variety of feedstocks. Biopolymers (polyhydroxyalkanoates), organic acids (lactic, succinic, and citric acid), and flavouring or aroma chemicals are a few examples. Precision fermentation and synthetic biology have made it feasible to use low-cost substrates, including industrial side streams or food processing by-products, as raw materials to produce valuable biomolecules (Böhm et al., 2023).

TABLE 2.3
Conventional Fermentation Techniques for Biogas, Biofuel, and Bioenergy Production

S. No.	Technique	Substrates Used	Microorganisms Involved	End Products	Key Features/ Limitations
1	**Anaerobic Digestion**	Agricultural residues, sewage sludge, animal manure, food waste	Anaerobic consortia (hydrolytic bacteria, acidogenic and methanogenic archaea)	Biogas (CH_4+ CO_2)	Widely used; produces renewable energy; limitations: slow process, sensitive to pH/ temperature.
2	**Alcoholic Fermentation**	Sugarcane, corn starch, molasses, lignocellulosic hydrolysates	*Saccharomyces cerevisiae*, *Zymomonasmobilis*	Bioethanol	Simple and established; limited by substrate cost and efficiency of lignocellulose hydrolysis.

(Continued)

TABLE 2.3 (*Continued*)
Conventional Fermentation Techniques for Biogas, Biofuel, and Bioenergy Production

S. No.	Technique	Substrates Used	Microorganisms Involved	End Products	Key Features/ Limitations
3	**Butanol Fermentation (ABE process)**	Starch, molasses, lignocellulose hydrolysates	*Clostridium acetobutylicum* and related spp.	Acetone, Butanol, Ethanol (ABE)	Produces valuable solvents and biofuel; low yield, product inhibition common.
4	**Lactic Acid Fermentation (for bio-based chemicals/fuel precursors)**	Sugars, food waste, lignocellulosic biomass hydrolysates	*Lactobacillus spp., Bacillus coagulans*	Lactic acid (precursor for polylactic acid bioplastics, biofuels)	Environmentally friendly; requires neutralization; cost-intensive substrates.
5	**Syngas Fermentation**	CO, CO_2, H_2 (from gasification of biomass/coal)	Acetogenic bacteria (*Clostridium ljungdahlii, Moorella spp.*)	Ethanol, acetic acid, biofuels	Converts waste gas streams; efficient use of carbon; expensive reactor design.
6	**Dark Fermentation (Hydrogen production)**	Carbohydrate-rich waste, starch, cellulose hydrolysates	Anaerobic bacteria (*Clostridium spp., Enterobacter spp.*)	Biohydrogen (H_2) with volatile fatty acids (VFAs)	Rapid H_2 production; low yield; often coupled with other processes for efficiency.

In the food industry, new fermented food and beverage items with improved functionality and distinctive sensory characteristics are being produced by fusing traditional knowledge with contemporary biotechnology. These include fermented protein substitutes, plant-based yoghurts enhanced with probiotics, and drinks that contain microbial strains that have been tailored to increase particular bioactive substances. Fermentation is frequently employed to customize nutritional and health-related qualities in addition to preservation and flavour development.

2.11 CULTURAL AND SAFETY ASPECTS

A number of significant safety and sociocultural issues are brought up when contemporary technology is incorporated into conventional fermentation methods. To guarantee microbial safety in altered processes, the application of engineered strains or multi-species consortia needs to be carefully assessed from a microbiological perspective. In order to rule out virulence factors, genes for antibiotic resistance, or undesired metabolic activities, strains employed in precision fermentation must either exhibit a proven track record of safe use or go through a thorough toxicological and genomic evaluation. Furthermore, the natural barriers that restrict pathogen growth during conventional batch fermentation may change with the introduction of novel substrates or continuous processing systems. As a result, improved risk assessment frameworks that consider both deliberate microbial mutations and process design changes have been used (Cassotta et al., 2022).

The cultural attitudes surrounding the shift from traditional artisanal fermentation to engineered or data-driven processes are equally significant. Fermented foods are associated with long-standing culinary traditions and frequently have distinct regional identities. Therefore, it may be argued that using synthetic biology or CRISPR-based methods conflicts with the cultural values ingrained in traditional cuisine. According to recent consumer research studies, new methods are more likely to be accepted when they are presented as ways to preserve or revitalize traditional items (such as stabilizing microbial starters obtained from heritage foods) as opposed to replacing them. Therefore, establishing trust and preventing the impression that cultural authenticity is being lost need a clear explanation of the goals and limitations of the technology interventions (Dai et al., 2023; Fang & Zhao, 2021).

Last but not least, innovative fermentation methods need to adhere to legal and labelling specifications. The use of synthetic biology techniques or genetically modified microbes necessitates obligatory pre-market authorization in many jurisdictions, along with traceability and labelling requirements. Regulatory bodies are increasingly using "process-based" criteria that may nonetheless identify the substance as genetically modified even when no recombinant DNA is left in the finished product. In the food industry, clear labelling is essential for promoting market acceptability and empowering consumers to make educated decisions (García-García et al., 2022). Bio-based chemicals and bioenergy products are subject to similar regulatory considerations, since certification programmes (such as those for sustainability or feedstock origin) are frequently needed to prove adherence to legal and industry requirements (Ghosh et al., 2022a). Therefore, it is crucial to include these cultural and regulatory factors in the design of contemporary fermentation systems in order to guarantee both technical success and public acceptance.

2.12 PROSPECTS FOR THE FUTURE

Recent developments in fermentation research point to a paradigm shift towards predictive biomanufacturing, where data-driven models are employed to both predict and track the behaviour of microbial communities during processing (Jakob et al., 2020). Future systems will employ real-time physiological data (metabolites, gene expression, and redox state) in place of static recipes to dynamically modify ambient factors and direct microbial metabolism in the direction of the intended result (Liu & Li, 2023). These systems will be able to predict changes in the structure of the microbial population, metabolic bottlenecks, or the emergence of undesirable by-products by combining multi-omics data with machine-learning algorithms. They will then be able to take corrective action on their own.

Additionally, fermentation is a good fit for new models of the circular economy. In contrast to many traditional industrial processes, fermentation can turn low-value by-products such as food waste, agricultural leftovers, or waste gases into valuable goods like protein-rich foods, organic acids, and biofuels (Mahato et al., 2019; Molnár et al., 2022). It is probable that fermentation systems of the future will be constructed as modular components that may be easily included into regional supply chains. For instance, nearby processing facilities may have micro-fermentation units that employ biomass that is readily available in the area to recycle value-added products back into the food or energy system. Regional bio-economies will benefit from this type of resource-efficient, decentralized design, which also lowers emissions associated with transportation (Vivian et al., 2021; Awasthi et al., 2021).

The creation of next-generation microbial engineering tools is another exciting field. Beyond CRISPR, novel techniques including whole genome recoding, transcriptional regulators, and base editing will enable the creation of microbial strains with customized metabolic roles and increased robustness (Gutiérrez-Ríos et al., 2022; Tsoi et al., 2019). The development of stable, multi-species consortia where individual strains carry out complementary tasks will also be made possible by advancements in synthetic ecology (Zhang et al., 2023; Derabli et al., 2022). These modified microbial communities could be employed to extend fermentation into completely new substrate spaces or customized to dynamic fermentation settings. When taken as a whole, these advancements suggest that fermentation will grow into a versatile, intelligent, and sustainable biomanufacturing platform that can support both cutting-edge industrial biotechnology and conventional food systems.

2.13 CONCLUSION

From an empirical, age-old method with roots in food preservation and cultural heritage, fermentation has developed into a very versatile platform for contemporary biomanufacturing. A remarkable range of fermented foods and beverages, many of which are still essential components of local diets and identities, was made possible by traditional techniques based on spontaneous microbial activity. Industrial-scale uses were made possible by the introduction of controlled starting cultures and process optimization techniques in response to growing demands for uniformity, safety, and scalability. Recent developments in data-driven modelling, process automation, synthetic biology, and microbiome research have made fermentation a potent tool for converting a variety of feedstocks into high-value biochemicals and bioenergy in addition to food. These innovative methods show how cutting-edge science and traditional wisdom may be combined to produce higher yields, faster processing times, and better sustainability. However, it's crucial that this advancement in technology doesn't ignore the customer confidence and cultural significance that support conventional fermentation methods. Therefore, future advancements will need to carefully balance incorporating advances that enable safe, efficient, and environmentally responsible production with maintaining the authenticity and practical benefits of legacy techniques. By preserving this equilibrium, fermentation will be able to make a significant contribution to both the developing bio-based economy and world food culture.

REFERENCES

Abdul-Wahid, B., & Peters, R. (2022). Machine learning applications in microbial fermentation: Process optimisation and prediction. *Biotechnology Reports*, *34*, e00731.

Adesina, A., & Nwobi, O. (2023). Valorisation of food-processing side streams through precision fermentation. *Current Opinion in Green and Sustainable Chemistry*, *39*, 100784.

Awasthi, M. K., Sarsaiya, S., & Liu, T. (2021). Bioconversion of agricultural waste into energy: Circular bioeconomy approach. *Renewable & Sustainable Energy Reviews*, *135*, 110226.

Ayyash, M., Al-Dhaheri, A., Kizhakkayil, J., & Al-Nabulsi, A. (2021). Bioactive peptide production by engineered *Lactobacillus* strains in dairy fermentation. *Journal of Dairy Science*, *104*(7), 7093–7102.

Böhm, V. M., Pellegrino, G., & Willemsen, P. (2023). Regulatory developments for genome-edited foods. *Nature Biotechnology*, *41*(8), 1125–1132.

Calero, P., & Nikel, P. I. (2019). Chasing bacterial chassis for metabolic engineering: A perspective review from classical to non-traditional organisms. *Microbial Biotechnology*, *12*(1), 98–124.

Cassotta, M., Avanzini, C., & Baldi, F. (2022). Microbial risks in novel fermentation processes and the role of safety assessment. *Food Control*, *139*, 109068.

Castillo, O., Melin, P., &Kacprzyk, J. (2022). Model predictive control enhanced by AI techniques in biotechnological processes. *Engineering Applications of Artificial Intelligence*, *112*, 104807.

Curiel, J. A., Taccari, M., &Gobbetti, M. (2022). Microbiome-based approaches for flavour and functionality improvement in fermented foods. *Comprehensive Reviews in Food Science and Food Safety*, *21*(1), 237–262.

Dai, Z., Ruan, J., & Li, Y. (2023). Consumer perception toward precision-fermented foods: A survey study. *Food Research International*, *170*, 113030.

de Souza, H. B., Niu, C., & Fang, X. (2022). Development of soft sensors for real-time monitoring of probiotic fermentation. *Sensors*, *22*(19), 7383.

Derabli, B., Nancib, A., Nancib, N., Aníbal, J., Raposo, S., Rodrigues, B., &Boudrant, J. (2022). Opuntia ficus indica waste as a cost-effective carbon source for lactic acid production by *Lactobacillus plantarum*. *Food chemistry*, *370*, 131005.

Dinh, T., Quang, M., & Kim, J. (2021). High-productivity continuous fermentation systems for biofuel production using engineered microorganisms. *Renewable & Sustainable Energy Reviews*, *148*, 111254.

Doudna, J. A., & Sternberg, S. H. (2022). The biology and application of CRISPR systems: Harnessing nature's toolkit. *Science*, *377*(6602), 840–847.

Fang, Z., & Zhao, X. (2021). Engineered microorganisms for sustainable bioenergy and biochemical production. *Biotechnology Advances*, *50*, 107769.

Fernandez-Nava, Y., Marañón, E., & Alonso-Fariñas, B. (2020). Process strategies for industrial-scale fermentation. *Journal of Bioprocess Engineering*, *45*(3), 232–245.

Fisher, D., Doyle, S., & Brewster, J. (2023). Artificial intelligence in fermentation: Opportunities and challenges. *Trends in Biotechnology*, *41*(6), 692–705.

García-Domínguez, A., Navarrete, P., & Leclercq, S. (2022). Kinetic modelling and optimisation strategies for precision fermentation. *Journal of Industrial Microbiology and Biotechnology*, *49*(3), 285–296.

García-García, M., Griffin, O., & Afzal, M. (2022). Biorefinery routes for second-generation bioethanol. *Renewable Energy*, *187*, 1007–1017.

Ghosh, C., Levy, R., & Glockner, F. (2022a). Novel fermented foods: Opportunities for functional benefits. *Trends in Food Science & Technology*, *119*, 407–418.

Ghosh, S., Chatterjee, T., & Jana, S. (2022b). Variability and reproducibility issues in spontaneous fermented food systems. *Applied Microbiology and Biotechnology*, *106*, 5397–5412.

González-Camejo, J., Aparicio-Trejo, H., & Garcia-Peña, E. I. (2022). Smart sensors and real-time monitoring in industrial fermentation. *Biochemical Engineering Journal*, *186*, 108554.

Gujar, G., Naik, S., & Pandit, A. (2023). Advances in continuous fermentation technologies for food and bioenergy applications. *Bioresource Technology Reports*, *22*, 101456.

Gutiérrez-Ríos, R. M., Romero-Campero, F., & Peña-Castro, J. M. (2022). Integration of multi-omics data and machine learning for predictive bioprocessing. *Current Opinion in Biotechnology*, *75*, 102702.

Huang, L., Xu, Y., & Liu, G. (2022). Microorganisms and metabolites in traditional fermented foods: Diversity and function. *Critical Reviews in Food Science and Nutrition*, *62*(22), 6166–6181.

Huang, X., & Zhang, T. (2022). Data-driven modelling approaches for optimisation of fermentation processes. *Journal of Industrial Microbiology & Biotechnology*, *49*(1), 1–12.

Hucka, M., Clancy, P., &Wipat, A. (2022). Synthetic biology tools and design strategies for microbial fermentation systems. *ACS Synthetic Biology*, *11*(4), 1325–1339.

Iqbal, H., Bello, M., & Hashem, A. (2021). Advances in metabolic engineering of microorganisms for sustainable fermentation. *Biotechnology Progress*, *37*(5), e3189.

Jakob, F., Vuillemin, M., & Schmid, A. (2020). Fermentation as a platform for novel plant-based food products. *Food Bioscience*, *37*, 100739.

Jia, H., Zhao, Y., & Li, D. (2024). Deep learning for in silico optimisation of lactic acid fermentation. *Bioprocess and Biosystems Engineering*, *47*(1), 31–44.

Jing, H., Zhang, Q., & Yu, H. (2020). Archaeological evidence and evolution of ancient fermentation technology in China. *Journal of Ethnic Foods*, *7*(1), 13.

Kim, H., Park, H. J., & Lee, H. J. (2021). Functional fermented foods and their health benefits. *Foods*, *10*(10), 2329.

Koontz, J. L., & Gupta, C. (2021). Precision fermentation and the future of functional food ingredients. *Trends in Biotechnology*, *39*(11), 1129–1133.

Koutinas, A. A., Khawaja, N., & Webb, C. (2019). Fermentation-based biorefinery models for sustainable systems. *Bioresource Technology*, *294*, 122234.

Kumar, V., & Ray, S. (2022). Oxygen transfer and respiration control in food fermentation. *Biochemical Engineering Journal*, *187*, 108567.

Kutyna, D. R., Curran, J., & Varela, C. (2022). Hybrid fermentation systems: Combining continuous and batch approaches for improved flavour development. *Journal of Food Engineering*, *330*, 111044.

Lee, J., & Saluja, A. (2022). Cross-cultural acceptance and expansion of traditional fermented foods. *International Journal of Gastronomy and Food Science*, *28*, 100542.

Liu, H., & Li, X. (2023). Evaluating microbial safety in precision fermentation systems. *Comprehensive Reviews in Food Science and Food Safety*, *22*(1), 465–482.

Liu, L., Bestel, S., Shi, J., Song, Y., & Chen, X. (2018). Paleolithic human foraging in the Levant: The Natufian beer hypothesis. *Journal of Archaeological Science*, *92*, 168–176.

Liu, Y., Liu, S., & Chen, X. (2020). Energy metabolism and flavour formation in anaerobic fermentation systems. *Food Research International*, *137*, 109715.

Lopes, M., Rodríguez, C., & Fernández-Arrojo, L. (2021). Fermentation of lignocellulosic residues for sustainable production of biochemicals and biofuels. *Sustainable Chemistry and Pharmacy*, *22*, 100469.

López-Garzón, C. S., & Straathof, A. J. J. (2022). Cell retention technologies for continuous fermentation. *Trends in Biotechnology*, *40*(3), 357–371.

Machado, R. R., & Rocha, I. (2021). Hybrid mechanistic–machine learning models in microbial biotechnology. *Current Opinion in Biotechnology*, *69*, 199–206.

Mahato, N., Sharma, K., Sinha, M., & Cho, M. H. (2019). Bio-based platform chemicals and value-added products via microbial fermentation. *Biotechnology for Biofuels*, *12*, 189.

Marco, M. L., Heeney, D., &Hutkins, R. (2021). A role for fermentation in global sustainable food production. *Nature Food*, *2*(7), 494–495.

Medina, M., Peres, J., & Martins, G. (2022). Application of precision fermentation in industrial-scale food production. *Food Biotechnology*, *36*(2), 223–238.

Molnár, E., Török, Á., & Kovács, S. (2022). Labelling requirements and consumer acceptance of precision-fermented food ingredients. *Food Control*, *141*, 109190.

Mukhopadhyay, A. (2022). Combining traditional fermentation and synthetic biology for biomanufacturing. *Nature Chemical Biology*, *18*(11), 1174–1178.

Navarrete, P., Rojas, V., & Simpson, R. (2021). Microbial succession and metabolic changes during cocoa fermentation. *Trends in Food Science & Technology*, *115*, 456–465.

Prajapati, J. B., & Nair, B. M. (2021). Fermented functional foods and their benefits. *Current Opinion in Food Science*, *42*, 75–81.

Qin, N., Xia, Y., & Lu, H. (2023). Dynamic control of microbial metabolism using CRISPR-based regulatory circuits. *Nature Communications*, *14*, 741.

Rezac, S., Kok, C. R., Heermann, M., &Hutkins, R. (2018). Fermented foods as a dietary source of live organisms. *Frontiers in Microbiology*, *9*, 1785.

Santos, R. D., Vieira, E., & Vicente, A. (2023). Selection of microbial strains for multi-phase fermentation processes. *Biotechnology Advances*, *63*, 108085.

Shi, L., & Zeng, X. (2021). Starter culture strategies for improved microbial interactions in food fermentation. *Trends in Food Science & Technology*, *112*, 327–338.

Skaar, I., Stenhammer, T., &Brendehaug, J. (2021). Monitoring and control of fermentation processes in the food industry. *Food Control*, *123*, 107737.

Soomro, S., Tareen, A., & Anwar, M. (2023). Substrate specificity and adaptability issues in traditional fermentation systems. *LWT - Food Science and Technology*, *171*, 114121.

Sun, J., & Alper, H. (2020). Engineering yeast for improved biofuel production from lignocellulosic biomass. *Current Opinion in Chemical Biology*, *59*, 69–76.

Tamang, J. P., Watanabe, K., & Holzapfel, W. H. (2020). Review: Diversity of microorganisms in global fermented foods and beverages. *Frontiers in Microbiology*, *11*, 592703.

Tran, M. D., Seo, H. J., & Kim, B. (2023). Real-time monitoring and control of continuous kimchi fermentation using redox and organic acid feedback. *Food Control*, *147*, 109568.

Tsoi, R., Dai, Z., & You, L. (2019). Emerging strategies for engineering microbial communities. *Biotechnology Journal*, *14*(8), 1800441.

Udugama, I. A., Lim, C., & Nielsen, L. K. (2020). Real-time bioprocess monitoring and control for smart continuous fermentation. *Trends in Food Science & Technology*, *106*, 301–313.

Vivian, A., Gibbons, W., &Vedantam, S. (2021). Certification requirements and market access for fermentation-derived biofuels. *Renewable and Sustainable Energy Reviews*, *148*, 60–84.

Wu, X., Suo, Y., & Zeng, J. (2019). Lactic acid production by lactic acid bacteria: Kinetic characteristics and regulation. *Journal of Biotechnology*, *302*, 70–77.

Wuyts, S., Vrancken, G., & De Vuyst, L. (2023). Selection of mixed starter cultures for improved vegetable fermentation. *Food Microbiology*, *110*, 104140.

Zhang, B., Li, T., & Gao, L. (2023). Synthetic microbial consortia for industrial biotechnological applications. *Trends in Biotechnology*, *41*(4), 409–421.

Zhang, D., & Chen, G. (2020). Time-dependent changes in microbial dynamics during spontaneous fermentation. *Journal of Food Engineering*, *281*, 109031.

Zhang, H., Chen, J., Xu, J., & Wang, X. (2024). Integrating synthetic biology and fermentation technology for the production of value-added bioproducts. *Biotechnology Advances*, *70*, 108162.

Zhou, J., He, Q., & Zhang, Y. (2021). Real-time optimisation of food fermentation processes. *Trends in Food Science & Technology*, *114*, 455–465.

3 Foundations of Precision Fermentation for Sustainable Bioenergy Production

G. B. Varshitha, N. Gnanasankaran, and Kanagaraj Rajalakshmi

3.1 INTRODUCTION

With the rise in global population, there is a requirement for the expansion of renewable energy, which has increased the demand for oil and gas resources. The exhaustion of fossil fuel reserves has resulted in a global energy crisis and growing environmental problems. When fossil fuels are used continuously, they can lead to air pollution, greenhouse gas emissions, and global warming. One alternative way is through the use of biofuels, which are environmentally friendly compared to conventional fossil fuels (Abubakar Musa Shuaibu et al., 2024).

The advanced technology uses microorganisms to make valuable products through precision fermentation, not only for the food industry, such as plant-based proteins, dairy substitutes, and even sustainable leather, but also in other industries. Precision fermentation helps manufacturers create products in customized way, which is needed, and it is done consistently to meet the specific needs for different fields. The advancements in biotechnology, like genetic engineering and synthetic biology, has made scientists to control the growth of microorganisms and the production that leads to precise and reliable outcomes.

Fermentation is a process through which microorganisms are produced on a large scale, to create specific products, by using cellular biomass or different metabolites. The substrates used in the fermentation process influence the final product, and the overall yield is determined by the selected fermentation method and reactor operation mode. Optimizing environmental and process parameters that favor microbial growth plays a crucial role in achieving the highest possible product yield (Soccol et al., 2010).

Sustainable bioenergy can be produced from biomass resources in multi-functional, integrated agriculture, forestry, fisheries, and aquaculture systems, along with food, feed, and/or bio-based products, from biogenic waste and residue streams, or as a co-product of ecosystem management.

 DOI: 10.1201/9781003684374-3

Biomass and derivatives from its bioenergy are versatile, storable, and dispatchable. They can replace fossil energy and complement variable renewables and other low-carbon options in transport, power, and heat production, industrial processes, and clean cooking, thereby enhancing resilience in the energy system.

Sustainable bioenergy plays a unique role in just and inclusive energy transitions and is especially important for sectors and regions where other decarbonization options are costly or not yet available. Sustainable bioenergy can contribute to energy security, clean energy access, rural development, increased agricultural productivity, improved farmer incomes, job creation, gender equality, responsible industrial development, poverty eradication, and climate change mitigation and adaptation strategies (Sharma et al., 2016).

3.2 PRINCIPLES OF PRECISION FERMENTATION

Precision fermentation is an essential process in microbiology and genetic engineering which is put into practice. The process starts by selecting a suitable host microorganism—often species like *Pichia pastoris* or *Saccharomyces cerevisiae* (baker's yeast) because they grow well and are genetically well-mapped. These organisms are picked for their ability to be engineered so they can manufacture a specific protein, fat, or other complex molecule that would otherwise be hard or costly to source.

Insertion of the new gene into microbes DNA is the central scientific step in encoding the ingredient that is required to be produced. This method is completely different from traditional fermentation used in bread or beer making, where microbes depend on their natural metabolism to turn sugars into alcohol and carbon dioxide. In contrast, precision fermentation is tightly controlled and highly targeted. In the bioreactor, the microbe consumes nutrient-rich feed, usually sugar-based, that helps to modify the metabolic pathway, and it generates the desired ingredient.

The advancement in technology has resulted in benefits, such as consistent quality, high purity, and the products are made identical to their natural products without any impact on the environment. Fermentation done in the lab involves careful tuning of every element, particularly from the genetic design to the fermentation conditions, to achieve the highest possible yield and efficiency.

Precision fermentation is a process of biotechnology that controls and influences the power of microorganisms, industrial-scale bioreactors, and modern gene engineering tools to produce high-value, specific products with high efficiency and purity. The key principles and technologies involved, including subsections on genetic engineering, CRISPR/Cas9, and other gene editing methods.

In industrial biotechnology, fermentation means growing a large number of microorganisms under controlled conditions inside a vessel called a fermenter or bioreactor. Depending on the process, microorganisms may grow with or without oxygen (aerobic or anaerobic conditions). Many industries, such as food, textiles, and transportation, make use of the products formed from fermentation.

The major role of the fermenter is to produce the desired product, and this is done by creating a suitable and controlled environment for the microorganisms to grow. A fermentation process includes several steps, like preparing the medium, sterilizing it, making the inoculum, running the fermentation, and finally, purifying the product.

Fermentation is the key stage, but at the same time, the steps before (upstream) and after (downstream) are also very important. There are parameters to be considered in the process to convert the substrates into products. Whether the process is done in batches or continuously, the major parameters in the process include temperature, pH, aeration, agitation, and careful control of oxygen transfer (Stanbury et al., 1995).

Scientists make specific changes in an organism's DNA by using CRISPR-Cas9. This works efficiently with two main parts: the guide RNA (gRNA), which helps find the exact spot on the DNA to be edited, and the Cas9 protein, which acts like molecular scissors to cut the DNA at this spot. The guide RNA binds to the target DNA next to a tiny DNA signature called the protospacer adjacent motif (PAM), and with the help of Cas9, both DNA strands are cut a few steps before the PAM, causing a break. In the process of fixing this break, the cell often makes the mistake of either adding or removing small bits of DNA, which can turn off the targeted gene (Mengstie & Wondimu, 2021) to fight off viruses and foreign genetic material, CRISPR-Cas9 was discovered as a defense system used by bacteria. In recent years, this technology has been used to edit genes in various organisms, through which crops, livestock, and advancements in medical research have been improved. Getting the CRISPR-Cas9 parts into the right cells efficiently is one of the major problems. Nanomaterials, such as specialized nanoparticles or delivery vehicles, are being used to protect and direct CRISPR components into the target cells, improve uptake, and help the components escape from cell compartments where they could be trapped or destroyed. These materials can also be planned to deliver the gene-editing tools to specific cells, making the process safer and accurate (Hossain, 2021).

A new phase of gene editing is done with the introduction of CRISPR-Cas9 technology, which enables genetic changes to be precise and efficient. Using special materials called nanomaterials can help solve problems with delivering CRISPR-Cas9 into cells, making the process better and safer. Research now focuses on innovative ways to use these nanomaterials to improve how CRISPR-Cas9 is given to cells, showing their promise for the future of genome editing (Lotfi et al., 2023).

A comparison was done on the three genome editing tools, namely CRISPR-Cas9, TALENs, and ZFNs, considering the important criteria like efficiency, specificity, ease of use, design flexibility, cost effectiveness, availability, and adoption. CRISPR-Cas9 is more efficient at making genetic changes than TALENs and ZFNs. It also has high specificity for its target site, and while TALENs and ZFNs are also specific, they are typically regarded as slightly less convenient to get adjusted for wide use comparatively. When it comes to practical application, CRISPR-Cas9 is relatively easy to use because only the guide RNA sequence needs to be redesigned for new targets, whereas TALENs require moderate effort, and ZFNs are complex and labor-intensive to design and apply. CRISPR-Cas9 also offers high design flexibility, allowing targeting of many different DNA sequences and enabling multiplex editing, while TALENs and ZFNs provide only moderate flexibility. Overall, CRISPR-Cas9 is more cost-effective than the other platforms and has become widely available and broadly adopted in research, while TALENs are relatively expensive with moderate current use, and ZFNs are the most expensive and now less commonly used. CRISPR-Cas9 has high efficiency, specificity, easy design, flexibility, affordability, and widespread use in genome editing (Aljabali et al., 2024) (Figure 3.1).

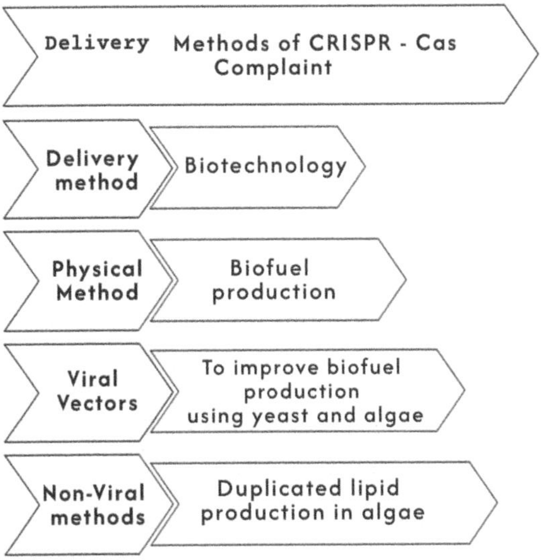

FIGURE 3.1 Delivery methods of CRISPR-Cas component.

3.3 METABOLIC ENGINEERING AND OPTIMIZATION OF STRAINS

Over the past century, industrial biotechnology has grown to focus on making fuels and food on a large scale. Metabolic engineering has given this field a strong boost by helping scientists change cells so they show useful traits, mainly by changing how substances flow through their metabolic pathways. In the post-genomic era, huge amounts of data on genes, proteins, and metabolites have made it possible to study cell metabolism more accurately and have also helped lower the cost of making DNA pieces. This has led to systems metabolic engineering, which uses new methods such as planned gene deletions, adding genes from other organisms, and using tools to finely control gene activity and cell behavior (Lee et al., 2012).

Using systems metabolic engineering in microbes like Escherichia coli and Saccharomyces cerevisiae allows high production of many secondary metabolites that are valuable for industry and medicine, such as biofuels, chemicals, and biopolymers. Plants are also rich sources of secondary metabolites; for example, they can produce tens of thousands of terpenoids, many of which can serve as anticancer drugs or protective compounds. Metabolic engineering in crops can also improve farming by increasing nitrogen use, nutrient content, photosynthetic efficiency, and resistance to disease. Together, metabolic engineering in microbes and plants supports faster and more cost-effective drug discovery by enabling the efficient production of molecules with pharmacological value (Lee et al., 2012).

To design these systems well, researchers must carefully study how cells work, often using high-throughput experiments on genomes and building mathematical models of cellular processes. A major challenge is that complete models for whole pathways, such as those involved in protein secretion, are still limited, even though smaller kinetic models exist. Full genome-scale models that link metabolism, signaling, gene regulation, and protein–protein interactions are only beginning to develop, but they will help scientists analyze complex networks and design better routes for increased protein production. Because of this, metabolic engineers rely on both computer models for prediction and screening, and lab experiments to test new ideas and confirm predictions (Otero-Muras & Carbonell, 2021; Paley, 2021).

Bringing large-scale "omics" data (genomics, transcriptomics, proteomics, metabolomics) into computer models makes predictions more accurate and helps identify good drug targets for complex diseases. Genomics can also reveal "silent" metabolic gene clusters that are not usually active, allowing these hidden pathways to be turned on and used, further linking genomics and metabolic engineering. Artificial intelligence (AI) is expected to become increasingly important here, helping to analyze big datasets, suggest new engineering strategies, and speed up the design–build–test cycle (Otero-Muras & Carbonell, 2021; Paley, 2021).

To understand and optimize how pathways behave over time, approaches such as inverse metabolic engineering and metabolic control analysis (MCA) are very useful. Inverse metabolic engineering starts from a desired trait and works backward to find which genes should be changed, often using high-throughput screening. MCA, in contrast, focuses on measuring how fluxes are distributed in a pathway and which steps have the most control. This review first looks at applications of metabolic engineering in microbes, plants, and drug discovery, then discusses goals, methods, and technologies, and finally considers how genomics, AI, and advanced modeling can guide the future development of modern metabolic engineering (Sanchez & Demain, 2002; Shiue & Prather, 2012; Yadav et al., 2018).

3.4 BIOENERGY PRODUCTION USING PRECISION FERMENTATION

Fermentation processes are classified into two main types: solid-state fermentation and submerged fermentation. Solid-state fermentation (SSF) refers to the cultivation of microorganisms on a solid, insoluble substrate with minimal or no free water present. In contrast, submerged fermentation is a liquid-based process in which microbial cells are dispersed throughout the growth medium, existing as individual cells or aggregates. A characteristic that distinguishes them from SSF is effective mixing in submerged fermentations, which is a homogeneous system.

Algae are aquatic in nature, and they require water for growth. Therefore, there are very few studies on SSF involving algal cultures. Hence, submerged or suspension culture systems are commonly used for algal cultivation.

The fermentation process includes several biochemical reactions, like hydrolysis, oxidation, reduction, polymerization, biosynthesis, and cell formation. Fermentation represents the metabolic activity of microorganisms utilizing carbon that contains substrates for energy and growth. According to the oxygen requirement, fermentation can be divided into two types:

1. Aerobic fermentation: The word "aerobic fermentation" is technically not accurate; the correct term is aerobic respiration. This process involves the breakdown of organic compounds into simpler molecules, releasing both

carbon dioxide (CO_2) and energy. Under anaerobic conditions, similar degradation can occur when alternative electron acceptors such as nitrate (NO_3^-) or sulfate (SO_4^{2-}) replace oxygen. The pathways of carbon cycling, therefore, differ between aerobic and anaerobic environments. When no external electron acceptor is available, organic matter is degraded through fermentation. Traditionally, fermentation method has actually been carried out aerobically in several industries. The aerobic processes depend on air to provide sufficient oxygen when the fermentation media is exposed. Modern aerobic fermentation uses sealed fermenters in which there is a continuous oxygen or air supply for submerged cultivation. Stirrers are used to mix adequately, through which sterilized air is introduced to maintain optimal aeration throughout the process.

2. Anaerobic fermentation: Anaerobic fermentation is a process in which algal biomass is converted into methane after lipid extraction. Fermentation processes are categorized based on their primary products, such as alcohol, lactate, acetate, or methane. Both microalgae and macroalgae act as renewable feedstocks for anaerobic digestion processes (ADP). Algal biomass includes both organic and inorganic components, with the organic fraction consisting of large, complex polymers like proteins, polysaccharides, lipids, and nucleic acids. These macromolecules are mostly present in particulate or colloidal forms. Through ADP, the complex organic constituents of algal biomass are converted into methane, carbon dioxide, new microbial biomass, and inorganic residues. One of the major advantages of algae is that their cell walls lack lignin, a recalcitrant structural polymer common in terrestrial plants, making them easier to biodegrade during ADP. Microalgae, which are rich in lipids and oils rather than sugars, can pose challenges for anaerobic digestion due to their high protein content. Macroalgae contain a higher proportion of fermentable carbohydrates and comparatively lower lipid levels. After lipid extraction, carbohydrates become rich in microalgae, which are suitable substrates for anaerobic digestion and efficient biogas production.

3.4.1 Bioethanol

The benefits to the environment have gained a lot of interest in bioethanol worldwide. It can be produced from various sources such as algae, lignocellulosic biomass (plant materials), and other types of plant matter. In the process of making bioethanol from algae, the algae are first collected and dried to remove most of the water. This makes it easier to handle the algae as a solid material. This drying step is important before extracting the fuel. Fermentation is carried out to convert the starches in algae into sugars, which are then used to produce bioethanol. Fermenters are special tanks in which the process of breaking down the biomass and the conversion of starch into sugars is done, which are then mixed with water and yeast. Yeast is used because it turns sugars into bioethanol. After fermentation, the ethanol is separated and purified by distillation, which removes water and impurities to get a concentrated liquid fuel (Abubakar Musa Shuaibu et al., 2024).

Traditional bioethanol sources produce more greenhouse gases than fossil fuels during production and usage. Similarly, the use of microalgae as a feedstock for bioethanol can help solve environmental problems. For example, waste sugar beetroot pulp has been studied as a raw material for bioethanol production. A plant processing around 17,000 tons of this waste each year can have its costs and productivity modeled to evaluate new technologies. Models can also predict how different parts of the process contribute to the total costs, and allow comparison of different scenarios for handling additional products that come from the bioethanol production process (Abubakar Musa Shuaibu et al., 2024).

3.4.2 Biodiesel

Biodiesel is a type of oxygen-rich fuel that burns completely and causes fewer harmful environmental effects while helping reduce emissions. A variety of raw materials can be used to produce biodiesel, including used cooking oils, animal fats, and oilseeds such as canola and oil palm. The most common method for making biodiesel is called transesterification. This process involves converting the long fatty acid chains in fats and oils, which are made of triglycerides (molecules formed by three fatty acids linked together), into smaller molecules. During transesterification, these triglycerides are broken apart into fatty acid ester molecules and glycerol, which can then be further processed to make biodiesel. This breakdown happens when the carbon–carbon bonds within the triglycerides are split, producing the molecules needed for biodiesel production (Abubakar Musa Shuaibu et al., 2024).

3.4.3 Biogas and Methane

Biogas refers to the mixture of gases produced through the anaerobic digestion of organic waste originating from human, animal, and plant sources. It forms due to oxygen deprivation and the metabolic activity of anaerobic microorganisms, particularly methanogenic bacteria. Biogas is naturally generated in environments in places where there is containment in swamps, marshlands, and landfill sites. Biogas production is done and controlled in a digestion tank or bioreactor, where anaerobic fermentation of biomass is conducted for practical and economical utilization (Jameel et al., 2024).

The composition of biogas consists majority of methane, which is 55–70% of the total gas mixture, and with 35–40% carbon dioxide, with trace amounts of nitrogen, hydrogen sulfide, and other gases. The proportion of the mix changes with the substrate type and digester temperature. Methane is a valuable constituent, because of which higher methane concentration enhances the energy quality and overall economic efficiency of biogas production (Jameel et al., 2024).

Biogas is a colorless, tasteless gas with a strong, rotten egg-like smell. Similar to carbon dioxide, it functions as a greenhouse gas but has approximately 25 times higher warming potential. Due to its high ignition capability, it burns with a blue flame at about 800°C and can form combustible mixtures with air in ratios from 1:1 to 1:20. For household uses like cooking, it operates best at a pressure of 5–20 cm of the water column. One cubic meter of biogas produces about 5,500–6,500 kcal of heat, which is enough to run a one-horsepower internal combustion engine for nearly two hours. Energy-wise, one cubic meter of biogas is equivalent to around 0.4 kg of diesel, 0.6 kg of oil, or 0.8 kg of coal (Jameel et al., 2024).

Therefore, it shows that by using biogas, there is less carbon in the air, cleaner air, healthier soil and ultimately a healthier life that is needed for human life.

3.4.4 Biohydrogen

In different metabolic processes, microorganisms produce biohydrogen as a by-product. Biohydrogen is an efficient and sustainable energy source, but it has a short lifespan. It is an alternative way to reduce dependence on limited energy resources. Biophotolysis is a method to produce biohydrogen where microalgae use water and sunlight to generate hydrogen, as well as anaerobic fermentation of carbohydrate-rich biomass. Monomers are the process of breaking complex polymeric sugars into simple sugars, as these units serve as the pioneer for biohydrogen generation. There is different pretreatment methods used to depolymerize these sugars and enhance hydrogen output from algae. Dark fermentation that is performed without light consumes more energy than the hydrogen it produces, leading to a negative energy balance. Therefore, it is necessary to improve the economic feasibility by integrating algal dark fermentation within a biorefinery system. In this process, fermentation by-products are changed into valuable proteins and other useful compounds. This integrated strategy boosts overall efficiency and enhances the sustainability of biohydrogen production (Abubakar Musa Shuaibu et al., 2024).

Biological hydrogen production is considered as an eco-friendly method for generating sustainable energy from biomass. Hydrogen is a clean fuel because its combustion in fuel cells does not release carbon dioxide (Park et al., 2009). Biohydrogen (bio-H_2) through dark fermentation or photo-fermentation using anaerobic microorganisms is converted from algal biomass, whether whole or post, oil and/or starch extraction. Biochemical compounds such as cellulose, alginates, sulfated fucans, and proteins are found in marine macroalgae particularly in brown seaweeds (Kloareg et al., 1986). Brown seaweeds are considered suitable substrates because of the carbohydrate content for hydrogen production via anaerobic fermentation. The following reactions show how anaerobic bacteria that produce hydrogen break down organic substrates.

$$\text{Glucose} + 2H_2O \rightarrow 2\,\text{Acetate} + 2CO_2 + 4H_2$$

$$\text{Glucose} \rightarrow \text{Butyrate} + 2CO_2 + 2H_2$$

During photochemical hydrogen formation in cyanobacteria and algae, water molecules are broken down into oxygen and hydrogen by hydrogenase enzymes (Adams, 1990). In cyanobacteria, nitrogenase supports light-dependent hydrogen generation, whereas hydrogenase becomes active during anaerobic darkness (Geetanjali Yadav et al., 2015).

The well-known hydrogen producers are Enterobacter and Clostridium species use various carbon substrates. In the same way, among different algal species tested, it was found that *Ulva lactuca*, *Porphyra tenera*, *Undaria pinnatifida*, and *Laminaria japonica*, the latter exhibited the highest hydrogen yield, producing about 4164 mL of hydrogen from 50 g/L of dry algal biomass over 50 hours under favorable conditions: 35°C, 5% substrate concentration, and initial pH 7.5 (Park et al., 2009).

Hydrogenase and nitrogenase are the two main enzymes responsible for hydrogen production.

1. Hydrogenase-mediated hydrogen production:
   Hydrogenases catalyze the reversible reduction of protons and electrons to form molecular hydrogen:

$$2H^+ + 2e^- \rightleftharpoons H_2$$

   Based on metal composition at active sites, hydrogenases are classified into three types: NiFe-, Fe-, and metal-free hydrogenases. *Scenedesmus* species have been reported to evolve hydrogen under light exposure following anaerobic and dark incubation (Park et al., 2009).
2. Nitrogenase-mediated hydrogen production:
   Nitrogenase, responsible for nitrogen fixation, is mainly found in prokaryotes such as Cyanobacteria but is absent in eukaryotic microalgae. It is a two-subunit enzyme complex requiring MgATP (2 ATP per electron) and low-potential electrons from reduced ferredoxin or flavodoxin to reduce substrates. During nitrogen fixation, molecular nitrogen is converted to ammonia as follows:

$$N_2 + 6H^+ + 6e^- \rightarrow 2NH_2$$

$$12\,ATP \rightarrow 12\,ADP + 12\,Pi$$

   In the absence of nitrogen gas (e.g., under argon), nitrogenase reduces protons to produce hydrogen:

$$2H^+ + 2e^- \rightarrow H_2$$

This hydrogen evolution occurs as a side reaction, typically at one-third to one-fourth the rate of nitrogen fixation, even under nitrogen-saturated conditions. Nitrogen reduction to ammonia by nitrogenase has been documented in *Anabaena cylindrica* (Park et al., 2009).

3.4.5 Algal Biofuels

Algal biomass is made up of both organic and inorganic materials. The organic part contains large complex molecules like proteins, polysaccharides, lipids, and nucleic acids. During anaerobic digestion (AD), this organic matter is broken down into final products like methane and carbon dioxide, along with new biomass and inorganic residues. AD involves four main stages: hydrolysis, acidogenesis (or fermentation), acetogenesis, and methanogenesis. In the first stage, bacteria break down the complex organic polymers (carbohydrates, lipids, proteins) into simpler molecules called monomers, which are then fermented into methane-rich biogas (about 50%–75% methane). Carbon dioxide makes up the remaining 25–50% of biogas and must be removed before the methane can be used for generating electricity.

1. Hydrolysis: In the first phase, the complex organic polymers in algal biomass are broken down into smaller components that anaerobic bacteria can access. In this process, the smaller molecules are dissolved in the solution.

2. Acidogenesis (Fermentation): During this phase, fermentative bacteria convert these smaller molecules, like sugars, amino acids, and fatty acids, into volatile fatty acids (VFAs) along with different alcohols.
3. Acetogenesis: In this phase, these VFAs and alcohols are digested by acetogenic bacteria into acetic acid, hydrogen, and carbon dioxide.
4. Methanogenesis: This final phase involves methanogens that use the products of the earlier stages to produce methane, carbon dioxide, and water. The result is biogas, a combustible gas mixture.

There are various factors that influence biogas production efficiency. When producing biogas from algae, two main options are producing biodiesel from microalgae and then using the leftover residues for biogas production, or directly digesting whole microalgae or macroalgae to generate biogas as the primary biofuel (Geetanjali Yadav et al., 2015) (Table 3.1).

TABLE 3.1
Different Biofuels with the Major Sources, Key Production Process and Optimization, and the Typical Yield

Biofuel Type	Major Sources	Key Production Process and Optimization	Typical Yield	Reference
Bioethanol	Algae, sugarcane, lignocellulosic biomass	Fermentation (using yeast); optimal sugar and nutrient content, controlled temperature, strain selection	Up to 0.51 kg/kg glucose	Abubakar Musa Shuaibu et al. (2024)
Biodiesel	Algal oil, used cooking oil, oilseeds (canola)	Transesterification of triglycerides; optimization of reaction temp, catalyst, molar ratio	Typically, 80–90% conversion of input oil	Abubakar Musa Shuaibu et al. (2024)
Biogas/ Methane	Organic wastes, algae, animal/ manure, residues	Anaerobic digestion; optimal substrate mix, temperature (35–45°C), pH, retention time	Methane: 0.25–0.30 m^3/kg VS (volatile solids)	Jameel et al. (2024)
Biohydrogen	Algal/ cyanobacterial biomass, brown seaweeds	Dark/photo fermentation; optimization of substrate (e.g., mannitol), temp (35°C), pH 7.5	~4164 mL H_2 from 50 g/L dry macroalgae in 50 h	Park et al. (2009)
Algal biofuels	Microalgae/ macroalgae	Hydrolysis, anaerobic digestion, sequential valorization of oil/carb residues; retention time, temp	Biogas: 0.3–0.4 m^3/ kg VS; Biodiesel: varies by lipid content	Geetanjali Yadav et al. (2015)

3.5 RECENT ADVANCES IN PRECISION FERMENTATION

Improving the productivity and efficiency of biofuel technologies requires optimizing recent developments in lignocellulosic biomass conversion, diversifying feedstocks to ensure viability within ecological and economic boundaries, and designing high-performance molecules that enhance fuel efficiency while reducing carbon emissions. Along with technological improvements, it is important to integrate social, economic, and environmental dimensions to achieve sustainable, consistent, and cost-effective biofuel production systems.

With the adoption of innovative technologies and biological approaches, biofuel generation from different biological resources is expanding globally. Technological innovations have advanced in the development of next-generation biofuels with improved compatibility and performance. The major focus lies in producing drop-in biofuels, as they are chemically identical to traditional fuels and can easily integrate with existing engines and infrastructures. With advanced refining and catalytic methods, new biofuel variants such as renewable diesel, bio-based jet fuels, and bio-derived chemicals are being developed as sustainable alternatives to petroleum-based fuels.

The biofuel processing technologies have been improved with the recent progress in fermentation and enzymatic hydrolysis methods. The revolution in biotechnology and genetic engineering production has improved feedstock properties and optimized transformation processes. Genetic modification can improve crop traits like biomass yield, stress resistance, and nutrient uptake, resulting in more vigorous and energy-rich feedstocks that are ideal for biofuel production. Advancements such as genetically engineered microbes and enzymes enable the efficient conversion of complex biomass into biofuels, contributing to both environmental sustainability and economic feasibility.

Biotechnology also provides facilities for the development of microbial strains capable of producing valuable biofuels and bioproducts from various sources like waste materials and lignocellulosic residues. Using genetic engineering and biotechnological techniques, scientists and the industry can overcome major limitations of traditional biofuel systems, such as low conversion efficiency and narrow substrate range, thus enabling biofuels to become clean, renewable, and scalable energy options. The biofuel sector is refined with technological breakthroughs in feedstock cultivation, conversion pathways, and product innovation. With advancements in precision agriculture, biotechnology, and sophisticated processing, biofuels are becoming increasingly efficient, sustainable, and competitive alternatives to fossil fuels, paving the way for a greener global energy *transition* (Sulis et al., 2025).

3.5.1 BIOREACTORS

The ancient civilizations such as the Egyptians, Mesopotamians, Romans, and Greeks used ceramic containers used by to ferment foods and beverages like bread, wine, beer, and mead which were the earliest bioreactors in human history. During that time, people were not aware about the microorganisms or fermentation science. The biotechnology field developed rapidly and steadily following the discovery of microorganisms in the 19th century. Louis Pasteur's discovery of microorganisms made a move in the focus to experimental microbiology. Microbial butanol was produced by Pasteur using a mixed bacterial culture that included at least one Clostridium species.

In the early 20th century, during World War I, Chaim Weizmann applied a *Clostridium* strain to generate acetone for manufacturing artillery propellants. The design of industrial-scale bioreactors was led after Alexander Fleming's discovery of penicillin created an urgent need for large-scale antibiotic production. As there were needs, by 1945, these systems were capable of producing up to seven trillion units of penicillin. By the latter half of the 20th century, the people witnessed transformative advancements in bioreactor technology, including innovations in sterilization techniques, aeration and agitation systems, and the development of multivessel configurations that paved way for the simultaneous production of multiple products. The modern reactors, fermenters, and processing units were built to handle significantly greater quantities of material at once. This shift toward large-scale systems improved industrial efficiency in several ways. Higher capacities allowed continuous processing of greater biomass volumes, reducing downtime and increasing overall productivity. Therefore, industries achieved higher yields, streamlined their production workflows, and made biofuel manufacturing more commercially effective (Palladino et al., 2024).

In 1980, bioreactor research expanded beyond microbial systems to the cultivation of animal and plant cells, evolving industrial needs to process different types of feedstocks, which led engineers and researchers to design entirely new types of reactors. These reactors were customized with different features such as optimized mixing patterns, enhanced temperature and pressure control, improved mass and heat transfer, and advanced monitoring systems. As industries introduced new processes and operated under more complex conditions, traditional reactor designs were no longer sufficient.

In the 21st century, the integration of automation, AI, and 3D printing has changed bioreactor design, making these systems more adaptive, precise, and efficient. Contemporary bioreactor designs focus on safe, cost-effective, and scalable production of biotechnology-derived products.

Bioreactors remain fundamental to the bioprocessing industry, serving as the critical interface between laboratory-scale research and industrial-scale production. Each bioreactor design offers distinct advantages and limitations, depending on its application (Palladino et al., 2024).

At present, bioreactors are indispensable instruments in modern biotechnology, enabling the cultivation of microbial, plant, and animal cells for the production of high-value bioproducts. To maintain optimal conditions for cell growth and productivity, the industries incorporated advanced control systems for parameters that include pH, temperature, aeration, agitation, nutrient supply, and product removal. The selection of an appropriate bioreactor type, with consideration of its advantages and drawbacks, is important for achieving process efficiency (Palladino et al., 2024).

3.5.2 Optimization Techniques

The bio-process engineering has improved along with the growth in AI and digital twins. This has shown renewed industrial focus on modeling and data-driven analytics to support innovation and technology pathway selection. Particularly, precision fermentation has involved machine learning (ML) and advanced statistical approaches such as high-throughput sequencing, omics-based profiling, and automated screening to refine metabolic pathways, optimize microbial strains, and boost process yields (Okolie, 2024).

To find complex links between experimental variables and performance outcomes ML algorithms, including Gaussian process regression, support vector machines, and neural networks, are particularly effective. To improve experimental design, researchers used ML tools to analyze data from published studies on important fermentation conditions in *Yarrowia lipolytica*. They found that Gaussian process models gave the most accurate predictions. Similarly, another team of researchers developed a system called PERISCOPE-Opt, which used methods such as XGBoost, random forests, and support vector machines to predict the best fermentation settings. Together, these studies showed the use ML in optimizing fermentation processes (Awogbemi & Kallon, 2023).

A recent technology focuses on applying AI-based techniques to make the work quick with detailed physics simulations. Though these methods are only beginning to gain traction in the fermentation field, but still using data-driven models and machine learning-enhanced computational fluid dynamics (CFD), researchers are now able to model larger and more complex systems with improved computational efficiency. These technologies show results for solving long-standing scale-up challenges, particularly problems related to gas-transfer limitations in the process (Okolie, 2024).

There is growing interest in using low-carbon materials and converting waste streams into valuable products to support circular bioprocessing, as renewable feedstocks have become important. Executing these systems is challenging because of the biological complexity and fluctuating composition of the feedstocks. To solve all these problems, some methods were implemented, such as advanced data-driven methods, better experimental design approaches like Bayesian optimization through which the productivity increased (Arif et al., 2025).

In the last five years, though there are still several experiments being done, engineering research has increasingly focused on creating predictive surrogate models for processes that are hard to understand or too time-consuming to simulate. Using these surrogate models in bigger optimization problems has now become common, that is not just in bioprocessing, but also for modeling complex physical systems. Recent progress in Bayesian optimization and physics-informed AI gives new ways to combine the strengths of traditional modeling with the speed and flexibility of modern data-driven methods (Okolie, 2024).

3.5.3 Process Control and Monitoring

The bioreactor technologies have recently progressed and have greatly enhanced fermentation efficiency through the integration of real-time monitoring and control systems. Modern bioreactors with advanced sensors are capable of continuously measuring key parameters such as pH, temperature, dissolved oxygen, and biomass concentration without requiring sample withdrawal. Hence, this method helps in minimizing contamination risks and improving batch uniformity (Niyigaba et al., 2025). There are also other new sensing approaches that include fluorescence-based and turbidimetric methods. This method has enabled direct quantification of cell mass, substrates, and product concentrations. Along with these approaches, portable and multiplexed thread-based microprobes have been developed for continuous tracking of pH and ammonium ions during fermentation (Niyigaba et al., 2025).

Joining sensor data with computational models has advanced process fingerprinting, smart control, and predictive capabilities in biopharmaceutical manufacturing. These innovations have fostered automation and digital transformation that led to improved insight and control in cell culture operations. For instance, sensor-integrated bioreactors markedly improved product homogeneity and minimized contamination risks. Advanced online monitoring enables real-time parameter adjustments, enhancing culture stability and reliability that are considered to be essential factors for optimal growth and contamination prevention in bioprocessing environments. These advancements contribute to higher process efficiency and the consistent production of quality biologics that align with industry priorities for safety and reliability (Niyigaba et al., 2025).

There are also smart packaging technologies which has revolutionized the food supply chain. This is done by integrating sensors, indicators, and communication systems for real-time quality and safety monitoring during transportation and storage (Niyigaba et al., 2025). The advantages of these technologies include enhanced traceability, reduced food waste, and greater consumer trust, particularly important for perishable or fermented products. Smart packaging systems using time-temperature indicators and gas sensors provide continuous feedback on product integrity, enabling early detection of spoilage or microbial growth. These materials monitor essential factors such as temperature, humidity, oxygen concentration, microbial activity, pH, and chemical composition throughout distribution. Active functions like ethylene and oxygen scavenging, humidity regulation, and freshness sensing in fruits and vegetables help mitigate deterioration during fermentation.

These types of intelligent interventions aim to lower food loss, extend shelf life, and uphold quality across the supply chain, a critical consideration for fermented foods with short shelf stability. But widespread implementation faces barriers like high costs, consumer perception, and regulatory compliance, which must be addressed for commercial viability (Niyigaba et al., 2025). Many advanced systems also incorporate oxygen scavengers, desiccants, and phase change materials to sustain ideal temperature and moisture levels. Further, this helps to improve consumer awareness of product freshness and expiration (Niyigaba et al., 2025). Although several prototype sensing systems exist, future research should focus on resolving safety, regulatory, and cost-related challenges to enable large-scale adoption of smart packaging solutions tailored to fermentation-based industries (Niyigaba et al., 2025).

3.6 CASE STUDIES

There are some case studies quoted to show that edible filamentous fungi such as Neurospora intermedia and Aspergillus oryzae can be integrated into sugar- or molasses-to-ethanol processes to valorize vinasse, a polluting by-product, by converting it into additional ethanol and protein-rich fungal biomass for animal feed. This biorefinery approach enhances ethanol production, reduces waste pollution, and produces valuable feed, improving overall process sustainability.

Similarly, pilot studies with lignocellulosic biomass like wheat straw show that combining dilute acid pretreatment with fungal fermentation and biogas production can maximize energy recovery and generate protein-rich feed. These integrated fungal biorefineries offer promising sustainable solutions for improving bioenergy efficiency and waste valorization in ethanol industries.

3.6.1 Commercial Examples for Bioenergy Production

The study investigated the integration of edible ascomycetes filamentous fungi, Neurospora intermedia and Aspergillus oryzae, into existing sugar- or molasses-to-ethanol processes to valorize vinasse, a major by-product of ethanol distillation. Vinasse, produced at 10–15 m^3 per m^3 of ethanol, is an acidic, high-organic-load waste (COD 50–150 g/L) that poses environmental challenges like soil pollution and greenhouse gas emissions when used as fertilizer. Traditional disposal methods, including AD, are hindered by sulfate content promoting sulfide formation and inhibits methane production. This approach uses vinasse as a substrate for fungal cultivation to recover energy as extra ethanol and generate protein-rich fungal biomass for animal feed (Ferreira et al., 2015).

Sugarcane and beet molasses are produced in large amounts—about 61 million tons per year—and are important raw materials for making bioethanol. Countries like Brazil and India often have combined sugar and ethanol factories. In this study, the researchers used two safe, food-grade fungi. *Neurospora intermedia*, which is used in Indonesian oncom, is good at fermenting ethanol from starch and plant materials. *Aspergillus oryzae*, known as the yellow koji mold, is widely used in Japanese foods like sake and miso and can produce many useful enzymes. The fungi were grown in shake flasks and airlift bioreactors using diluted vinasse, which is the leftover liquid from ethanol production. The goal was to find the best conditions for producing biomass and ethanol while also lowering COD and reducing the thickness (viscosity) of the vinasse (Ferreira et al., 2015).

The results showed that both fungi were able to use the glycerol and sugars in vinasse very well. *N. intermedia* produced 202.4 g/L of dry biomass and 11.8 g/L of ethanol, while *A. oryzae* produced 222.8 g/L of biomass and 10.2 g/L of ethanol. The vinasse also became cleaner, with COD reduced by 34% and viscosity reduced by 21%, which helps in treating the waste. For a large ethanol plant producing 100,000 m^3 of ethanol per year, adding this fungal process could create 200,000–250,000 tons of dry fungal biomass each year. This biomass contains 40–45% protein, making it a valuable animal-feed source. The process would also generate 8,800–12,600 m^3 of extra ethanol, increasing total ethanol output by 8.8–12.6%. This makes the system useful for both waste treatment and producing high-protein feed during a time of rising global protein needs (Nair & Taherzadeh, 2016).

These results show that adding this biorefinery process turns vinasse from a waste problem into a useful resource. It increases ethanol production, creates nutritious fungal biomass for feed, and reduces pollution at the same time. This approach fits well with circular economy ideas in the sugar–ethanol industry. The fungi *N. intermedia* and *A. oryzae* both worked well with industrial waste (Nair & Taherzadeh, 2016).

Another study looked at using olive mill wastewater (OMW) as a material for producing single-cell oil with Zygomycetes fungi. OMW is a very polluting by-product from olive oil production in the Mediterranean. The researchers tested whether certain oil-producing fungi could grow on this waste and turn it into useful lipids.

They found that fungi such as *Thamnidium elegans* and *Zygorhynchus moelleri* grew well even when the medium contained up to 50% OMW. These fungi produced good amounts of biomass (3.5–4.4 g/L) and had a high lipid content of about 60%. The lipids mainly included oleic and palmitic acids. *Z. moelleri* also produced gamma-linolenic acid (up to 17.7%), which is valuable for nutraceutical and pharmaceutical uses.

The process also removed many phenolic compounds from the wastewater, helping reduce its toxicity and environmental impact. Because OMW has a high organic load and is difficult to dispose of safely, converting it into single-cell oils provides a useful way to treat the waste while creating valuable bioactive lipids.

The study's kinetic modeling showed that the fungi used the substrate efficiently and produced lipids effectively. The work demonstrates that Zygomycetes fungi can be good candidates for sustainable OMW treatment and the production of valuable bioproducts (Bellou et al., 2014).

3.6.2 Pilot Studies and Practical Implications

There are successful pilot and demonstration-scale technologies showing ethanol production from lignocellulose, but commercial viability remains challenging. Therefore, economic feasibility can be enhanced by adopting a biorefinery approach. This integration in process of production of biofuels, chemicals, heat, or electricity from lignocellulosic waste to maximize value extraction depending on market conditions. Wheat straw is an important lignocellulosic feedstock due to its low cost and abundance as an agricultural residue the significance has been reduced globally.

The integrated biorefinery concept combining ethanol and biogas production from wheat straw has been less studied, despite wheat straw's suitable availability. Pretreatment is important because of wheat straw's complex cellulose, hemicellulose, and lignin composition. Traditional sulfur-based pretreatments can inhibit biogas quality by promoting sulfate-reducing bacteria that produce corrosive hydrogen sulfide. Dilute phosphoric acid pretreatment is a promising alternative, improving both ethanol yield (through nutrient benefits) and biogas production.

A study demonstrated the integrated process at demonstration scale using dilute phosphoric acid pretreatment, followed by fermentation with the edible fungus Neurospora intermedia, and biogas production from fermentation residues. This approach also produces protein-rich fungal biomass that is used as animal or fish feed, addressing global feed demand. Methane production from thin stillage, a by-product of wheat starch ethanol processing, was also evaluated to complete the integrated biorefinery concept.

The study showed that milder pretreatment favored ethanol and fungal biomass yields. The study also revealed that harsher pretreatment enhanced biogas production. Co-digestion of fermentation syrup and thin stillage further boosted methane generation. The fungal biomass had high nutritional value, providing a valuable alternative protein feed source. The integrated process achieves near-theoretical ethanol yields, significant fungal biomass generation, and enhanced methane production, optimizing energy and economic returns from wheat straw.

This integrated wheat straw biorefinery model, balancing ethanol, biogas, and feed production via tailored pretreatment severity, offers a novel route for lignocellulosic biofuel commercialization. Further techno-economic and lifecycle assessments are necessary for industrial-scale implementation and sustainable adoption, and this was discussed in the study (Nair et al., 2018).

3.7 CHALLENGES

Though precision fermentation has advanced technologies there are still some challenges in the field of application. One of the primary challenges in precision fermentation is scaling up from laboratory to industrial production. Scaling precision fermentation introduces complex technical challenges such as maintaining culture purity, ensuring stable environmental conditions, and replicating small-scale process efficiencies on a much larger scale. Innovations in bioreactor design, automation, and process optimization are crucial to overcoming these hurdles. However, the difficulty of replicating lab-scale conditions in massive industrial bioreactors leads to increased production costs and risks of inconsistent product quality. When it comes to industrial production, it becomes difficult to maintain consistent and optimized strains for reliable yields. The industry faces a significant capacity gap that must be bridged to meet growing demand, requiring substantial investments in infrastructure and technology. Iterative prototyping and pilot-scale testing are essential for a seamless transition to commercial-scale manufacturing. Regulatory approval processes for novel foods and ingredients are also challenging. This process tends to be lengthy, underscoring the need for early and proactive engagement with regulatory agencies to expedite product approvals and market entry. The other challenge is that transparency with consumers about production practices and early regulatory engagement remain key to accelerating adoption and building confidence in precision fermentation products. Consumer concerns regarding the safety of fermentation-derived products require transparent communication strategies that emphasize environmental benefits to foster trust and acceptance. Adding more to the above challenges, the high production costs compared to traditional fermentation pose significant barriers; thus, demonstrating superior product quality, functionality, and sustainability becomes crucial for differentiation in the market (Augustin et al., 2024). The other challenges come from the market. Establishing market presence in a competitive environment is often a challenge, and that is aided by strategic partnerships with established brands and industry players, which help extend reach and credibility.

Maintaining product quality during scale-up is analogous to navigating turbulent conditions, necessitating robust quality control frameworks and precise process validation. Collaborative efforts, continuous research, and automation advancements collectively pave the way for sustainable and scalable precision fermentation (Verma et al., 2025). These are the various challenges in the precision fermentation technology, and the same has been depicted in Figure 3.2.

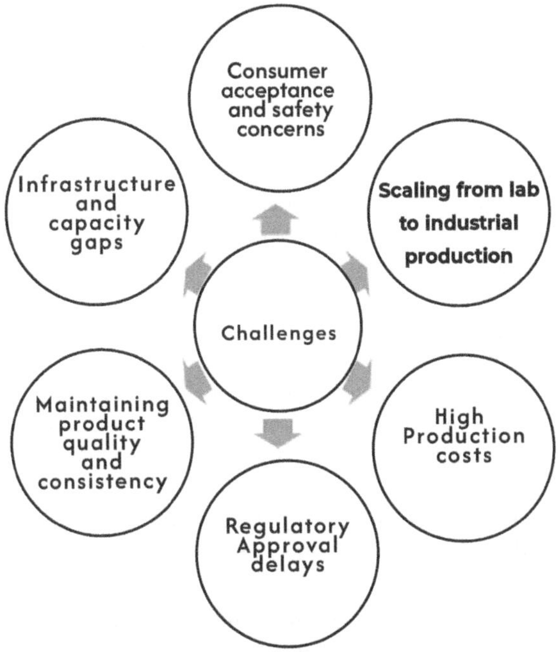

FIGURE 3.2 Challenges in precision fermentation.

3.8 FUTURE DIRECTIONS

The future direction of precision fermentation includes the recent developments in smart fermentation systems using ML and Internet of Things (IoT) technologies to increase the efficiency and control of fermentation processes. ML is applied in metabolic engineering. Using advanced algorithms to fine-tune parameters like temperature, humidity, and microbial composition to facilitate strain development and optimize metabolic fluxes. IoT supports real-time monitoring and remote automation of fermentation, allowing dynamic adjustments to critical variables. By integrating constraint-based modeling with ML, it further improves predictability and performance in fermentation analysis.

AI and ML in food fermentation have advanced quality control and process optimization, which can enable improved operational efficiency, sales forecasting, and food safety. For instance, applying ML to Raman spectroscopy data in yeast fermentation significantly reduced model training time and enhanced the accuracy of predicting key metabolites and cell concentration. In beer production, AI combined with sensor networks like electronic noses and robotics achieves high accuracy in identifying fermentation types and forecasting consumer acceptability, providing brewers with precise, cost-effective quality assessment tools.

Forecasting microbial growth, aiding timely interventions, and optimizing processing can be done using predictive microbiology models integrated with sensor technologies have enhanced food safety by accurately. Real-time environmental sensors coupled with predictive models improve supply chain management, although challenges remain in modeling dynamic microbial interactions. In future, AI will play a key role in managing microbial communities in fermentation, fostering beneficial microbes while controlling pathogens. Therefore, reliance on AI-driven automation must be balanced with human expertise for quality control.

Recent studies show that ML works well for improving fermentation yields, spotting contamination, and predicting molecular properties. This is done by using better parameter tuning and advanced data analysis. Such approaches reduce resource-intensive experimentation and enhance process robustness in biotechnological and food fermentation applications (Niyigaba et al., 2025; Packiam et al., 2022).

3.9 CONCLUSION

Precision fermentation is a new technology invented to produce clean energy and products by using the microbes and chemicals produced from plants. At present, scientists make use of different advanced special tools, biological information, and computer programs to have control over production in the industries. Precision fermentation has made the production more efficient and at the same time, less expensive. There are different methods in precision fermentation, particularly algae are being used to make fuels, like bioethanol, biodiesel, biogas, and biohydrogen, because algae are good, renewable resources that can be used in many ways.

The latest advancement in the fermentation process using smart systems and designs has resulted in the best results. Even though there are challenges like growing the process to the industrial scale, legal rules, customer trust, and high costs, ongoing research and teamwork in the industry are helping to solve these problems. To conclude, precision fermentation provides a different method for producing renewable energy and useful products that benefit both the environment and the economy.

REFERENCES

Adams, M. W. W. (1990). The structure and mechanism of iron-hydrogenases. *Biochimica et Biophysica Acta (BBA) - Bioenergetics*, *1020*(2), 115–145. https://doi.org/10.1016/0005-2728(90)90044-5

Aljabali, A. A. A., El-Tanani, M., & Tambuwala, M. M. (2024). Principles of CRISPR-Cas9 technology: Advancements in genome editing and emerging trends in drug delivery. *Journal of Drug Delivery Science and Technology*, *92*, 105338. https://doi.org/10.1016/j.jddst.2024.105338

Arif, M., Alalawy, A. I., Zheng, Y., Koutb, M., Kareri, T., Salama, E. S., & Li, X. (2025). Artificial intelligence and machine learning models application in biodiesel optimization process and fuel properties prediction. *Sustainable Energy Technologies and Assessments*, *73*, 104097. https://doi.org/10.1016/J.SETA.2024.104097

Augustin, M. A., Hartley, C. J., Maloney, G., & Tyndall, S. (2024). Innovation in precision fermentation for food ingredients. *Critical Reviews in Food Science and Nutrition*, *64*(18), 6218–6238. https://doi.org/10.1080/10408398.2023.2166014

Awogbemi, O., & Kallon, D. V. Von. (2023). Application of machine learning technologies in biodiesel production process—A review. *Frontiers in Energy Research*, *11*. https://doi.org/10.3389/fenrg.2023.1122638

Bellou, S., Makri, A., Sarris, D., Michos, K., Rentoumi, P., Celik, A., Papanikolaou, S., & Aggelis, G. (2014). The olive mill wastewater as substrate for single cell oil production by Zygomycetes. *Journal of Biotechnology*, *170*(1), 50–59. https://doi.org/10.1016/J.JBIOTEC.2013.11.015

Ferreira, J. A., Lennartsson, P. R., & Taherzadeh, M. J. (2015). Production of ethanol and biomass from thin stillage by *Neurospora intermedia*: A pilot study for process diversification. *Engineering in Life Sciences*, *15*(8), 751–759. https://doi.org/10.1002/elsc.201400213

Hossain, A. M. (2021). *CRISPR-Cas9: A Fascinating Journey from Bacterial Immune System to Human Gene Editing* (pp. 63–83). https://doi.org/10.1016/bs.pmbts.2021.01.001

Jameel, M. K., Mustafa, M. A., Ahmed, H. S., Mohammed, A. J., Ghazy, H., Shakir, M. N., Lawas, A. M., Mohammed, S. K., Idan, A. H., Mahmoud, Z. H., Sayadi, H., & Kianfar, E. (2024). Biogas: Production, properties, applications, economic and challenges: A review. *Results in Chemistry*, *7*, 101549. https://doi.org/10.1016/J.RECHEM.2024.101549

Kloareg, B., Demarty, M., & Mabeau, S. (1986). Polyanionic characteristics of purified sulphated homofucans from brown algae. *International Journal of Biological Macromolecules*, *8*(6), 380–386. https://doi.org/10.1016/0141-8130(86)90060-7

Lee, S. Y., Mattanovich, D., & Villaverde, A. (2012). Systems metabolic engineering, industrial biotechnology and microbial cell factories. *Microbial Cell Factories*, *11*, 156. https://doi.org/10.1186/1475-2859-11-156

Lotfi, M., Morshedi Rad, D., Mashhadi, S. S., Ashouri, A., Mojarrad, M., Mozaffari-Jovin, S., Farrokhi, S., Hashemi, M., Lotfi, M., Ebrahimi Warkiani, M., & Abbaszadegan, M. R. (2023). Recent advances in CRISPR/Cas9 delivery approaches for therapeutic gene editing of stem cells. *Stem Cell Reviews and Reports*, *19*(8), 2576–2596. https://doi.org/10.1007/s12015-023-10585-3

Mengstie, M. A., & Wondimu, B. Z. (2021). Mechanism and applications of CRISPR/Cas-9-mediated genome editing. *Biologics: Targets and Therapy*, *15*, 353–361 . https://doi.org/10.2147/BTT.S326422

Nair, R. B., Kabir, M. M., Lennartsson, P. R., Taherzadeh, M. J., & Horváth, I. S. (2018). Integrated process for ethanol, biogas, and edible filamentous fungi-based animal feed production from dilute phosphoric acid-pretreated wheat straw. *Applied Biochemistry and Biotechnology*, *184*(1), 48–62. https://doi.org/10.1007/s12010-017-2525-1

Nair, R. B., & Taherzadeh, M. J. (2016). Valorization of sugar-to-ethanol process waste vinasse: A novel biorefinery approach using edible ascomycetes filamentous fungi. *Bioresource Technology*, *221*, 469–476. https://doi.org/10.1016/J.BIORTECH.2016.09.074

Niyigaba, T., Küçükgöz, K., Kołożyn-Krajewska, D., Królikowski, T., & Trząskowska, M. (2025). Advances in fermentation technology: A focus on health and safety. *Applied Sciences*, *15*(6), 3001. https://doi.org/10.3390/app15063001

Okolie, J. A. (2024). Introduction of machine learning and artificial intelligence in biofuel technology. *Current Opinion in Green and Sustainable Chemistry*, *47*, 100928. https://doi.org/10.1016/J.COGSC.2024.100928

Otero-Muras, I., & Carbonell, P. (2021). Automated engineering of synthetic metabolic pathways for efficient biomanufacturing. *Metabolic Engineering*, *63*, 61–80. https://doi.org/10.1016/j.ymben.2020.11.012

Packiam, K. A. R., Ooi, C. W., Li, F., Mei, S., Tey, B. T., Ong, H. F., Song, J., & Ramanan, R. N. (2022). PERISCOPE-Opt: Machine learning-based prediction of optimal fermentation conditions and yields of recombinant periplasmic protein expressed in Escherichia coli. *Computational and Structural Biotechnology Journal*, *20*, 2909–2920. https://doi.org/10.1016/j.csbj.2022.06.006

Paley, E. L. (2021). Microbial shikimate pathway in diseases. *Microbial Metabolism and Disease*, 67–76. https://doi.org/10.1016/B978-0-323-88445-7.00007-5

Palladino, F., Marcelino, P. R. F., Schlogl, A. E., José, Á. H. M., Rodrigues, R. de C. L. B., Fabrino, D. L., Santos, I. J. B., & Rosa, C. A. (2024). Bioreactors: Applications and Innovations for a Sustainable and Healthy Future—A Critical Review. *Applied Sciences, 14*(20), 9346. https://doi.org/10.3390/app14209346

Park, J.-I., Lee, J., Sim, S. J., & Lee, J.-H. (2009). Production of hydrogen from marine macroalgae biomass using anaerobic sewage sludge microflora. *Biotechnology and Bioprocess Engineering, 14*(3), 307–315. https://doi.org/10.1007/s12257-008-0241-y

Sanchez, S., & Demain, A. L. (2002). Metabolic regulation of fermentation processes. *Enzyme and Microbial Technology, 31*(7), 895–906. https://doi.org/10.1016/S0141-0229(02)00172-2

Sharma, N., Bohra, B., Pragya, N., Ciannella, R., Dobie, P., & Lehmann, S. (2016). Bioenergy from agroforestry can lead to improved food security, climate change, soil quality, and rural development. *Food and Energy Security, 5*(3), 165–183. https://doi.org/10.1002/fes3.87

Shiue, E., & Prather, K. L. J. (2012). Synthetic biology devices as tools for metabolic engineering. *Biochemical Engineering Journal, 65*, 82–89. https://doi.org/10.1016/j.bej.2012.04.006

Shuaibu, A. M., Alhassan, U. H., Shafi'I, A. M., Kolere, M. S., Sharma, R., & Abdurrashid, I. (2024). Biofuel: A sustainable and clean alternative to fossil fuel. *GSC Advanced Research and Reviews, 21*(2), 204–222.

Soccol, C. R., Vandenberghe, L. P. de S., Medeiros, A. B. P., Karp, S. G., Buckeridge, M., Ramos, L. P., Pitarelo, A. P., Ferreira-Leitão, V., Gottschalk, L. M. F., Ferrara, M. A., Silva Bon, E. P. da, Moraes, L. M. P. de, Araújo, J. de A., & Torres, F. A. G. (2010). Bioethanol from lignocelluloses: Status and perspectives in Brazil. *Bioresource Technology, 101*(13), 4820–4825. https://doi.org/10.1016/J.BIORTECH.2009.11.067

Stanbury, P. F., Whitaker, A., & Hall, S. J. (1995). *Principles of Fermentation Technology* (2nd ed.). Butterworth-Heinemann Publications.

Sulis, D. B., Lavoine, N., Sederoff, H., Jiang, X., Marques, B. M., Lan, K., Cofre-Vega, C., Barrangou, R., & Wang, J. P. (2025). Advances in lignocellulosic feedstocks for bioenergy and bioproducts. *Nature Communications, 16*(1), 1244. https://doi.org/10.1038/s41467-025-56472-y

Verma, K., Duhan, P., Pal, D., Verma, P., & Bansal, P. (2025). Precision fermentation for the next generation of food ingredients: Opportunities and challenges. *Future Foods, 12*, 100750. https://doi.org/10.1016/J.FUFO.2025.100750

Yadav, D., Tanveer, A., Malviya, N., & Yadav, S. (2018). Overview and principles of bioengineering: The drivers of omics technologies. *Omics Technologies and Bio-Engineering: Towards Improving Quality of Life, 1*, 3–23. https://doi.org/10.1016/B978-0-12-804659-3.00001-4

Yadav, G., Kumar, R., & Sen, R. (2015). Algal photobioreactors. In *Marine Bioenergy* (pp. 100–129). CRC Press. https://doi.org/10.1201/b18494-10

4 Selection of Microbial Population for Fermentation Process

Lekhavani Ramesh, Suguna Krishnamoorthy, Kumaraguru Sivakumar, Vinayagam Mohanavel, Ramalakshmi Subbarayalu, and Rajesh Pandiyan

4.1 INTRODUCTION

Fermentation is a biochemical process in which the sugars are converted into a new product by chemical reactions that are performed by microbes. Humans have been using fermentation, the natural process, from time immemorial to produce a range of products, some of them being foods, drugs, and fuels (Taveira et al., 2021). Fermentation enables the preservation of perishable foods due to their short shelf life and enhances the sensory qualities of the final product along with preservation (Yee et al., 2025). The food constituents' enzymatic or microbial processes are accountable for fermenting food to produce the desired biochemical transformations responsible for large-scale food change. Spontaneous fermentation was initially performed with the involvement of microbes occurring intrinsically within the feedstock. The microflora group undertaking fermentation reactions in this instance, however, remained unknown (Sionek et al., 2023).

Pasteur uncovered during the mid-19th century that microbial metabolism governs fermentative transformations, and this was the start of intentional selection of environmental microbes to maximise fermentation efficacy (Praveen & Brogi, 2025). In 1890, pure yeast cultures were used in wine production, whereas, in the early 20th century, pasteurisation techniques enhanced the regulation of microbial populations. These developments underscore the pivotal functional significance of microorganisms in modulating the physicochemical characteristics associated with fermented foods. Microbe-mediated fermentation is an intrinsically anaerobic process in which specific carbohydrates, such as those from grains or sugars, are broken down into acids, alcohols, or gases (Okoye et al., 2025). Thus, fermentation methods today still base much of their practice relying on thoroughly characterised microbial cultures to provide functionality, quality, and consistency in food manufacturing, both industrial and conventional.

Microbial fermentation refers to the anaerobic breakdown of nutrients, primarily sugars and other carbohydrates, by microbial enzymes (Sun et al., 2022). In general,

DOI: 10.1201/9781003684374-4

diverse microorganisms participate in the fermentation process, each producing unique enzymes that generate a complex array of by-products responsible for the characteristic aroma and flavour of fermented foods (R. Sharma et al., 2020). In microbial fermentation, bacteria, yeasts, and moulds are the most commonly used microorganisms. LAB encompasses *L. acidophilus* (H. Gao et al., 2022), *L. casei* (Qin et al., 2022), *L. bulgaricus* (Lee et al., 2024), *L. lactis* (Kondrotiene et al., 2023), *L. cremoris* (Poudel et al., 2022), etc., and are used in fermentation industries due to their symbiotic attributes. Whereas AAB includes *Saccharibacter* (He et al., 2022), *Acetobacter cerevisiae*, *Komagataeibacter europaeus* (Han et al., 2024), *Acetobacter pasteurianus*, *Acetobacter malorum*, and *Gluconoacetobacterentanii* (He et al., 2022) that are used in organic acid production due to their high acid-fermenting capacity. Yeasts, particularly *Saccharomyces cerevisiae* (Q. Yang et al., 2022) and *Geotrichumcandidum*, moulds including *Rhizopus oryzae*, *Aspergillus oryzae*, *Penicillium* spp, *P. camemberti*, and *P. roqueforti* are extensively studied for their role in alcoholic fermentation in beer and wine production, by metabolising sugars to yield alcohol and carbon dioxide (Praveen & Brogi, 2025).

Microbial selection is vital in fermentation because the specific microorganisms chosen directly influence the efficiency, yield, safety, and quality of the end product. Careful selection ensures that the microbes can efficiently utilise the substrate, tolerate process conditions, consistently produce the desired compounds, and outcompete undesirable or pathogenic organisms (Ji et al., 2023). The chosen microbes influence the aroma, flavour, nutrient, and texture content of the fermented commodity, as well as its repeatability between production lots (K. Zhang et al., 2023). Effective fermentation requires a capable microbial strain to be employed to metabolise the chosen substrate or feedstock. Various strains possess unique capacities for the metabolism of individual sugars, lignocellulosic substrates, or lipids. It is important to select a strain that can efficiently convert the target substrate into the desired target biofuel product (Siddiqui et al., 2023).

Effective selection of strains also provides control over safety by minimising the risk of contamination and spoilage; this is accomplished through the inhibition of unwanted or harmful microbes and preventing the production of toxins or off-flavours (Praveen & Brogi, 2025). Microbes also differ in their tolerance to temperature, pH, oxygen, and other environmental factors, so selecting strains that thrive under specific fermentation conditions ensures a robust process and reduces the risk of process failure (Ojeda-Linares et al., 2022). Product specificity is another key consideration; certain microbes are more suited for producing specific products like alcohol, organic acids, or various fermented foods, so microbial selection targets the most appropriate strains for the desired outcome (Foti et al., 2025). Genetic stability is important as well, since stable strains reduce variability in product characteristics, making industrial production more reliable and predictable over time (Duperray et al., 2024). Microbial selection can involve traditional starter cultures, selective breeding, or advanced genetic engineering, further to optimise strain performance for industrial or artisanal applications. There is limited understanding of interactions in mixed microbial communities, insufficient knowledge of stress tolerance and genetic stability, and a need for better tools to predict and optimise strain performance, especially for non-food applications (N. Sharma et al., 2025).

The chapter discusses these points by highlighting the research gap in understanding complex interactions within microbial consortia, particularly in mixed and co-culture fermentations, and notes that most current approaches focus on single-strain optimisation, often overlooking interspecies relationships that impact yield, quality, and flavour. The chapter covers principles and methods for selecting and optimising microbial populations in fermentation, discusses challenges, and highlights both traditional and advanced techniques for food, enzyme, and biofuel production.

4.2 CRITERIA FOR SELECTING MICROBIAL STRAINS

Microbial strain selection is a crucial step in fermentation. It serves a critical function in regulating the efficiency, productivity, and the integrity of the fermentation process. The microbial strain selected should be able to efficiently convert the substrate or feedstock available for the fermentation process.

4.2.1 Productivity

The yield and productivity of the chosen microbial strain are among the most significant issues to be addressed. High conversion efficiency strains can maximise the output from a fixed volume of feedstock (Mannan et al., 2025). The microorganism selected for production possesses favourable characteristics that directly impact the efficiency of fermentation. The capability to grow in minimal media lowers the cost of inputs and supports large-scale production. A higher growth rate may shorten fermentation time and increase productivity. This may recover end products, such as secretion into medium in extractable form, which reduces downstream processing demands. These features enhance both economic and practical feasibility of microbial production systems (Olsson et al., 2022). Overall, these criteria regulate the development and selection of microbial strains, which leads the fermentation process to produce maximum yield with minimal resources.

4.2.2 Substrate Utilisation

The substrate and substrate composition have a critical influence on fermentation properties and, indirectly, final product attributes (Zannini et al., 2022). Utilisation of mixed substrates may improve the capabilities of microbial growth. Introducing mixed cultures into mixed sugar substrates may help to produce high yields from microbes, as a single sugar or carbon substrate used for fermentation only improves the growth of a monoculture, which consumes only a single type of sugar or carbon (Rafieenia et al., 2022). Strains that tolerate substrate variability can maintain stable production, even when feedstock composition fluctuates, supporting scalable industrial processes. Inefficient substrate utilisation leads to incomplete fermentation, accumulation of intermediates, lower yields, and higher costs, making this criterion a primary driver in strain selection for optimal fermentation outcomes.

4.2.3 Environmental Tolerance

Environmental tolerance is the major criterion for fermentation, as it determines the microbial stability to withstand environmental and other physico-chemical conditions like pH, temperature, ethanolic concentration, humidity, and other stressors encountered during industrial fermentation (Mengesha et al., 2022). Microorganisms used in fermentation have their respective optimum conditions, which affect their growth rate and metabolism. Industrial fermentations often require microbial strains that can withstand fluctuations in these environmental factors to maintain consistent production at scale (Ye et al., 2024). Microbial environmental stressors influence the aroma, flavour, and quality of the final product, especially in food and beverage fermentation, by enhancing metabolic pathways under certain conditions.

4.2.4 pH

When selecting microbial strains for fermentation based on pH, consider a strain's tolerance and optimal growth range at the specific pH of the fermentation process, alongside other factors like substrate utilisation, product yield, and safety (Vulart et al., 2025). The chosen strain must be able to survive and function effectively within the fermentation's pH range. pH control determines the production of the end product during fermentation (Reale et al., 2020). Uncontrollable pH may accumulate toxic or non-dissociated organic acids that may collapse microbial energy and inactivate biomass. Maintaining pH at optimal conditions enhances microbial metabolism to sustain high ATP levels and enzymatic activity, translating into better yields (Vargas-Luna et al., 2025).

4.2.5 Temperature

Microbial strains for fermentation are selected based on their thermotolerance, optimal growth temperature, and productivity at specific temperatures used in the process. It should withstand a high temperature range during fermentation without losing efficiency or viability (Liszkowska & Berlowska, 2021). Selected strains should maintain metabolic stability and high enzymatic activity during the process, ensuring product yield and efficient substrate conversion. Strains tolerating high temperature retain genetic stability for large-scale fermentation under desirable temperature conditions (Zhang et al., 2021). The effects of temperature directly impact the production of volatile compounds and by-products during fermentation. Higher temperature influences the formation of volatile compounds, including acetaldehyde, ethyl esters, acetate esters, and higher alcohols. Whereas lower temperature enhances flavour, aroma, and complexity in fermented foods (Zhao et al., 2022). Thus, temperature plays a vital role in determining microbial performance and the overall success of the fermentation process.

4.2.6 Humidity

Selecting microbial strains for fermentation based on humidity involves assessing their ability to tolerate and adapt to moisture variations in the fermentation environment, as humidity affects microbial growth, metabolism, and product development

(Groff et al., 2024). Strains should be resistant to moisture content or relative humidity in the fermentation substrate or environment, maintaining viability and metabolic activity. Few strains tolerate low moisture levels without losing their metabolic activity, which is a critical function during the solid-state fermentation process. In the initial fermentation process, higher moisture content accelerates the growth of bacterial communities (Zhu et al., 2022). Humidity also impacts metabolic pathways, causing fluctuations in succession speed and community function. The microbial population may vary with humidity changes, so maintaining beneficial or dominant activity despite humidity variation is important. Resistant or effective strains grow and ferment in sustainable conditions across the expected humidity range of the process (Terefe et al., 2021).

4.2.7 Genetic Stability

Genetic integrity and metabolic robustness of production strains are key demands for industrial manufacturing of bio-based products through microbe-mediated fermentation. Instability of the producer strain is by no means accounted for owing to the fact that the reason could arise from genetic and/or non-genetic factors, whose interactions add complexity to the study and regulation of strain instability (Olsson et al., 2022). Large-scale biomanufacturing is frequently limited by challenges in maintaining the long-term stability of the production phenotype. Under a prolonged fermentation process, cells may undergo genetic instability due to environmental stressors. This decreases the amount of final product after fermentation. This may increase cost and impact the overall process and make long-term sustainable production difficult (Rugbjerg & Olsson, 2020).

4.2.8 Safety and Regulatory Compliance

Safety and Regulatory Compliance are fundamental criteria, especially in the food, drink, and fermentation industry, which sets the standard and the overall quality of the final product during production. This mainly focuses on GRAS (Generally Recognised as Safe) status for food and other by-product applications (Ashaolu, 2020). This designation applies to microorganisms that do not produce toxins during production. The Food and Drug Administration (FDA) authorises the usage of non-pathogenic microorganisms employed as legal microbial starter culture in food fermentation, ensuring consumer protection and product safety (Capozzi et al., 2020). Apart from the FDA, Good Manufacturing Practices (GMP), Hazard Analysis and Critical Control Points (HACCP) are also followed for safe and hygienic products from production, which ensures safety compliance (Awuchi, 2023).

4.3 SOURCES OF MICROBIAL STRAINS

Microbial strains for the food industry can be sourced from various locations, including natural sources, culture collection banks, and food processing environments. Additionally, starter cultures, often containing specific strains of bacteria, yeasts, or moulds, are a crucial source for controlled fermentation processes in food production.

4.3.1 Natural Sources

Naturally fermented food items and liquor in most nations and certain parts of the world continue to remain prepared at home in routine parameters. Foods undergoing natural fermentation liquids carry functional as well as non-viable microorganisms. Microorganisms from natural sources are isolated from soil, water, plants, and gut regions (Siddiqui et al., 2023). These diverse types of organisms can be isolated and cultured for the fermentation process. Microorganisms from natural sources adapt to certain substrates and environmental conditions, which improve aroma, flavour, textures, and health benefits (Suman et al., 2022).

LAB, particularly *Lactobacillus* and *Bifidobacterium* species, are commonly considered as probiotic microorganisms, which are involved in fermented dairy product production. *Saccharomyces sp*., *Candida sp*., *Rhizopus*, and *Pichia sp*. are other types of beneficial microorganisms from natural sources used in beverage and other enzymatic fermentation processes. The gastrointestinal tract also plays a major role in isolating microorganisms for fermentation, serving as a natural source of beneficial microorganisms. The gut microbiota contributes microbes that ferment dietary fibres and other carbohydrates (Jahn et al., 2023). The human gut microbiota is primarily composed of five major phyla, namely Bacteroidetes, Proteobacteria, Verrucomicrobia, Firmicutes, and Actinobacteria representing 90% of the gut microbiota (P. Zhang, 2022).

4.3.2 Culture Collections

Certified culture collections are organisations that preserve microbial culture collections and supply microbial materials. They serve as microbial strain repositories in the form of patent deposits, proprietary services to conserve precious organisms for research, industry, and society. There are various recognised certified bacterial strains, e.g., American Type Culture Collection (ATCC) strains, National Collection of Type Cultures (NCTC), Microbial Type Culture Collection and Gene Bank (MTCC), China Centre of Industrial Culture Collection (CICC), which have thousands of microbial strains with classifications (Stark, 2025). These culture banks have a reliable source of standardised and safe microbial strains which can be used in food fermentation. They support microbial quality control, regulatory compliance and strain improvement that facilitate the development of novel fermented products. They also ensure intellectual property protection and patent-related depositions for microbial innovations (Kim et al., 2024).

4.3.3 Mutant and Engineered Strains

Mutant and engineered microbial strains are essential in fermentation for enhancing the production of desired products and improving process efficiency. These strains, either derived from traditional methods like mutagenesis or through modern genetic engineering techniques, offer advantages over wild-type strains in terms of yield, substrate utilisation, tolerance to inhibitors, and other desirable traits. Production output can be optimised by improving the fermentation process, either through the use of genetically modified microorganisms (GMM) strains or by producing recombinant enzymes (Deckers et al., 2020). This can be achieved by recombinant technology and other genetic engineering procedures. For example, in enzyme production, genetically modified fungi and bacteria are used to increase yield and consistency. Whereas engineered microbes are used in precision fermentation to produce animal-free proteins and bioactive ingredients for novel food product production (Mirsalami & Mirsalami, 2025).

Scientific advancements in genetic engineering and molecular biology provided the route to the utilisation of GMM to overproduce carbon metabolism products, including carboxylic acids and amino acids. Whole pathways, for instance, of plant origin, were successfully carried into microorganisms, and this can potentially tap a yet broader set of products available (Baptista et al., 2024). Random mutagenesis occurred in *Wickerhamomycesanomalus* strain YWP1-3 using UV and ethyl methane sulfonate (EMS) in UV22-2 and UV22-3, which exhibit enhanced amylase and pectinase activities. These enzymatic improvements advance the quality of Arabica coffee flavour (Meeampun et al., 2024). According to Wu et al. (2024), *Actinobacillus succinogenes* CICC11014 is used as a high-stress-resistant strain for succinic acid production from corn straw. EF-G (fusA, S539G substitution), NagK (T298P in N-acetylglucosamine kinase), and RNase G (R392H substitution) make the strain more inhibitor-tolerant and physiologically stable under stress, which enhances the succinic acid production after mutation. *Streptococcus thermophilus* MAU80285Y was produced by treating *S. thermophilus* IMAU80285 with the chemical mutagen for rerouting of the galactose metabolism pathways, and provides excellent characteristics for dairy application (Dan et al., 2023).

In summary, mutant and engineered microbial strains are invaluable tools in modern fermentation, enabling enhanced productivity, safety, and innovation in food and industrial biotechnology. Their application combines traditional fermentation with cutting-edge genetic and engineering technologies, always under strict regulatory oversight to ensure product safety and efficacy.

4.4 SCREENING AND CHARACTERISATION

Screening and characterisation of microorganisms in fermentation involves identifying and studying microbes with desirable traits for industrial uses, particularly for the production of specific metabolites. The importance of microbial screening lies in its ability to optimise production processes, improve product quality, and reduce costs.

4.4.1 Primary and Secondary Screening

Primary screening can be utilised to suggest the isolation and identification of the target microorganism based on its qualitative capability to create the product of interest, i.e., an antibiotic, amino acid, or enzyme. In doing so, desired microorganism is usually isolated from a natural origin like soil containing various other species (Olicón-Hernández et al., 2022). Sometimes, the desired microorganism must be isolated from a large population of various microorganism species. Primary screening techniques include the crowded plate technique, indicator dye method, enrichment culture technique, hydrolysis halos, and inhibition halos (Z. Zhang et al., 2021) (Figure 4.1).

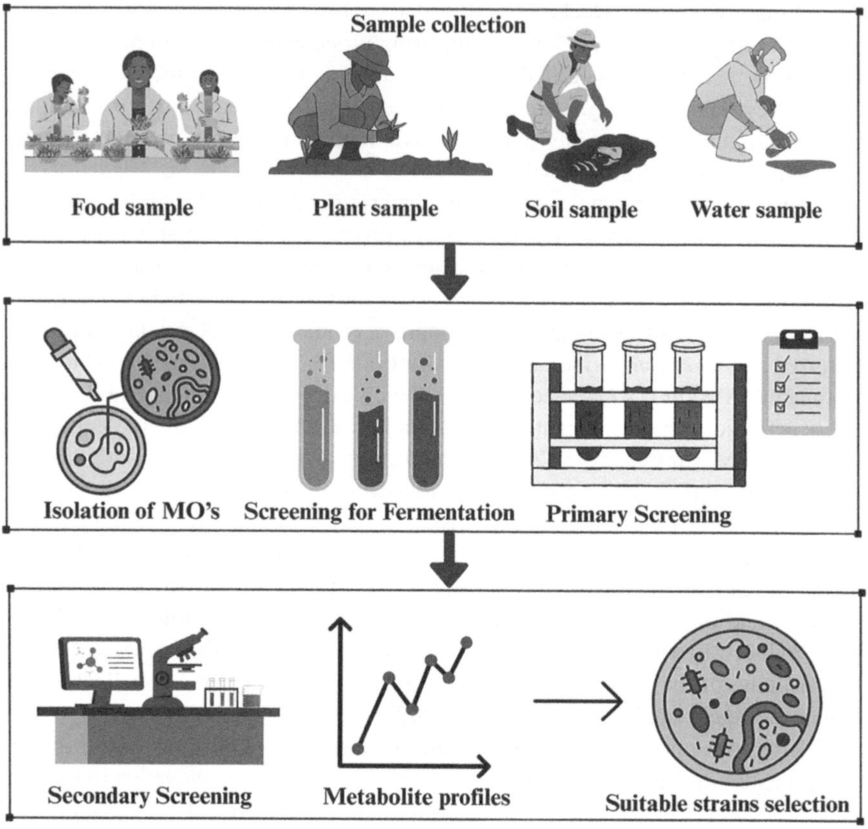

FIGURE 4.1 The workflow from sample collection, microbial isolation, and fermentation screening to the selection of suitable strains.

While the initial screening tends to employ fast tests modified for a substantial number of samples, the subsequent screening phase may serve even with enhanced sensitivity and specificity. Consequently, it employs even more accurate and even more chromatographic methods for direct quantification or for spectrophotometric analysis. The success of the initial selection would rely on the isolation source, culture type, and isolation method, among others (Nekoval et al., 2023). Nonetheless, secondary screening relies upon primary screening for the efficacy of isolation, and thus, the primary selection criteria must exhibit high precision and sensitivity enough not to produce spurious positive detections. The primary screening criteria are formulated for analysing only the detection or lack of the target activity (Awe et al., 2023).

In quantitatively defined secondary selection criteria, methods for the target activity, i.e., enzyme-mediated activities, measurement of the quantity of the desired metabolite is carried out. In certain situations, primary screening can be bypassed, depending on the target activity, the number of samples to be analysed, available resources, and other relevant considerations (Urbelienė et al., 2025). In secondary screening, the main objective is to sort, evaluate, and verify the exact properties and value of microorganisms, in terms of the food and dairy industry. Qualitative secondary screening determines range and spectrum of microbial activity. Quantitative screening determines the amount of product produced by microbes under different conditions. Secondary screening may also find novel compounds through methods like chromatographic and spectroscopic methods. It also simulates production in small flasks to large fermenters for a better yield and economic viability. At last, secondary screening is a comprehensive, in-depth stage that determines whether isolated microorganisms have real, measurable, and stable potential for industrial application, using a range of advanced analytical and fermentation methods to quantify, compare, and characterise their products and performance.

4.4.2 Physiological and Biochemical Profiling

Physiological and biochemical profiling of microbes involves analysing metabolic and functional characteristics to identify and differentiate microorganisms (Wang et al., 2022). Physiological profiling includes analysis of metabolic activity, growth kinetics, nutrient utilisation, enzyme production, and other physiological traits of microorganisms. Biochemical characterisation determines the metabolism of macromolecules (carbohydrates, amino acids, fats) and the production of microbial secondary metabolites (Seydametova & Zainol, 2021). They assess the production of specific metabolites like dextran, acetoin, or other compounds relevant to the fermentation process (Kong et al., 2025).

Physiological and biochemical profiling play a major role in fermentation as they provide insights into the industrial suitability of microbial strains. This helps in identifying robust strains capable of successful fermentation conditions (Khushboo et al., 2023). These screening methods also determine stress responses and substrate assimilation patterns, which include nitrogen and carbon utilisation methods (Y. Zhang et al., 2024). Correlating this, biochemical profiling maintains metabolic and enzymatic activities that include oxidase, catalase, nitrate reduction, protein, and starch hydrolysis. They also assess to enhance homo and hetero fermentative pathways for performing sugar fermentation assays (Masi et al., 2021). Together, these profiles allow differentiation of microbial species, prediction of fermentation efficiency, and optimisation of culture conditions to maximise yield and stability.

4.4.3 Molecular Identification

Molecular identification is considered a critical and vital screening method in fermentation, especially in the food industry, for accurate strain selection. This reveals strain selection, functionality, and purity of microbial strain to avoid toxicity in the fermentation process (Beyene et al., 2020). Molecular techniques allow accurate and rapid determination of microbes for selecting starter culture utilisation in fermentation. In fermentation, only known genetic and metabolic traits should be used for suitable intended process. This prevents contamination and assures consistent fermentation by accurate monitoring and controlled microbial populations (Olatunde et al., 2018).

Methods such as 16S rRNA gene sequencing, or internal transcribed spacer sequencing, which is a specific technique for yeast and fungal isolates, are used to identify species involved in fermentation processes (Srinivas et al., 2025). Other molecular typing techniques, such as quantitative PCR (qPCR), Polymerase Chain Reaction (PCR), amplified fragment length polymorphism (AFLP), random amplified polymorphic DNA (RAPD), and restriction fragment length polymorphism (RFLP), reveal strain differentiation and allow monitoring microbial dynamics in complex fermentation consortia (Adesulu-Dahunsi et al., 2020). In contrast to the aforementioned methods, whole genome sequencing and bioinformatics tools (KEGG – Kyoto Encyclopedia of Genes and Genomes, BLAST – Basic Local Alignment Search Tool, RAST – Rapid Annotation using Subsystem Technology, etc.) provide insights into stress tolerance mechanisms and metabolic networks for genetic engineering for designing genetically modified organisms (Stergiou et al., 2021). Compared to conventional methods, molecular identification ensures rapid, accurate, and reproducible screening of microbes, which is crucial for optimising fermentation processes, ensuring product quality, and supporting the development of novel bio-based products.

4.4.4 Optimisation of Selected Microbial Population

Optimising a selected microbial population involves strategically manipulating environmental conditions and/or genetic characteristics to enhance desirable traits and improve overall performance in a specific application. This can involve regulating the growth of undesirable microbes, enhancing the stability of desired ones, and utilising computational tools to streamline the selection process (Chai et al., 2022). Microbial community optimisation can be managed through bio-stimulation by supplying nutrients specific to the microbial community, bioaugmentation via the introduction of dominant or efficient hydrogen-producing species, and online process control to maintain community stability.

The efficiency of the fermentation process is not always consistent based on specific strain selection, but also based on the optimisation of their functionality under industrial conditions. Optimisation strategies comprise adaptation of microbes to stresses, tuning media nutrient formulations and exploiting synergistic interactions with co-cultures to enhance robustness. Optimisation of the production medium is required to maximise the metabolite yield. This acts as the ultimate source in fermentation, as it is solely responsible for the high yield through microbial fermentation. Carbon and nitrogen are the key elements in media formulation, where carbon sources like starch, glucose, fructose, and glycerol act as energy suppliers, whereas nitrogen sources like yeast extract, ammonium salts, peptones, and corn steep liquor assist in protein and enzymatic synthesis. Additionally, other sources like vitamins, minerals, surfactants, growth factors, or precursors may influence metabolic pathways and inhibit product reduction (Du & Cai, 2025).

Similarly, microbial consortia benefit fermentation by providing complementary metabolic activities with different types of final products. Since the single-strain culture has limitations, these are overcome by microbial consortia. For instance, one may hydrolyse complex polysaccharides into simple sugars, while the other may ferment sugar efficiently and produce ethanol or organic acids (Xin & Qiao, 2025). Microbial consortia also have interactions between them, such as mutualism, commensalism, syntrophy, or competitive exclusion, which broadly enhance efficiency, stability, and diversity of the product. For instance, LAB and yeasts can enhance the conversion of flavour substrates to stimulate the redolence and flavour in fermented foods and foodborne pathogens and moulds, and pathogenic factors, such as nitrite, can be effectively inhibited (W. Liu et al., 2025).

The benefits of consortia include higher yields, faster processing, stress tolerance, and the potential to generate novel metabolites; however, challenges remain in maintaining population balance, preventing contamination, and scaling up for industrial use. With advances in synthetic biology, systems biology, and metabolic modelling, engineered microbial consortia with predictable interactions are being developed, paving the way for more efficient, robust, and sustainable fermentation processes (Table 4.1).

4.5 ADVANCED STRATEGIES FOR MICROBIAL STRAIN IMPROVEMENT IN INDUSTRIAL FERMENTATION

4.5.1 Omics Technologies

The emergence of omics-based techniques revolutionised the approach to evaluating microbial communities by specifying microbial interactions to choose strains during fermentation. In the context of molecular interactions between organisms, omics technologies encourage a holistic perspective based on integrating more than one source of data at a large scale (Bhimani et al., 2024). Application of omics technologies makes optimisation apparent, driving microbial manufacturing performance, and expressing substantial economic implications and societal value (Figure 4.2).

With the rapid advance of omics technologies, optimisation of microbial production is increasingly informed by integrated omics datasets—spanning genomes, transcripts, proteins, metabolites, and the standing metabolic network model derived from these data (Wan et al., 2023). By capturing the interactions between transcription, translation, and metabolic flux, these investigative platforms provide a holistic snapshot of biological systems.

High-throughput platforms for metaproteomics and metabolomics have matured in tandem, fostering consensus that deliberate fusion of all omics levels is now essential (Wen et al., 2023). Successful deployments of integrated multi-omics workflows, bolstered by bioinformatics pipelines, have generated a coherent corpus of evidence that elucidates fermentation, scaling multi-resolution profiles of microbial and molecular constituents macro- and micro-heterogeneities. Consequently, the multi-omics approach surpasses single-layer analyses, cementing its capacity to capture fermentation operative principles. Recent multi-omics initiatives have characterised the structural, regulatory, and expression landscapes of functional genes in conventional fermented products. Ab initio macrogenomics and companion metabolomics co-map the functional gene repertoire with their metabolic end products, delineating community function and product performance in a unified framework. By amalgamating multi-resolution data tiers, the multi-omics paradigm acts synergistically,

TABLE 4.1
Advanced Approaches for Strain Selection

S. No.	Approaches	Description	Suitable Strains and Applications	References
1	Genomics	Studies the entire set of genetic instructions (DNA) of an organism. It involves sequencing and analysing DNA to understand genes, their organisation, and their variations.	It is a breeding strategy to predict the genotypic merits of individuals for selection. E.g., *Saccharomyces cerevisiae* and *Kluyveromyces marxianus*	Lee et al. (2022)
2	Transcriptomics	Examines the complete set of RNA transcripts (the transcriptome) produced by a cell or organism. This includes mRNA (which carries genetic information from DNA to ribosomes), tRNA, rRNA, and other non-coding RNAs.	Used to analyse gene expression patterns to identify traits of interest and flavour formation gene response pathways. E.g., *Streptococcus thermophilus*	(Fu et al., 2024)
3	Proteomics	Focuses on the complete set of proteins (the proteome) expressed by a cell or organism. Proteins are the workhorses of the cell, carrying out most biological functions. Proteomics studies protein structure, function, interactions, and modifications.	Provides a detailed protein level insight into dairy fermentation and stress-related proteins. E.g., *Lactococcus cremoris*	van Olst et al. (2024)
4	Metabolomics	Deals with the complete set of small molecule metabolites (the metabolome) found within a biological sample. Metabolites are the end products of cellular metabolic pathways and include sugars, amino acids, lipids, and other molecules.	This includes understanding flavour and aroma, optimising fermentation conditions, improving nutritional value, ensuring product authenticity, identifying bioactive compounds, and monitoring-specific metabolites in microbial metabolism. E.g., *Aspergillus oryzae*, *Bacillus amyloliquefaciens*	Lee et al. (2016)
5	CRISPR/Cas Editing	Enables precise genetic modifications in microorganisms used for fermentation, allowing for enhanced production of desired compounds, improved fermentation efficiency, and targeted removal of undesirable microbes.	Engineering food-grade microbes for higher biofuel or food ingredient yields, knocking out undesirable pathways. E.g., *Lactobacillus sp., E.coli*	Mu et al. (2022)

6	Metabolic Engineering	Enhances microbial fermentation by strategically altering a microorganism's metabolic pathways to increase the yield of desired products, such as lactic acid or biofuels, or to produce novel compounds.	Redesigning microbial metabolic pathways to enhance product yield, minimise by-products, improve stress tolerance, and enable efficient utilisation of diverse substrates for sustainable bioprocessing. E.g., *Lactococcus lactis*	Luo et al. (2023)
7	Adaptive Laboratory Evolution (ALE)	Technique that applies Darwinian principles in the lab to select for microbial strains with improved phenotypes, such as enhanced tolerance to stressors, increased growth rates, or higher yields of target products like alcohols or carotenoids.	ALE can be used to activate dormant metabolic pathways and optimise a strain's ability to produce specific metabolites or compounds needed for industrial applications. E.g., *Levilactobacillus brevis*	(Lee & Kim, 2020)
8	Synthetic Biology	Leverages engineering and molecular biology to redesign or create novel biological systems, optimise metabolic pathways within these engineered microbes to enhance high production of desired molecules.	Enables the conversion of plant sugars into specific compounds for use in food, beverages, cosmetics, and fragrances, such as the sweetener products by redesigning and constructing novel genetic circuits and metabolic pathways in microorganisms. E.g., *Clostridium sp., Pichia pastoris*	Walker and Pretorius (2018)
9	High-throughput Screening	Automated, rapid method for evaluating large numbers of microorganism strains and fermentation conditions to identify those that perform best at a commercial scale.	Used to rapidly evaluate large microbial strain libraries and identify high-performing producers and optimisation of metabolic traits. E.g., *Saccharomyces cerevisiae, Cyberlindnera fabianii*	Vong et al. (2024)
10	Bioinformatics and AI	Revolutionise fermentation by enabling the optimisation of complex bioprocesses, leading to enhanced yields, product quality, and efficiency.	Used to analyse multi-omics data, metabolic network, and predict microbial interactions. E.g., lactic acid bacteria and Bacillus spp., and filamentous fungi	Yuan et al. (2025)

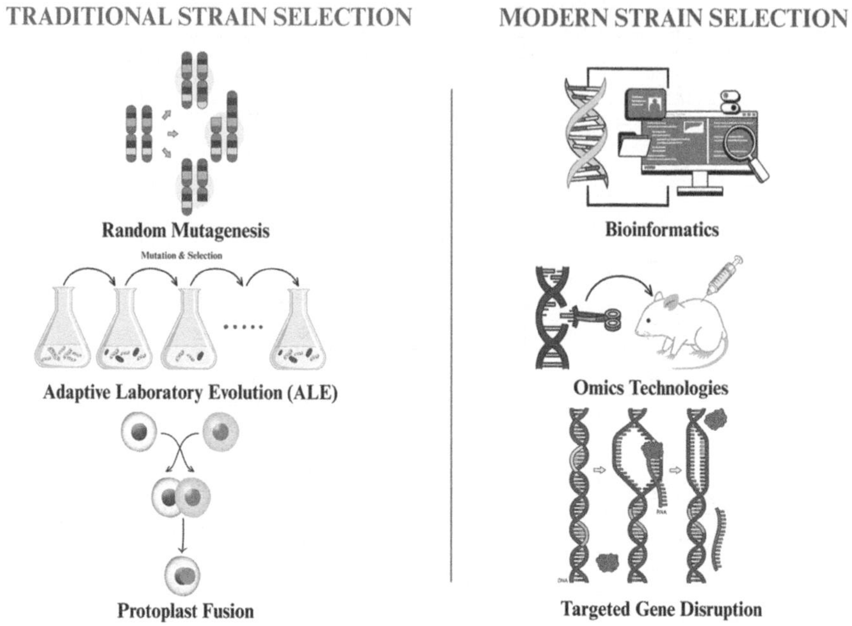

FIGURE 4.2 Compares traditional and modern strain selection methods, including mutagenesis, adaptive laboratory evolution, protoplast fusion, bioinformatics, omics technologies, and targeted gene disruption.

multiplying leverage for screening novel strains, monitoring sequence dynamics, and mining high-value functional genes for subsequent metabolic engineering.

4.5.2 Bioinformatics and Metabolic Modelling

Bioinformatics and metabolic modelling are powerful tools in fermentation research, offering insights into microbial metabolism and guiding strain development for improved fermentation processes. Metabolic models, especially genome-scale models, help predict metabolic fluxes, understand microbial interactions, and design strategies for enhanced production of desired fermentation products (Somerville et al., 2021). This strategy has been used in fermentative production of bio-based industrial chemicals and fuels from alternative carbon sources. Bioinformatics and metabolic modelling also help in microbial selection for fermentation through system-level understanding and rational design of microbial strains (Walsh et al., 2022).

Genome-scale metabolic models (GEMs) are mathematical representations of all known metabolic reactions within an organism, encompassing genes, proteins, and metabolites. They are used to study cellular metabolism, predict metabolic fluxes, and integrate various types of biological data (Passi et al., 2021). GEMs have applications in understanding genotype-phenotype relationships, identifying therapeutic targets, and guiding metabolic engineering strategies. GEMs are valuable tools for understanding and optimising fermentation processes (Pelicaen et al., 2020).

They provide a comprehensive, in silico representation of an organism's metabolism, enabling researchers to simulate and predict the impact of genetic and environmental changes on fermentation outcomes. By integrating genomic, transcriptomic, and metabolomic data, GEMs can be used to design and engineer microbial strains for improved production of desired fermentation products (Y. Q. Lee et al., 2025).

4.5.3 CRISPR and Synthetic Biology for Targeted Improvements

Genome editing techniques, such as CRISPR-Cas9, leverage this information to introduce specific mutations or delete undesirable genes, thereby enhancing desirable traits. For instance, modifying genes involved in metabolic pathways can increase the yield of a target compound, while removing regulatory bottlenecks can relieve constraints on production (Cui & Qu, 2024). CRISPR technologies enable site-specific, successful, and multiplexed editing of microbial genomes, enhancing stress tolerance, high substrate utilisation, and improved metabolite yields (Husaini et al., 2023).

Production of exopolysaccharides (EPSs) in *S. thermophilus* and lactic acid by *Pediococcus acidilactici* has been enhanced using CRISPR-Cas-based genetic tools. CRISPR-Cas9 allows the insertion and deletion of genes that release waste products and the construction of metabolic pathways: Production of lactic acid and flavour compounds. Lactic acid production is increased in LAB species, such as *Lactococcus lactis* and *Lactobacillus plantarum* (Dong et al., 2022). These deletions in *C. acetobutylicum* also showed increased stability of solvent formation during prolonged fermentation runs (Cheng et al., 2019). Taken together, these advances highlight the transformative role of CRISPR-based genome editing in fine-tuning microbial metabolism for fermentation. By enabling targeted pathway optimisation, by-product minimisation, and enhanced stress tolerance, CRISPR technologies not only improve current production efficiencies but also open new avenues for designing next-generation microbial cell factories. When integrated with synthetic biology and systems-level modelling, these tools hold immense potential to accelerate industrial-scale biomanufacturing of fuels, chemicals, and high-value bioproducts.

4.6 APPLICATIONS AND CASE STUDIES

4.6.1 Food and Beverage Fermentation

Microbial strain selection in the food and beverage industry not only enhances consistent production but also promotes innovative approaches in health-promoting and functional foods. It also determines the safety, efficiency, and quality of fermented food products (Praveen & Brogi, 2025). Appropriate starter cultures are included to commence and control fermentation in the production of fermented food products, enzyme, and beverage production. Yeast and moulds such as *Zygosaccharomyces rouxii*, *Saccharomyces cerevisiae*, *Neurospora crassa*, *Rhizopus oryzae*, and *Aspergillus oryzae* are used in the production of soy sauce, miso, tempeh, rice vinegar, and koji (Zhao, Ding et al., 2018; Zheng, Shi et al., 2021). (Figure 4.3).

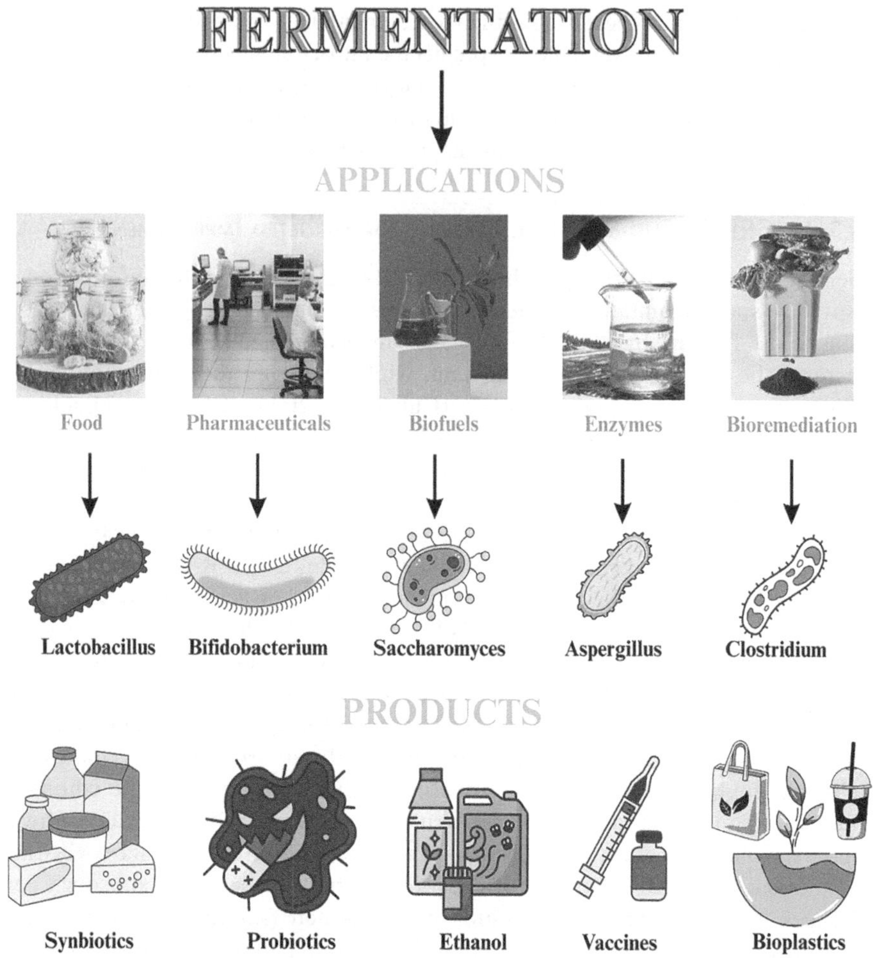

FIGURE 4.3 Overall fermentation process, highlighting key applications in food, pharmaceuticals, biofuels, enzymes, and bioremediation, along with representative microorganisms and their derived products.

Bacterial species, including *Lactobacillus spp* (Wang et al., 2021), *Lactococcus lactis subsp. Lactis* (Li et al., 2020), *Propionibacterium acidipropionici* (de Assis, Machado et al., 2022), *Acetobacter aceti* (Haghshenas et al., 2015), and *Leuconostocmesenteroides* (Jideani et al., 2021), are involved in probiotic product manufacturing. Advances in CRISPR and synthetic biology are improving the importance and scope of strain selection for enhanced flavour, nutritional value, and industrial scalability.

4.6.2 Biofuel Production

As biofuel is considered a sustainable product, microbial selection serves a key function in making efficient manufacturing and elevated yield, such as biobutanol, bioethanol, and biodiesel. Microorganisms rich in lipid content ferment biofuel from substrates (Ramamurthy et al., 2021). The selection of microbial strain influences the overall yield, stress tolerance and substrate range. Notably, compared to a single type of strain, microbial consortia enhance the overall conversion rate and increase in product yield, *Clostridium* sp., *Saccharomyces* sp., *Trichoderma* sp., and *Kluyveromyces* sp. They have been extensively utilised to achieve high yields of simple sugars, reduced levels of inhibitory compounds, and enhanced biofuel production (Kumari et al., 2022). *S. cerevisiae* is considered the main source to produce bioethanol through anaerobic fermentation (de Melo et al., 2020). Due to the obligate anaerobic nature of *Clostridium* and the limited number of species producing commercially viable amounts of butanol, the butanol-producing genes have been transferred into the more manageable, facultatively anaerobic *E. coli* for industrial applications (Lapuerta et al., 2017).

4.6.3 Pharmaceutical Production

Microbial selection is central to pharmaceutical production because the choice of microorganism directly determines the type, yield, and quality of bioactive compounds that can be manufactured (Sadanov et al., 2025). Many antibiotics, anticancer agents, immunosuppressants, enzymes, and vitamins are of microbial origin, with species such as *Streptomyces*, *Penicillium*, and *Aspergillus* serving as classical producers (Chunduri et al., 2022). Selecting strains with naturally high biosynthetic potential or engineering them to optimise secondary metabolite pathways ensures consistent and scalable drug yields. Moreover, microbial selection is crucial for identifying organisms with desirable industrial traits, such as rapid growth, tolerance to stress conditions, and amenability to genetic modification (Zeng et al., 2023). With the aid of modern tools such as CRISPR-Cas, synthetic biology, and high-throughput screening, novel strains can be developed or improved to enhance metabolite diversity, reduce by-product formation, and lower production costs. Thus, effective microbial selection not only underpins the discovery of new therapeutic molecules but also drives the large-scale, sustainable, and economically viable production of pharmaceuticals (Mayorga-Ramos et al., 2023).

4.6.4 Challenges and Future Directions

Despite significant advances in microbial selection and engineering for fermentation, several challenges remain. Contamination and strain deterioration can compromise fermentation efficiency and product consistency, necessitating robust monitoring and preservation strategies (Praveen & Brogi, 2025). Scale-up issues and process variability between laboratory and industrial settings often hinder predictable outcomes, requiring careful optimisation of bioprocess parameters (Elhalis et al., 2023). Additionally, regulatory and consumer acceptance concerns, particularly with respect to genetically modified organisms (GMOs) and synthetic biology-based interventions, pose hurdles to the widespread adoption of engineered strains (Singh & Kumar, 2025).

Looking ahead, emerging trends hold promise for overcoming these barriers. Precision fermentation aims to integrate genome editing, metabolic modelling, and synthetic circuits to fine-tune microbial performance with greater accuracy (Mudoor Sooresh et al., 2023). The application of artificial intelligence (AI) and machine learning in strain selection and process optimisation will accelerate the identification of robust microbial candidates (Cheng et al., 2023). Furthermore, greater emphasis on sustainability, including the use of renewable feedstocks, waste valorisation, and circular bioprocessing, will define the future direction of microbial selection in industrial fermentation.

4.7 CONCLUSION

Microbial selection remains a cornerstone of successful fermentation processes, directly influencing yield, product quality, and economic feasibility. From biofuels to pharmaceuticals and food-grade metabolites, the careful identification and improvement of microbial strains determine the scalability and efficiency of industrial bioprocesses. Advances in CRISPR-Cas genome editing, synthetic biology, and adaptive laboratory evolution have transformed strain development from trial-and-error approaches into targeted, predictable strategies. Looking forward, the integration of precision fermentation, AI-driven strain discovery, and sustainable bioprocessing frameworks will shape the next generation of microbial applications. By addressing current challenges and aligning innovations with regulatory and societal expectations, microbial selection will continue to drive safe, scalable, and environmentally responsible fermentation technologies.

REFERENCES

Adesulu-Dahunsi, A., Sanni, A., & Jeyaram, K. (2020). Diversity and technological characterization of *Pediococcuspentosaceus* strains isolated from Nigerian traditional fermented foods. *LWT*, *140*, 110697.

Ashaolu, T. J. (2020). Safety and quality of bacterially fermented functional foods and beverages: A mini review. *Food Quality and Safety*, *4*(3), 123–127.

Awe, S., Ajiboye, A. E., & Agboola, F. O. (2023). Screening of microorganisms producing polymer (PHB) from dump sites soil in Ilorin Metropolis, Kwara State, Nigeria. *Journal of Biochemistry Microbiology and Biotechnology*, *11*(1), 26–30.

Awuchi, C. G. (2023). HACCP, quality, and food safety management in food and agricultural systems. *Cogent Food & Agriculture*, *9*(1) 93–111.

Baptista, M., Costa, C. E., & Domingues, L. (2024). Engineering thermotolerant *Kluyveromycesmarxianus* strains for resveratrol production by simultaneous saccharification and fermentation of whole slurry corn cob. *Bioresource Technology*, *416*, 131755.

Beyene, E., Tefera, A. T., Muleta, D., Fantahun, S. K., & Wessel, G. M. (2020). Molecular identification and performance evaluation of wild yeasts from different Ethiopian fermented products. *Journal of Food Science and Technology*, *57*(9), 3436–3444.

Bhimani, P., Mahavar, P., Rajguru, B., Bhatt, V. D., Nathani, N. M., & Shri, M. (2024). Unveiling the green dialogue: Advancements in omics technologies for deciphering plant–microbe interactions in soil. *Discover Plants*, *1*(1) pp. 4–10.

Capozzi, V., Fragasso, M., & Russo, P. (2020). Microbiological safety and the management of microbial resources in artisanal foods and beverages: The need for a transdisciplinary assessment to conciliate actual trends and risks avoidance. *Microorganisms*, *8*(2), 306.

Chai, W. Y., Teo, K. T. K., Tan, M. K., & Tham, H. J. (2022). Fermentation process control and optimization. *Chemical Engineering & Technology*, *45*(10), 1731–1747.

Cheng, C., et al. (2019). Engineering *Clostridium* for improved solvent production: Recent progress and perspective. *Applied Microbiology and Biotechnology*, *103*(14), 5549–5566.

Cheng, Y., et al. (2023). Artificial intelligence technologies in bioprocess: Opportunities and challenges. *Bioresource Technology*, *369*, 128451.

Chunduri, A., et al. (2022). Role of microbes in the pharmaceutical industry. In *Role of Microbes in Industrial Products and Processes* (pp. 1–34).

Cui, Y., & Qu, X. (2024). CRISPR-Cas systems of lactic acid bacteria and applications in food science. *Biotechnology Advances*, *71*, 108323.

Dan, T., Hu, H., Tai, J., He, Y., & He, B. (2023). Production and evaluation of a mutant galactose-utilizing strain of *Streptococcus thermophilus* for application in milk fermentation. *LWT*, *187*, 115284.

de Assis, D. A., et al. (2022). High cell density culture of dairy *Propionibacterium* sp. and *Acidipropionibacterium* sp.: A review for food industry applications. *Food and Bioprocess Technology*, *15*(4), 734–749.

de Melo, A. H. F., et al. (2020). Evolutionary engineering of two robust Brazilian industrial yeast strains for thermotolerance and second-generation biofuels. *Industrial Biotechnology*, *16*(2), 91–98.

Deckers, M., Deforce, D., Fraiture, M., & Roosens, N. H. (2020). Genetically modified microorganisms for industrial food enzyme production: An overview. *Foods*, *9*(3), 326.

Dong, H., et al. (2022). CRISPR/Cas tools for enhancing the biopreservation ability of lactic acid bacteria in aquatic products. *Frontiers in Bioengineering and Biotechnology*, *10*, 1114588.

Duperray, M., Delvenne, M., François, J. M., Delvigne, F., & Capp, J. (2024). Genomic and metabolic instability during long-term fermentation of an industrial *Saccharomyces cerevisiae* strain engineered for C5 sugar utilization. *Frontiers in Bioengineering and Biotechnology*, *12*, 1357671.

Elhalis, H., et al. (2023). The potentials and challenges of using fermentation to improve the sensory quality of plant-based meat analogs. *Frontiers in Microbiology*, *14*, 1267227.

Foti, P., Caggia, C., & Romeo, F. V. (2025). New insight into microbial exploitation to produce bioactive molecules from agrifood and by-products' fermentation. *Foods*, *14*(8), 1439.

Gao, H., Li, X., Chen, X., Hai, D., Wei, C., Zhang, L., & Li, P. (2022). The functional roles of *Lactobacillus acidophilus* in different physiological and pathological processes. *Journal of Microbiology and Biotechnology*, *32*(10), 1226–1233.

Groff, M. C., Noriega, S. E., Meglioli, M. E. D., Rodríguez, L., Kuchen, B., & Scaglia, G. (2024). Determination of variable humidity profile for lactic acid maximization in fungal solid-state fermentation. *Fermentation*, *10*(8), 406.

Han, N. R., Yu, S., Byun, J. A., Yun, E. J., Cheon, S., Song, S., Shim, S., Choi, I. G., Lee, S. H., & Kim, K. H. (2024) Evolution-aided improvement of the acid tolerance of *Levilactobacillus brevis* and its application in sourdough fermentation. *Food Research International*, *190*, 114584.

He, Y., Xie, Z., Zhang, H., Liebl, W., Toyama, H., & Chen, F. (2022). Oxidative fermentation of acetic acid bacteria and their products. *Frontiers in Microbiology*, *13*, 879246.

Husaini, C. U. N., et al. (2023). The CRISPR technology: A promising strategy for improving dark fermentative biohydrogen production using *Clostridium* spp. *International Journal of Hydrogen Energy*, *48*(61), 23498–23515.

Jahn, L. J., Rekdal, V. M., & Sommer, M. O. (2023). Microbial foods for improving human and planetary health. *Cell*, *186*(3), 469–478.

Ji, X., Zhang, L., Yu, X., Chen, F., Guo, F., Wu, Q., & Xu, Y. (2023). Selection of initial microbial community for the alcoholic fermentation of sesame flavor-type baijiu. *Food Research International*, *172*, 113141.

Jideani, V. A., et al. (2021). *Leuconostocmesenteroides* and *Pediococcuspentosaceus* non-alcoholic pearl millet beverage enriched with *Moringa oleifera* leaf powder: Nutritional and sensory characteristics. *Processes*, *9*(12), 2125.

Khushboo, N., Karnwal, A., & Malik, T. (2023). Characterization and selection of probiotic lactic acid bacteria from different dietary sources for development of functional foods. *Frontiers in Microbiology*, *14*, 1170725.

Kim, S. Y., Ahn, Y. W., Kim, J. K., Lee, J. H., & Choi, Y. S. (2024). Equivalence analysis for five Korean NCCP microorganism strains as ATCC reference strains alternative in Korean Pharmacopoeia microbiological tests. *YakhakHoeji*, *68*(3), 168–209.

Kondrotiene, K., Zavistanaviciute, P., Aksomaitiene, J., Novoslavskij, A., &Malakauskas, M. (2023). *Lactococcus lactis* in dairy fermentation—Health-promoting and probiotic properties. *Fermentation*, *10*(1), 16.

Kong, X., Zhang, J., Shen, H., Shi, N., Zhou, H., Li, Y., Guo, Y., Luo, H., & Yu, L. (2025). Screening, identification, and fermentation characteristics of lactic acid bacteria from pickled potherb mustard and potential applications. *Foods*, *14*(8), 1431.

Kumari, A., et al. (2022). Role of microorganisms in production of biofuels. In A. Guldhe & B. Singh (Eds.), *Novel Feedstocks for Biofuels Production* (pp. 65–116). Springer.

Lapuerta, M., et al. (2017). Modeling viscosity of butanol and ethanol blends with diesel and biodiesel fuels. *Fuel*, *199*, 332–338.

Lee, D. E., et al. (2016). Metabolomic profiles of *Aspergillus oryzae* and *Bacillus amyloliquefaciens* during rice koji fermentation. *Molecules*, *21*(6), 773.

Lee, M., Bang, W., Lee, H., Yang, S., Lee, K., Kang, H., Hong, S., & Yang, J. (2024). Safety assessment and evaluation of probiotic potential of *Lactobacillus bulgaricus*. *Microorganisms*, *12*(3), 456.

Lee, Y.-G., et al. (2022). Genome-edited *Saccharomyces cerevisiae* strains for improving quality, safety, and flavor of fermented foods. *Food Microbiology*, *104*, 103971.

Lee, Y. Q., Choi, Y., Park, S., Kim, S., Lee, M., Kim, D., Koduru, L., Lakshmanan, M., Jung, S., Kim, M. J., Choe, Y. H., & Lee, D. (2025). Genome-scale metabolic model-guided systematic framework for designing customized live biotherapeutic products. *NPJ Systems Biology and Applications*, *11*(1) 73.

Li, W., et al. (2020). Fermentation characteristics of Lactococcus lactis subsp. lactis isolated from naturally fermented dairy products and screening of potential starter isolates. *Frontiers in Microbiology*, *11*, 2020.

Liszkowska, W., & Berlowska, J. (2021). Yeast fermentation at low temperatures: Adaptation to changing environmental conditions and formation of volatile compounds. *Molecules*, *26*(4), 1035.

Liu, W., Tang, Y., Zhang, J., Bai, J., Zhu, Y., Zhu, L., Zhao, Y., Daglia, M., Xiao, X., & He, Y. (2025). Microbial interactions in food fermentation: Interactions, analysis strategies, and quality enhancement. *Foods*, *14*(14), 2515.

Luo, Q., et al. (2023). Metabolic engineering of microorganisms to produce pyruvate and derived compounds. *Molecules*, *28*(3), 1418.

Mannan, A. A., Darlington, A. P. S., Tanaka, R. J., & Bates, D. G. (2025). Design principles for engineering bacteria to maximise chemical production from batch cultures. *Nature Communications*, *16*(1) 279.

Masi, C., Gemechu, G., & Tafesse, M. (2021). Isolation, screening, characterization, and identification of alkaline protease-producing bacteria from leather industry effluent. *Annals of Microbiology*, 71(1), 1631.

Mayorga-Ramos, A., et al. (2023). CRISPR-Cas-based antimicrobials: Design, challenges, and bacterial mechanisms of resistance. *ACS Infectious Diseases*, *9*(7), 1283–1302.

Meeampun, Y., Panyachanakul, T., Samosorn, S., Dolsophon, K., Jiamjariyatam, R., Lorliam, W., Arnthong, J., Suwannarangsee, S., Tantayotai, P., & Krajangsang, S. (2024). Characterization of yeast mutant strains for starter culture in Arabica coffee fermentation. *Scientific Reports*, *14*(1) 6069,.

Mengesha, Y., Tebeje, A., & Tilahun, B. (2022). A review on factors influencing the fermentation process of teff (*Eragrostis teff*) and other cereal-based Ethiopian injera. *International Journal of Food Science*, *2022*, 4419955.

Mirsalami, S. M., & Mirsalami, M. (2025). Advances in genetically engineered microorganisms: Transforming food production through precision fermentation and synthetic biology. *Future Foods*, 100601.

Mu, Y., et al. (2022). Development and applications of CRISPR/Cas9-based genome editing in *Lactobacillus*. *International Journal of Molecular Sciences*, *23*(21), 12852.

Mudoor Sooresh, M., et al. (2023). Opportunities and challenges of understanding community assembly in spontaneous food fermentation. *Foods*, *12*(3), 673.

Nekoval, S. N., Churikova, A. K., Chernyakovich, M. N., & Pridannikov, M. V. (2023). Primary screening of microorganisms against *Meloidogyne hapla* (Chitwood, 1949) under the conditions of laboratory and vegetative tests on tomato. *Plants*, *12*(18), 3323.

Ojeda-Linares, C. I., Solís-García, I. A., & Casas, A. (2022). Constructing micro-landscapes: Management and selection practices on microbial communities in a traditional fermented beverage. *Frontiers in Ecology and Evolution*, *10*, 821268.

Okoye, C. O., Ezenwanne, B. C., & Olasoji, O. (2025). Chap. 2: Mechanisms of preservation by lactic acid bacteria in food fermentation. In R. Tang (Ed.), *IntechOpen eBooks* (pp. 338–345).

Olatunde, O. O., Obadina, A. O., Omemu, A. M., et al. (2018). Screening and molecular identification of potential probiotic lactic acid bacteria in effluents generated during ogi production. *Annals of Microbiology*, *68*, 433–443.

Olicón-Hernández, D. R., Guerra-Sánchez, G., Porta, C. J., Santoyo-Tepole, F., Hernández-Cortez, C., Tapia-García, E. Y., & Chávez-Camarillo, G. M. (2022). Fundaments and concepts on screening of microorganisms for biotechnological applications: Mini review. *Current Microbiology*, *79*(12) 373, 1–13

Olsson, L., Rugbjerg, P., Pianale, L. T., & Trivellin, C. (2022). Robustness: Linking strain design to viable bioprocesses. *Trends in Biotechnology*, *40*(8), 918–931.

Passi, A., Tibocha-Bonilla, J. D., Kumar, M., Tec-Campos, D., Zengler, K., & Zuniga, C. (2021). Genome-scale metabolic modeling enables in-depth understanding of big data. *Metabolites*, *12*(1), 14.

Pelicaen, R., Gonze, D., De Vuyst, L., & Weckx, S. (2020). Genome-scale metabolic modeling of *Acetobacter pasteurianus* 386B reveals its metabolic adaptation to cocoa fermentation conditions. *Food Microbiology*, *92*, 103597.

Poudel, R., Thunell, R. K., Oberg, C. J., Overbeck, S., Lefevre, M., Oberg, T. S., & McMahon, D. J. (2022). Comparison of growth and survival of single strains of *Lactococcus lactis* and *Lactococcus cremoris* during Cheddar cheese manufacture. *Journal of Dairy Science*, *105*(3), 2069–2081.

Praveen, M., & Brogi, S. (2025). Microbial fermentation in food and beverage industries: Innovations, challenges, and opportunities. *Foods*, *14*(1), 114.

Qin, D., Ma, Y., Wang, Y., Hou, X., & Yu, L. (2022). Contribution of lactobacilli on intestinal mucosal barrier and diseases: Perspectives and challenges of *Lactobacillus casei*. *Life*, *12*(11), 1910.

Rafieenia, R., Atkinson, E., & Ledesma-Amaro, R. (2022). Division of labor for substrate utilization in natural and synthetic microbial communities. *Current Opinion in Biotechnology*, *75*, 102706.

Ramamurthy, P. C., et al. (2021). Microbial biotechnological approaches: Renewable bioprocessing for the future energy systems. *Microbial Cell Factories*, *20*(1), 55.

Reale, A., Zotta, T., Ianniello, R. G., Mamone, G., & Di Renzo, T. (2020). Selection criteria of lactic acid bacteria to be used as starter for sweet and salty leavened baked products. *LWT*, *133*, 110092.

Rugbjerg, P., & Olsson, L. (2020). The future of self-selecting and stable fermentations. *Journal of Industrial Microbiology & Biotechnology*, *47*(11), 993–1004.

Sadanov, A. K., et al. (2025). Engineering useful microbial species for pharmaceutical applications. *Microorganisms*, *13*(3), 599.

Seydametova, E., & Zainol, N. (2021). Morphological, physiological, biochemical and molecular characterization of statin-producing *Penicillium*microfungi isolated from little-explored tropical ecosystems. *Current Research in Microbial Sciences*, *2*, 100044.

Sharma, N., Mahawar, L., Mishra, A., & Albrectsen, B. R. (2025). Microbial contributions to plant growth and stress tolerance: Mechanisms for sustainable plant production. *Plant Stress, 2025*, 100966.

Sharma, R., Garg, P., Kumar, P., Bhatia, S. K., & Kulshrestha, S. (2020). Microbial fermentation and its role in the quality improvement of fermented foods. *Fermentation, 6*(4), 106.

Siddiqui, S. A., Erol, Z., Rugji, J., et al. (2023). An overview of fermentation in the food industry: Looking back from a new perspective. *Bioresources and Bioprocessing, 10*, 85.

Singh, A., & Kumar, S. (2025). Exploring the functionality of microbes in fermented foods: Technological advancements and future directions. *Fermentation, 11*(6), 300.

Sionek, B., Szydłowska, A., Küçükgöz, K., & Kołożyn-Krajewska, D. (2023). Traditional and new microorganisms in the lactic acid fermentation of food. *Fermentation, 9*(12), 1019.

Somerville, V., Grigaitis, P., Battjes, J., Moro, F., & Teusink, B. (2021). Use and limitations of genome-scale metabolic models in food microbiology. *Current Opinion in Food Science, 43*, 225–231.

Srinivas, M., Walsh, C. J., Crispie, F., et al. (2025). Evaluating the efficiency of 16S-ITS-23S operon sequencing for species-level resolution in microbial communities. *Scientific Reports, 15*, 2822.

Stark, J. F. (2025). Imperial microbiology: The National Collection of Type Cultures and the management of microorganisms, 1916–1922. *Journal of the History of Biology*.

Stergiou, O. S., Tegopoulos, K., Kiousi, D. E., Tsifintaris, M., Papageorgiou, A. C., Tassou, C. C., Chorianopoulos, N., Kolovos, P., & Galanis, A. (2021). Whole-genome sequencing, phylogenetic and genomic analysis of *Lactiplantibacilluspentosus* L33, a potential probiotic strain isolated from fermented sausages. *Frontiers in Microbiology, 12*:746659.

Suman, J., Rakshit, A., Ogireddy, S. D., Singh, S., Gupta, C., & Chandrakala, J. (2022). Microbiome as a key player in sustainable agriculture and human health. *Frontiers in Soil Science, 2*, 821589.

Sun, W., Shahrajabian, M. H., & Lin, M. (2022). Research progress of fermented functional foods and protein factory-microbial fermentation technology. *Fermentation, 8*(12), 688.

Taveira, I. C., Nogueira, K. M. V., De Oliveira, D. L. G., & Silva, R. D. N. (2021). Fermentation: humanity's oldest biotechnological tool. *Frontiers for Young Minds, 9*, 568656.

Terefe, Z. K., Omwamba, M. N., & Nduko, J. M. (2021). Effect of solid state fermentation on proximate composition, antinutritional factors and in vitro protein digestibility of maize flour. *Food Science & Nutrition, 9*(11), 6343–6352.

Urbelienė, N., Gasparavičiūtė, R., Vaitekūnas, J., Meškienė, R., Valantinaitė, U., Kruopis, P., Gudiukaitė, R., & Meškys, R. (2025). A screening method for polyester films-degrading microorganisms and enzymes. *Journal of Hazardous Materials, 487*, 137177.

van Olst, B., et al. (2024). Bacterial proteome adaptation during fermentation in dairy environments. *Food Microbiology, 121*, 104514.

Vargas-Luna, C., Godoy, L., Benavides, S., De Lecco, C. C., Urtubia, A., & Franco, W. (2025). Screening and selection of native lactic acid bacteria isolated from Chilean grapes. *Foods, 14*(1), 143.

Vong, W. C., et al. (2024). High-throughput screening for aroma production in food fermentations. *Food Research International, 177*, 113902.

Vulart, L., Estévez, Á., Peiro, E., Gòdia, F., & Ganigué, R. (2025). Combined effect of feeding regime and pH on carboxylic acids production in thermophilic mixed-culture fermentation. *Journal of Environmental Chemical Engineering, 2025*, 117818.

Walker, R. S., & Pretorius, I. S. (2018). Applications of yeast synthetic biology geared towards the production of biopharmaceuticals. *Genes, 9*(7), 340.

Wan, S., Liu, X., Sun, W., Lv, B., & Li, C. (2023). Current advances for omics-guided process optimization of microbial manufacturing. *Bioresources and Bioprocessing, 10*(1).

Wang, X., Liu, Y., & Sun, J. (2022). Physiological and biochemical characterization of isolated bacteria from a coccolithophore *Chrysotila dentata* (Prymnesiophyceae) culture. *Diversity, 14*(1), 2.

Wang, Y., et al. (2021). Metabolism characteristics of lactic acid bacteria and the expanding applications in food industry. *Frontiers in Bioengineering and Biotechnology*, *9*, 612285.

Wen, L., Yang, L., Chen, C., Li, J., Fu, J., Liu, G., Kan, Q., Ho, C., Huang, Q., Lan, Y., & Cao, Y. (2023). Applications of multi-omics techniques to unravel the fermentation process and the flavor formation mechanism in fermented foods. *Critical Reviews in Food Science and Nutrition*, *64*(23), 8367–8383.

Wu, J., Li, Y., Yin, J., Wang, C., Qi, X., Zhou, Y., Liu, H., Wu, P., & Zhang, J. (2024). Mutation breeding of high-stress resistant strains for succinic acid production from corn straw. *Applied Microbiology and Biotechnology*, *108*(1), 278, 1–14

Xin, Y., & Qiao, M. (2025). Towards microbial consortia in fermented foods for metabolic engineering and synthetic biology. *Food Research International*, *201*, 115677.

Yang, Q., Yao, H., Liu, S., & Mao, J. (2022). Interaction and application of molds and yeasts in Chinese fermented foods. *Frontiers in Microbiology*, *12*, 664850.

Ye, X., Yu, Y., Liu, J., Zhu, Y., Yu, Z., Liu, P., Wang, Y., & Wang, K. (2024). Seasonal environmental factors drive microbial community succession and flavor quality during acetic acid fermentation of Zhenjiang aromatic vinegar. *Frontiers in Microbiology*, *15*, 1442604.

Yee, C. S., Zahia-Azizan, N. A., Rahim, M. H. A., Zaini, N. A. M., Raja-Razali, R. B., Ushidee-Radzi, M. A., Ilham, Z., & Wan-Mohtar, W. A. A. Q. I. (2025). Smart fermentation technologies: Microbial process control in traditional fermented foods. *Fermentation*, *11*(6), 323.

Yuan, J., et al. (2025). A review and prospects: Multi-omics and artificial intelligence-based approaches to understanding the effects of lactic acid bacteria and yeast interactions on fermented foods. *Innovative Food Science & Emerging Technologies*, *99*, 103874.

Zannini, E., Lynch, K. M., Nyhan, L., Sahin, A. W., Riordan, P. O., Luk, D., & Arendt, E. K. (2022). Influence of substrate on the fermentation characteristics and culture-dependent microbial composition of water kefir. *Fermentation*, *9*(1), 28.

Zeng, J., et al. (2023). Microbial production of medicinally important agents. *Frontiers Media SA*, *14*, 1274087.

Zhang, K., Zhang, T., Guo, R., Ye, Q., Zhao, H., & Huang, X. (2023). The regulation of key flavor of traditional fermented food by microbial metabolism: A review. *Food Chemistry X*, *19*, 100871.

Zhang, P. (2022). Influence of foods and nutrition on the gut microbiome and implications for intestinal health. *International Journal of Molecular Sciences*, *23*(17), 9588.

Zhang, Y., Sun, Q., Liu, X., Basit, R. A., Ma, J., Fu, Z., Cheng, L., Fan, G., & Teng, C. (2024). Screening, identification, and fermentation condition optimization of a high-yield 3-methylthiopropanol yeast and its aroma-producing characteristics. *Foods*, *13*(3), 418.

Zhang, Z., Shah, A. M., Mohamed, H., Tsiklauri, N., & Song, Y. (2021). Isolation and screening of microorganisms for the effective pretreatment of lignocellulosic agricultural wastes. *BioMed Research International*, *2021*, 1–16.

Zhao, G., et al. (2018). Extracellular proteome analysis and flavor formation during soy sauce fermentation. *Frontiers in Microbiology*, *9*, 1872.

Zhao, Y., Liu, S., Han, X., Zhou, Z., & Mao, J. (2022). Combined effects of fermentation temperature and *Saccharomyces cerevisiae* strains on free amino acids, flavor substances, and undesirable secondary metabolites in huangjiu fermentation. *Food Microbiology*, *108*, 104091.

Zheng, X., et al. (2021). A review on the general cheese processing technology, flavor biochemical pathways and the influence of yeasts in cheese. *Frontiers in Microbiology*, *12*, 703284.

Zhu, M., Zheng, J., Xie, J., Zhao, D., Qiao, Z., Huang, D., & Luo, H. (2022). Effects of environmental factors on the microbial community changes during medium-high temperature Daqu manufacturing. *Food Research International*, *153*, 110955.

5 An Introduction to Artificial Intelligence (AI) Tools for Bioprocess Optimization

Arivumalar Chandrasekaran, Panneerselvam Theivendren, Leena Ajay Sawaikar, Swati Mayur Keny, Celina Nazareth, Malarkodi Velraj, and Raja Veerapandian

5.1 INTRODUCTION

The artificial intelligence (AI) represents a much-needed aspect of contemporary biotechnology, especially optimization of bioprocesses (Asin-Garcia et al., 2025; Malabadi et al., 2023). The recent and unprecedented changes in AI and machine learning (ML) technologies have resulted in a paradigm shift in numerous spheres, such as healthcare, agriculture, and environmental sciences. Within the framework of bioprocess optimization, AI offers a new potential for efficiency improvement, cost reduction, and supply of bioproducts of stable quality (Cheng et al., 2023; Sharma & Singh, 2025).

Typically, the systems in bioprocess optimization are complex, dynamic, e.g., fermentation, cell culture, enzyme catalysis, and protein production (Liu, 2020; Rathore et al., 2021). Historically, to a great extent, these processes were subject to the rule of trial-and-error techniques and scientific approaches that were predominantly based on empirical means. But when the bioprocesses become more complex, it is not possible to make any changes manually (Helleckes et al., 2023; Narayanan et al., 2020). The analysis of learning through vast volumes of data produced by these processes and modifying its trends as well as introducing real-time adjustments can be a formidable solution through the application of AI tools, especially ML algorithms, to help optimize productivity.

Figure 5.1 illustrates the transformative role of AI in bioprocess optimization. It highlights key applications, including enhancing efficiency, automating fermentation, and protein optimization, enabling real-time data analysis and fostering sustainable innovations in biotechnology.

 DOI: 10.1201/9781003684374-5

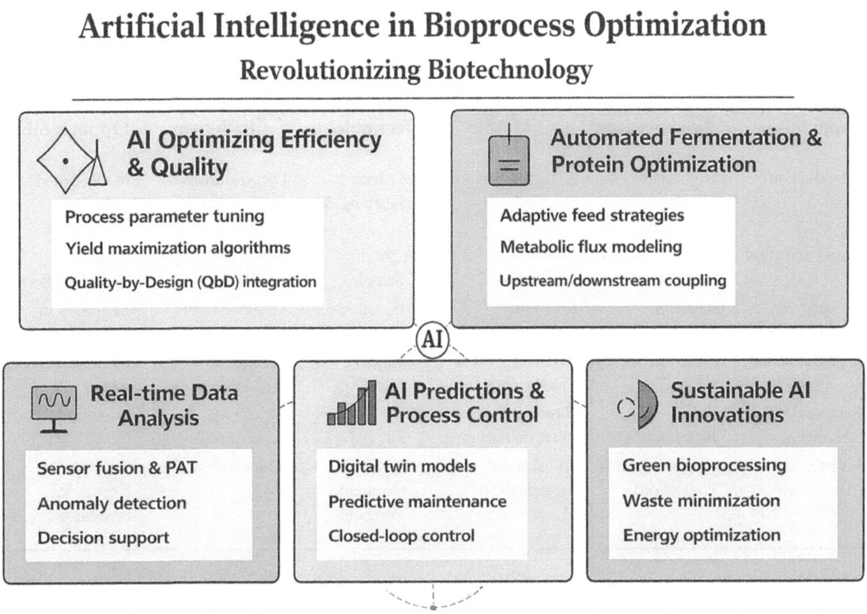

FIGURE 5.1 Transformative role of artificial intelligence in bioprocess optimization.

Fundamentally, AI involves the utilization of algorithms to handle and profile huge data volumes and forecast the future. The AI models used in bioprocess optimization are being trained on the past data of bioreactors, sensors, and process parameters (Butean et al., 2025; Mondal et al., 2023). They can then be used to predict the behaviour of a biological system to enable the scientists to optimize the process to give the best results.

Some of the applications of AI in bioprocess optimization would be monitoring and control of fermentation conditions, optimization of protein expression systems, and enhancing efficiency of enzyme catalysis (Vinestock et al., 2024). Also, AI applications can be employed to anticipate product quality, improve the reproducibility of any bioprocess, and accelerate the creation of new biopharmaceutical products. The possibility of AI automation of data analysis and decision-making in bioprocessing is changing the industry and enabling it to develop faster, at less cost, and with stronger production mechanisms (Butean et al., 2025; Yang et al., 2023).

The use of AI in the optimization of bioprocesses has a number of challenges despite the advantages. These are the necessity of high-quality data, how complex biological systems are, and the ways AI tools can be implemented alongside the current industrial infrastructure (Holzinger et al., 2023; Yetgin, 2025). However, the possibilities of AI to transform bioprocessing are enormous, especially with the development of new algorithms and methods to manage information.

The purpose of the review is to obtain a detailed insight into the use of AI tools in optimization of bioprocesses. In this sphere, we will discuss the current reality of AI technologies, the applications of AI to bioprocess optimization in a variety of ways,

TABLE 5.1
Key Aspects of AI in Bioprocess Optimization

Aspect	Bioprocesses	AI Role	Technologies	Challenges	Opportunities
Introduction	Fermentation, Protein Production	AI for efficiency and quality	Machine Learning (ML)	Data quality and integration	Enhanced product yield
Data Utilization	Enzyme Catalysis, Cell Culture	Data-driven predictions	Algorithms, Sensors	Data complexity and noise	Improved decision-making
Process Optimization	Bioreactor Monitoring	Real-time adjustments	ML and AI tools	Legacy systems	Automation in biomanufacturing
Cost Reduction	Bioproduct Supply	Reducing process inefficiencies	Predictive Modelling	Scaling AI implementation	Lower operational costs
Personalized Medicine	Genetic Bioprocessing	Customizing biomanufacturing	Deep Learning, AI	Regulatory concerns	Customized therapies
Future Directions	Biopharmaceutical Production	AI-enhanced sustainability	Blockchain, Quantum Computing	AI-tool integration	AI-powered bioprocess innovation

challenges and opportunities of such innovations (Ali et al., 2024; Helleckes et al., 2023). We will also discuss the perspectives of the future of the development of AI using tools in bioprocessing, especially within the realm of personalised medicine and sustainable biotechnology practices.

Table 5.1 summarizes the key aspects of AI in bioprocess optimization, highlighting its role in enhancing process efficiency, addressing challenges, and exploring opportunities for cost reduction, personalized medicine, and sustainable biotechnology practices.

5.2 OVERVIEW OF AI IN BIOPROCESS OPTIMIZATION

5.2.1 Evolution of Artificial Intelligence in Biotechnology

The development of AI in the biological technology has been characterized by the growing amount of incorporating computational methods, in order to deal with intricate biological issues (Saraf et al., 2024). Originally, biotechnology was mainly based on trial and error, and experimental procedures in the optimization of bioprocesses. Nevertheless, with the increase in computational power and the development of data analysis methods, AI applications, such as ML and artificial neural networks (ANNs), could be used as a viable solution to modelling and optimization of biotechnological systems. In the beginning, AI was used in biotechnology in the area of bioinformatics, and it was used to analyse genomic, transcriptomic, and proteomic data (Ogunjobi et al., 2024). These tools over time were expanded to drug discovery, protein folding, and optimization of bioprocesses. The ability of AI to deal with large quantities of data has facilitated greater accuracy in modelling biological mechanisms that enhance predictability and efficiency of production systems.

The greater the use of AI in biotechnology, the higher its capability to deal with the complexity of complex bioprocesses (Ali et al., 2024). The ML algorithms, supervised and unsupervised learning, have been useful in streamlining production processes like the fermentation and the enzyme catalysis. Due to the presence of big datasets, which are produced through bioreactors and sensors, nowadays the AI systems can process and analyse the real-time data and make predictions, improve models, and suggest optimization. This development has stimulated innovations in the field of biomanufacturing, enabling enhanced process control and improved scalability (Subramanian, 2021). In the present times, AI is on the edge of biotechnological development, and it is significant in making the biotechnological uses more sustainable, efficient, and precise.

5.2.2 Bioprocess Optimization Is Important in Biotechnology

The bioprocess optimization is an important element in biotechnology industry especially when biopharmaceuticals, biofuels, as well as other bio-based products are being produced (Bähner et al., 2021). The necessity to streamline bioprocesses is based on the fact that biological systems are complex systems and a minor change in the parameters (temperature, pH, oxygen, and nutrient) may have a profound effect on the efficiency, the final product yield, and its quality. Bioprocessing optimization in the biopharmaceutical industry has been used to make sure that production of biologics like monoclonal antibodies, enzymes, and vaccines is economical as well as scalable (Behera, 2023). These processes are vital in optimizing to meet the increasing demand in the world and at the same time ensuring the safety and consistency of the products.

In addition, optimization of bioprocesses can be critical towards enhancing the environmental sustainability of biotechnological processes (Mendes et al., 2024). With the improvement of its efficiency, AI-based optimization can reduce the energy consumption, decrease the use of raw materials, and decrease the quantity of waste. This is particularly valuable when it comes to high-scale production where any efficiency gains can translate to a great environmental and economic advantage. This is because, in cases where biotechnological processes have become more complex, the old methods of optimization can no longer properly manage the demands of the new, large-scale processes of production (Fasim et al., 2021). The solutions to this issue are AI tools that allow adjusting the processes in real time, predictive modelling, and increasing the accuracy in controlling the variables. This is the point of optimization that is important to enhance the profitability and sustainability of biotechnological manufacturing (Stalidzans & Dace, 2021).

Figure 5.2 illustrates AI tools in bioprocess optimization, showcasing their impact on process accuracy, scalability, real-time adjustments for sustainability, and the challenges of data quality and integration in biotechnology.

5.2.3 AI Tools: A Quick Overview

ANNs, ML, and deep learning have become the focus of optimization of bioprocesses in biotechnology through the use of AI tools (Yang et al., 2023). These tools utilize algorithms that work through a large amount of data to determine trends, predict results, and make decisions using past data. When it comes to

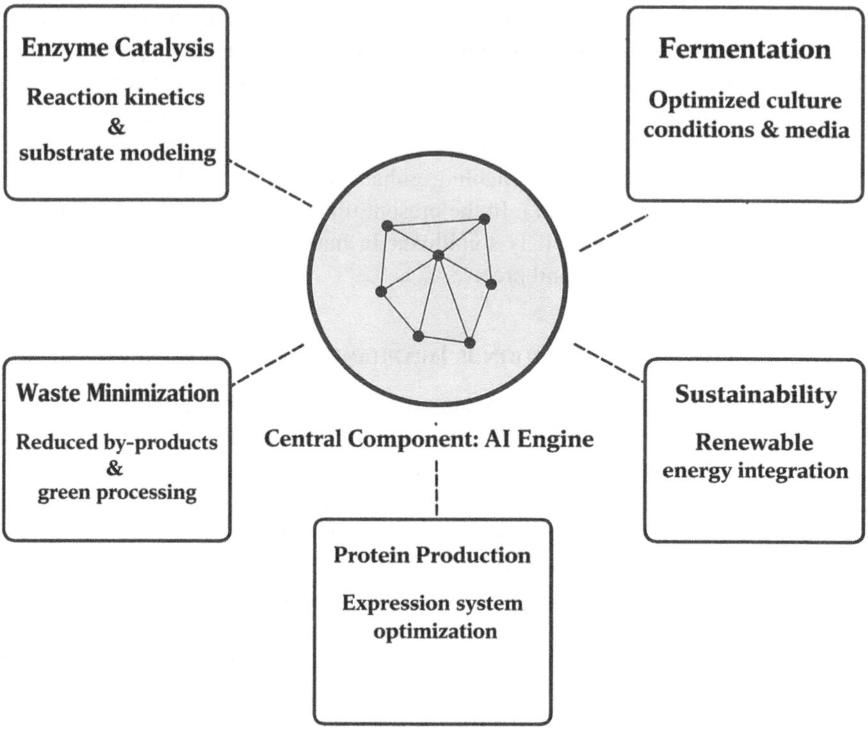

FIGURE 5.2 AI tools in bioprocess optimization.

bioprocess optimization, AI is able to use data of different types, including sensors, bioreactors, and laboratory experiments (Borgosz & Dikicioglu, 2024). It is through identifying trends and connexions in this data that AI systems can predict how biological systems would act in various circumstances and prescribe ideal parameters of the process. The ability to predict gives AI a priceless application in streamlining complex procedures like fermentation, cell culture, and protein production (Ali et al., 2024).

Supervised and unsupervised ML algorithms have been widely used to design predictive models based on the data that has been gathered during bioprocesses (Singh & Singhal, 2023). Such models can be used to predict the results of a particular change in the parameters of a process to have more efficient and target interventions. Also, ANNs that are based on the human brain structure can make predictions that are non-linear and enhance bioprocess optimization forecasts. Another branch of AI is reinforcement learning that is used to make the best decision in a dynamic environment, where changes in the processes should be continuous. All of these AI tools are used to make bioprocesses optimization more automated, precise, and efficient, which is why they cannot be ignored in contemporary biotechnology (Pandey et al., 2024).

5.2.4 Advantages and Problems of Using AI in Bioprocess Optimization

Application of AI in optimization of bioprocesses has a number of advantages particularly relating to improved efficiency of processes, accuracy, and scalability (Helleckes et al., 2023). Among the primary benefits, one can distinguish the possibility to analyse a significant amount of data to obtain valuable information that human beings would not easily have identified. AI is capable of determining the trends of the data measured in bioreactors, sensors, and laboratory experiments, which can be used to optimize the parameters of bioprocesses in real time (Yang et al., 2023). This results in increased yield, quality of products, and minimized wastage. Moreover, AI systems are capable of anticipating the possible problems in advance, which allows making a proactive change that will help avoid costly downtime or product discrepancies. Human error is also minimized as well as operational costs because routine tasks, including process monitoring and cleaning or processing of data, are automated (Biswas & Dutta, 2020).

Although these are the advantages, AI integration in the optimization of bioprocesses is associated with a number of challenges. A major challenge is that the quality and consistency of data required is high because the AI algorithms are much dependent on great volumes of data to make correct predictions (Zha et al., 2025). The information obtained with the help of the bioreactors and sensors should be of high-quality to guarantee the accuracy of the AI models. Also, AI models need to be constantly fed with new data to enable them to remain accurate over time. The complexity of the system of integration of AI tools with the already developed biotechnological systems is another difficulty that might need specific knowledge and experience (Kumar et al., 2025). Smaller biotechnology companies might also be hindered by the cost of development and maintenance of AI systems. However, with the further development of AI technology and its availability, the advantages of the implementation can far surpass these pitfalls, so it is a necessity in the future of biotechnology.

Table 5.2 summarizes AI tools used in bioprocess optimization, detailing their applications, advantages, challenges, data requirements, and integration issues. It highlights the need for high-quality data and the complexities involved in AI model integration with existing biotechnology systems.

TABLE 5.2
AI Tools Used in Bioprocess Optimization, Detailing Their Applications, Advantages, Challenges, Data Requirements, and Integration Issues

AI Tool	Application	Advantages	Challenges	Data Requirements	Technological Integration
Artificial Neural Networks	Predictive modelling for process optimization	Enhanced prediction accuracy	Data quality and consistency issues	High-quality, clean data required	Integration with legacy systems
Machine Learning (ML)	Real-time optimization of bioreactor systems	Improved process efficiency	Need for continuous data updates	Large datasets from bioreactors	Complex system integration

(*Continued*)

TABLE 5.2 (*Continued*)
AI Tools Used in Bioprocess Optimization, Detailing Their Applications, Advantages, Challenges, Data Requirements, and Integration Issues

AI Tool	Application	Advantages	Challenges	Data Requirements	Technological Integration
Deep Learning	Optimization of complex biological systems	Higher precision in biomanufacturing	Cost and maintenance of AI systems	Data from sensors and lab experiments	Compatibility with current infrastructure
Supervised Learning	Fermentation control and protein synthesis	More targeted interventions	AI model training complexities	Data from real-time sensors	Need for specialized knowledge
Unsupervised Learning	Modelling unknown trends in bioprocesses	Discovering new optimization patterns	Challenges in training without labelled data	Data from multiple sources (lab, sensors)	Lack of flexibility in older systems
Reinforcement Learning	Decision-making in dynamic environments	Automation of complex decision processes	High computational demands	Continuous data flow required	Requires interdisciplinary teams

5.3 APPLICATIONS OF AI TOOLS IN BIOPROCESS OPTIMIZATION

5.3.1 AI in Fermentation Process Control

The bioproducts that rely on fermentation to produce enzymes, antibiotics, and biofuels are among the critical bioprocesses (Gao et al., 2020). It is a complicated microbial growth, metabolic activities, and environmental conditions, and all of them need to be carefully managed to ensure a good yield and the quality of the products. The conventional systems of controlling the fermentation processes have been dependent on manual manipulations and trial and error (Yee et al., 2025). Nevertheless, more advanced solutions based on AI, especially ML algorithms, can deliver continuous analysis of real-time sensor data placed in bioreactors. The conditions in fermentation, like temperature, pH, oxygen, and nutrient levels, can be optimized by these models without the interference of man (Asiedu et al., 2024). AI tools can also be used to predict the most effective parameters to be used in future fermentation batches based on the large volumes of data on the past that can be processed to generate the desired product concentration and purity, thereby increasing productivity and reducing wastage.

In addition, fermentation control using AI can offer dynamic changes which can react to changes in the biological system which results in greater consistency and repeatability (Vinestock et al., 2024). Efficient AI models, such as deep learning, could model and forecast microorganisms behaviour in diverse situations and thus predictive control approaches could be implemented. This helps the system to reallocate itself in real-time to any unforeseen events like the depletion of the nutrients or contamination by the microbes (Wlodkowic & Karpiński, 2021). As well, AI is able to spot minor patterns and associations that may be missed through a conventional approach and provide an insight that can be applied to enhance subsequent fermentation cycles. With the further development of AI, it is bound to change the optimization of the fermentation process, and biomanufacturing will be more efficient, sustainable, and cost-effective (Ayub et al., 2025).

5.3.2 AI Is Utilized in Cell Culture Optimization

Cell culture can be described as a backbone of the biopharmaceutical manufacturing, especially in the production of monoclonal antibodies, vaccines, and recombinant proteins (Hu, 2020). Streamlining of cell culture parameters, including the media, temperature, and gaseous exchange, plays an important role in ensuring the highest yield and minimum variability. Conventionally, this has been a lengthy process that has frequently consumed huge doses of organizations and time. This is made much simpler by AI tools and, especially, ML algorithms, which automate the optimization of culture conditions (Hesami & Jones, 2020). The ML models may be used to analyse both historical and real-time data on cell cultures, such as the rate of cell growth, the intake of nutrients, and the formation of the products to identify the most desirable environment to be produced in the most productive way.

The behaviour of the cell culture system can also be predicted using AI models that allow deviations to be identified at an early stage and prevent the possible impact on the process (Park et al., 2023). AI is able to control the key parameters of cell growth by regulating the variables of the environment, including pH, oxygen, and the concentration of the metabolites in the air, in real time. With this ongoing observation and optimization, the yield of the bioproducts will not only be improved but it will also be the case with the consistency and quality of the bioproducts (Ögmundarson et al., 2020). Also, the large datasets can be incorporated with the help of AI, and the insights will be more comprehensive and valuable in the process of scaling up. As AI is increasingly becoming part of the cell culture systems, it promises to provide the prospect of more personalized, efficient, and cost-effective biopharmaceutical production (Neudeck, 2024).

5.3.3 AI in Drug Production and Purification

Proteins production is a crucial part of biopharmaceutical production, in which the effective expression and separation of proteins are essential in the production of therapeutic agents (Castro et al., 2021). Optimization of protein production is a process that entails complicated aspects such as the expression system selection, the media composition, and the induction condition. The algorithms of AI and deep learning have made a groundbreaking contribution to this process by using extensive datasets of previous experiments to forecast the best conditions to express proteins. These AI applications assist biotechnologists in rapid strain, culture media, and growth condition selection, which saves a lot of time and resources that are normally wasted on empirical optimization.

Protein production is followed by the next obstacle, the purification. Conventional way of purifying proteins is a series of chromatographic processes, which take up a lot of time and usually consume resources (Carta & Jungbauer, 2020). The purification protocols used can be optimized to be the most efficient based on the physicochemical properties of the target proteins which can be done with the help of AI. As an example, AI models can propose optimal resin and conditions of chromatography that will ensure maximum yield and lowest costs. Also, the purification strategy simulation that AI enables can be used to determine possible bottlenecks at the start of the process so

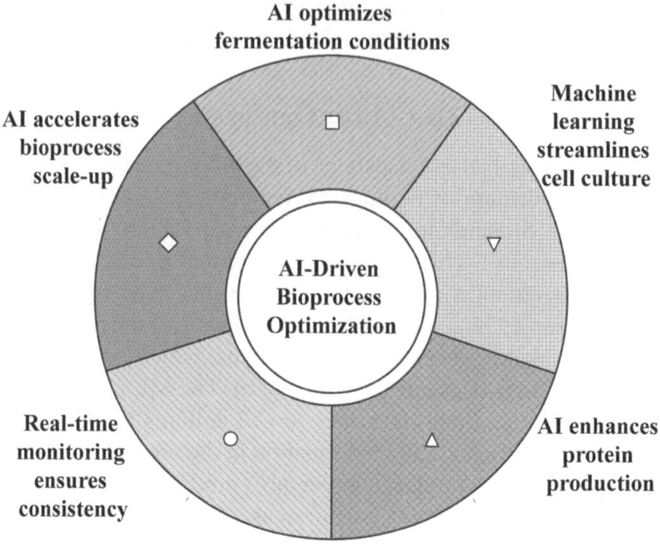

FIGURE 5.3 Applications of AI in bioprocess optimization.

that more efficient and cost-effective pipelines to produce proteins can be developed (Sharma et al., 2024). With the current developments of AI, it is expected to further streamline the process of protein production and purification, thus streamlining the production of biopharmaceuticals and making them efficient and scalable.

Figure 5.3 Applications of AI in bioprocess optimization, showcasing how AI improves fermentation conditions, accelerates bioprocess scaling, streamlines cell culture processes, enhances protein production, and ensures consistency through real-time monitoring for optimized production.

5.3.4 AI-Assisted Process Monitoring and Control

Monitoring and control in bioprocessing is important to ensure uniformity and quality in the process (Wasalathanthri et al., 2020). Conventionally, bioprocess monitoring has been carried out by gathering data from different sensors and instruments and then making adjustments manually by the operators through experience and intuition. Nonetheless, this strategy is prone to inefficiency and inefficient performance. AI-assisted process monitoring involves the automation of bioprocesses control through the use of real-time data of sensors, ML algorithms, and predictive models (Cheng et al., 2023). Those AI systems are able to process large volumes of data gathered by numerous sources, such as temperature, pH, dissolved oxygen, and pressure, and determine trends and forecast the most appropriate process conditions (Shekhar et al., 2025).

The control systems based on AI not only optimize the parameters of the process but also offer an early warning system in case of a possible deviation or failure (Luo et al., 2025). As an example, AI could identify subtle changes in data, which may indicate an upcoming problem, e.g., microbial contamination or nutrient imbalance, and take corrective measures, which will postpone the development of problems. Moreover, the AI tools may give the opportunity to predict the future tendencies of the process based on the previous data and implement proactive changes to the process, increasing efficiency and making the quality of the products more regular (Qudus, 2025). With the consistent optimization of the conditions of bioprocesses with the help of AI, bioprocesses can become more predictable, scalable, sustainable, and next-generation biomanufacturing, which will have a lesser amount of human involvement and better results, is possible.

5.3.5 AI in Bioprocess Scale-Up

There are many problems in scaling up a bioprocess that is developed in laboratory to the industrial-scale production, including the process consistency and effectiveness. Conventional scale-up practices tend to make use of trial-and-error practices and assumptions which are founded on past experiments on small scale (Memeti, 2025). Nonetheless, AI tools provide a statistical method of addressing these issues by predicting the behaviour of a bioprocess at bigger scales by relying on ML models. The models examine the data of small-scale bioreactors as well as large-scale bioreactors and, therefore, aid in determining the key parameters that regulate scaling success, including mixing, oxygen transfer, and nutrient diffusion (de Mello et al., 2024).

Optimization of the lab-scale to production-scale can also be achieved by AI, which can model the various scaling-up conditions in different scenarios and forecast the problems that can occur in scale (shear stress or nutrient gradients) at scale. ML algorithms are capable of determining the best scaling parameters so as to guarantee that the bioprocess gets stronger, efficient, and reproducible as it scales (Puranik et al., 2022). Besides, the scale-up aided by AI can greatly speed up and decrease the costs of the time-consuming scaling experiments. Through prediction and optimization of scaling behaviour with the help of AI, bio-manufacturers can experience a less bumpy ride to commercial production, less trial and error, and enhanced process robustness (Maharjan et al., 2023). This is possible not only to enhance productivity, but also to guarantee that the end product has the necessary quality standards regardless of the scale.

Table 5.3 summarizes the applications of AI in optimizing various bioprocesses, highlighting how AI tools enhance fermentation, cell culture, protein production, process monitoring, and bioprocess scaling, ensuring higher efficiency and improved product quality.

TABLE 5.3
AI Applications in Fermentation, Cell Culture, and Bioprocess Scaling

Process	AI Application	Benefit	Tools/Tech	Optimization Focus	Impact on Production
Fermentation Control	AI models to monitor temperature, pH, nutrients	Increases yield and product quality	Machine Learning, Deep Learning	Process consistency and repeatability	Higher productivity, lower wastage
Cell Culture Optimization	Machine learning for culture condition analysis	Optimizes growth rate, nutrient uptake	AI, Real-time Data Monitoring	Temperature, pH, gas exchange	Improved yield, reduced variability

(Continued)

TABLE 5.3 (*Continued*)
AI Applications in Fermentation, Cell Culture, and Bioprocess Scaling

Process	AI Application	Benefit	Tools/Tech	Optimization Focus	Impact on Production
Protein Production and Purification	AI models for strain and media optimization	Faster strain and media selection	Deep Learning, Data Modelling	Protein expression and separation	Cost savings, efficient purification
Process Monitoring and Control	Real-time sensor data analysis for process control	Predicts deviations like microbial contamination	Machine Learning, AI Algorithms	pH, oxygen, pressure control	Consistent product quality
Bioprocess Scale-Up	AI models predicting behaviour at larger scales	Ensures successful scaling from lab to industry	Machine Learning, Predictive Models	Mixing, oxygen transfer, nutrient diffusion	Faster scale-up, cost-effective
AI for System Integration	Modular AI tools for bioprocessing integration	Streamlines integration with existing systems	AI, Cloud Computing, Digital Twins	Legacy system compatibility	Improved system efficiency

5.4 CHALLENGES, FUTURE DIRECTIONS, AND CONCLUSION

5.4.1 Data-Related Problems in the Implementation of AI

The issue of data quality and availability is one of the most important obstacles to the implementation of AI tools in the optimization of bioprocess (Shoshi et al., 2024). To provide meaningful insights and draw accurate predictions, AI systems work under the condition of large quantities of quality data. In bioprocessing, data may be distributed in various sources, such as sensors and other laboratory devices and even in manual records, hence the problem of data consistency and completeness (Borgosz & Dikicioglu, 2024). Also, the information gathered is not necessarily applicable to the behaviour of the whole system, which can influence the accuracy of AI models. In addition, bioprocess data usually lacks cleanliness or is noisy, which makes it difficult to train ML algorithms efficiently that require large datasets to be clean (Helleckes et al., 2023).

To counter such problems, there is a need to have strong data management systems integrated (Aldoseri et al., 2023). Such systems have to guarantee the standardization, cleaning, and pre-processing of data and then pumping it into AI models. Also, the issue of data privacy and security should be considered, especially in the cases of proprietary or sensitive data in producing biopharmaceuticals (Alosert et al., 2022). It is important to have a more detailed data strategy that incorporates the use of IoT sensors to monitor data in real time and data fusion techniques implementation. Besides, the continuous refinements of AI predictions with real experimental or production outputs should also be performed to generate confidence and enhance the accuracy of the models with time (Liang et al., 2022).

5.4.2 AI Implications in Bulk Biotech

Another issue in the optimization of bioprocesses is the implementation of the AI tools in combination with the currently available biotechnological systems (Butean et al., 2025). The use of legacy systems by many biotechnological companies cannot be particularly compatible with the current AI technologies. These old systems are generally not flexible and scalable to embrace AI tools; thus, the integration process is hectic and expensive (Jain, 2022). Moreover, the inability to connect AI models with the existing bioprocessing devices of bioreactors or chromatography may result in inefficiency and slow adoption.

In order to address these difficulties, the combined efforts of AI specialists and biotechnologists should be involved (Ali et al., 2024). The gap in the knowledge between AI and the requirements of bioprocess optimization can be addressed with the help of interdisciplinary teams (Sharma & Singh, 2025). It is also important to have a modular and flexible integration approach so that AI tools can complement and supplement the existing systems without completely changing the current infrastructure. Moreover, cloud-based systems and digital twins can be the possible solution in conquering the problem of compatibility as they enable the sharing of data in real time and optimization of a system in different platforms. This type of integration might simplify the process, lower the implementation expenses, and make the optimization of bioprocesses using AI a faster process (Pan, 2023).

Figure 5.4 This infographic illustrates the key challenges and future directions in AI bioprocess optimization, highlighting data quality, legacy system integration, emerging technologies, AI's transformative role, and its application in personalized medicine.

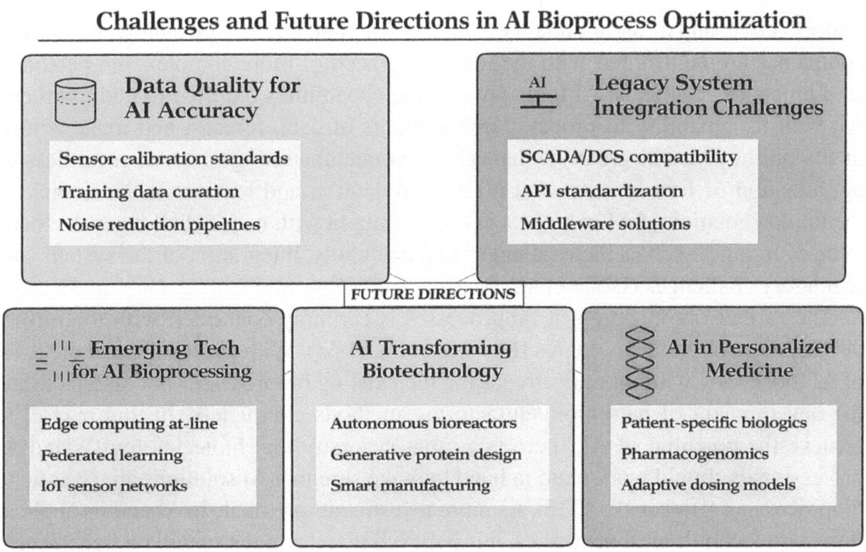

FIGURE 5.4 Key challenges and future directions in AI bioprocess optimization.

5.4.3 The Future of AI in the Optimization of Bioprocesses

The potential given by AI and ML, especially deep learning and real-time monitoring systems, is enormous, and the future of AI in bioprocess optimization cannot be ignored (Pan, 2023). It can be stated that AI will take a central role in personalized medicine in the coming years, where more efficient, patient-specific bio-manufacturing processes will be designed. Bioprocesses can be optimized to generate biologics specific to an individual genetic composition by taking advantage of the predictive nature of AI to enhance higher efficacy and little side effects (Živanović & Filipović, 2024). Complex bioprocesses such as gene therapy and tissue engineering will also proceed in the optimization of the processes with the help of AI-powered models, which will allow the development of future-generation biopharmaceutical products.

Also, AI will be integrated with other emerging technologies, including blockchain and quantum computing, which will revolutionize bioprocessing. Blockchain will also provide better security and transparency of data and quantum computing will find the answers to the complex optimization, which is impossible to solve with classical computing now (Krishnakumar, 2020). It can be expected that with the advancement of AI algorithms, they will be able to make much more accurate predictions and real-time readjustments and, thus, bioprocesses will become highly efficient, automated, and sustainable. The further rise in the development of AI models, as well as the availability of more data and the ability to compute, will speed up the large-scale implementation of AI-based solutions in biomanufacturing, changing the biotechnology environment (Bettanti et al., 2024).

5.4.4 Conclusion: AI in Transforming Bioprocessing

To sum up, AI can transform bioprocess optimization and make it more efficient, cost-effective, and of higher quality (Singh & Singhal, 2023). Since biotechnological companies are confronted with the increasing need of more complex and personalized biopharmaceuticals, AI tools provide the capabilities required to address them. AI, with its capability to process large amounts of data, forecast and make adjustments on the fly, will greatly enhance the biomanufacturing process, regardless of optimization of fermentation and protein production and expansion. Nevertheless, the implementation of AI in bioprocessing is fraught with major challenges that need to be overcome, such as the challenges in data quality, integration of the system, and regulatory challenges (Isoko et al., 2024).

The further optimization of bioprocesses is certainly connected with the further development of AI technologies (Rathore et al., 2023). With the increase in the level of AI tools, they will not only streamline the existing bioprocesses but also facilitate the development of new biomanufacturing methods completely. In this regard, to achieve the potential of AI, there is a great necessity that biotechnologists and AI professionals should work hand in hand to make sure that AI solutions are specific to bioprocessing (Bivi et al., 2025). As more investments are made in AI and its application in biotechnological processes, it is believed that the industry will be transformed so that it produces biopharmaceuticals that are faster, sustainable, and less costly.

Table 5.4 outlines the challenges, solutions, and future directions in AI bioprocess optimization, focusing on data quality, legacy system integration, emerging

TABLE 5.4
Challenges, Solutions, and Future Directions in AI Bioprocess Optimization

Challenge/Issue	Description	Impact	Solution/Strategy	Technologies Involved	Future Directions
Data Quality	Data consistency, completeness, and noise affect AI model accuracy.	Poor model predictions, inefficiency.	Implement strong data management systems, cleaning, and real-time data monitoring.	IoT sensors, AI models	Continuous improvement with real data.
Legacy System Integration	Legacy systems are not compatible with current AI tools.	Slow AI adoption, inefficiency.	Use modular, flexible integration with cloud-based systems and digital twins.	Cloud computing, digital twins	Accelerated integration of AI systems.
Personalized Medicine	AI can enhance patient-specific biomanufacturing processes.	Improved efficacy and minimal side effects.	Develop AI tools for more personalized, predictive processes.	AI, biomanufacturing, genomics	Widespread adoption in personalized therapies.
Emerging Technologies	Blockchain, quantum computing, and deep learning could revolutionize AI.	Enhanced data security, better predictions.	Integrate new technologies for AI system efficiency and security.	Blockchain, quantum computing, AI	Full integration in bioprocess optimization.
AI Model Complexity	Complex models need high-quality data for better predictions.	Limited model scalability and accuracy.	Ongoing refinement of AI models, feeding real experimental outputs.	Machine learning, AI models	Improved AI algorithms for better predictions.
Regulatory Challenges	AI implementation faces regulatory hurdles in biopharmaceuticals.	Delay in AI adoption in bioprocessing.	Develop regulations that accommodate AI tools in bioprocessing.	AI, regulatory compliance	Adaptation of AI within regulatory frameworks.

technologies, personalized medicine, and regulatory challenges. It emphasizes strategies for overcoming issues and the role of new technologies in transforming bioprocessing.

5.5 FUTURE DIRECTIONS

The future of AI in bioprocess optimization is bright, and the further development of ML models, integration of data, and computational resources is going on. It is likely that AI will allow increased personalization of bioprocesses, especially in the case of biopharmaceuticals, where custom ways of manufacture would make the treatment more effective. Also, the integration of the AI and real-time monitoring technology will further enhance the prediction and control of the outcomes of bioprocesses and minimize human control and enhance its accuracy.

With the further development of AI technologies, the field is expected to transition towards the integration of AI with more sophisticated technologies, such as blockchain to store data safely and quantum computing to solve more complex solutions. Such developments will open the way to more sustainable, efficient, and cost-effective process of bioprocessing.

5.6 DISCUSSION

Bioprocessing optimization is being revamped by AI that promises to provide a radical solution to the problems that biotechnology industry has been grappling with since time immemorial. Although the integration of AI tools into bioprocesses has considerable positive sides, such as higher efficiency and lower costs, there are still issues arising during the management of data, the integration of the system, and algorithm development. The question of AI technologies and classical biotechnological processes is not that simple, and their implementation on an industrial scale may have to pass significant challenges regarding infrastructure and staff preparedness.

5.7 CONCLUSION

Conclusively, AI tools can highly promote the efficiency and scalability of bioprocess optimization and thus can be considered as an essential part of the biotechnology future. With the use of AI capabilities in data analysis, predicting, and optimization of the processes, the biopharmaceutical industry will have an advantage in a shorter time of development, reduced costs, and the possibility of creating safer bioproducts. Although the adoption of AI tools is still problematic at the industrial level, the future of bioprocessing is enormous, as the number of new developments associated with AI and machine learning algorithms grows, and the quantity of raw data with high-quality increases accordingly. These tools are also undergoing continuous development; hence, they will be at the core of the next generation of biotechnology.

REFERENCES

Aldoseri, A., Al-Khalifa, K. N., & Hamouda, A. M. (2023). Re-thinking data strategy and integration for artificial intelligence: Concepts, opportunities, and challenges. *Applied Sciences*, *13*(12), 7082.

Ali, M., Shabbir, K., Ali, S., Mohsin, M., Kumar, A., Aziz, M., Zubair, M., & Sultan, H. M. (2024). A new era of discovery: How artificial intelligence has revolutionized the biotechnology. *Nepal Journal of Biotechnology, 12*(1), 1–11.

Alosert, H., Savery, J., Rheaume, J., Cheeks, M., Turner, R., Spencer, C., Farid, S., & Goldrick, S. (2022). Data integrity within the biopharmaceutical sector in the era of Industry 4.0. *Biotechnology Journal, 17*(6), 2100609.

Asiedu, N. Y., Bamaalabong, P. P., Johnson, J. E., Bonsu, J. K., & Addo, A. (2024). Modeling, simulations, and Simulink developments in the analysis of optimal control of temperature and pH in a batch ethanol fermentation process. *Journal of Engineering and Applied Science, 71*(1), 195.

Asin-Garcia, E., Fawcett, J. D., Batianis, C., & Dos Santos, V. A. M. (2025). A snapshot of biomanufacturing and the need for enabling research infrastructure. *Trends in Biotechnology, 43*(5), 1000–1014.

Ayub, A., Singh, R., Wani, A. K., Xie, R., Ahmad, F., Gorish, B. M. T., Sun, J., Nurhayati, N., Pangestu, R., & Hashemizadeh, A. (2025). Role of artificial intelligence (AI) in sustainable biofuel production from lignocellulosic biomass. In *Artificial Intelligence Applications for a Sustainable Environment* (pp. 329–348). Springer.

Bähner, F. D., Prado-Rubio, O. A., & Huusom, J. K. (2021). Challenges in optimization and control of biobased process systems: An industrial-academic perspective. *Industrial & Engineering Chemistry Research, 60*(42), 14985–15003.

Behera, B. K. (2023). *Conceptual Development of Industrial Biotechnology for Commercial Production of Vaccines and Biopharmaceuticals*. Elsevier.

Bettanti, A., Beccari, A. R., & Biccarino, M. (2024). Exploring the future of biopharmaceutical drug discovery: Can advanced AI platforms overcome current challenges? *Discover Artificial Intelligence, 4*(1), 102.

Biswas, A., & Dutta, P. K. (2020). Novel approach of automation to risk management: The reduction in human errors. *International Conference on Mobile Computing and Sustainable Informatics.*

Bivi, N., Graham, D., Joglekar, L., McGuire, K., Stoop, J., Zoghbi, J., Baker, B., Bandukwala, A., Bond, S., & Buoninfante, A. (2025). 2024 White paper on recent issues in bioanalysis: Impact of LDT in US and IVDR in EU; AI/ML for High Parameter Flow Cytometry; The rise of Olink Technology; CDx for AAV Gene Therapies; Integrative Bioanalysis by Multiple Platforms; Super Sensitive ADA/NAb LBA (PART 2A–Recommendations on Advanced Strategies for Biomarkers, IVD/CDx Assays (BAV), Cell Based Assays (CBA), and Ligand-Binding Assays (LBA) PART 2B–Regulatory Agencies' Input on Biomarkers, IVD/CDx, and Biomarker Assay Validation). *Bioanalysis, 17*(4), 211–248.

Borgosz, L., & Dikicioglu, D. (2024). Industrial internet of things: What does it mean for the bioprocess industries? *Biochemical Engineering Journal, 201*, 109122.

Butean, A., Cutean, I., Barbero, R., Enriquez, J., & Matei, A. (2025). A review of artificial intelligence applications for biorefineries and bioprocessing: From data-driven processes to optimization strategies and real-time control. *Processes, 13*(8), 2544.

Carta, G., & Jungbauer, A. (2020). *Protein Chromatography: Process Development and Scale-up*. John Wiley & Sons.

Castro, L. S., Lobo, G. S., Pereira, P., Freire, M. G., Neves, M. C., & Pedro, A. Q. (2021). Interferon-based biopharmaceuticals: Overview on the production, purification, and formulation. *Vaccines, 9*(4), 328.

Cheng, Y., Bi, X., Xu, Y., Liu, Y., Li, J., Du, G., Lv, X., & Liu, L. (2023). Artificial intelligence technologies in bioprocess: Opportunities and challenges. *Bioresource Technology, 369*, 128451.

de Mello, A. F. M., de Souza Vandenberghe, L.P., Herrmann, L. W., Letti, L. A. J., Burgos, W. J. M., Scapini, T., Manzoki, M. C., de Oliveira, P. Z., & Soccol, C. R. (2024). Strategies and engineering aspects on the scale-up of bioreactors for different bioprocesses. *Systems Microbiology and Biomanufacturing, 4*(2), 365–385.

Fasim, A., More, V. S., & More, S. S. (2021). Large-scale production of enzymes for biotechnology uses. *Current Opinion in Biotechnology, 69*, 68–76.

Gao, S., Song, W., & Guo, M. (2020). The integral role of bioproducts in the growing bioeconomy. *Industrial Biotechnology*, *16*(1), 13–25.

Helleckes, L. M., Hemmerich, J., Wiechert, W., von Lieres, E., & Grünberger, A. (2023). Machine learning in bioprocess development: From promise to practice. *Trends in Biotechnology*, *41*(6), 817–835.

Hesami, M., & Jones, A. M. P. (2020). Application of artificial intelligence models and optimization algorithms in plant cell and tissue culture. *Applied Microbiology and Biotechnology*, *104*(22), 9449–9485.

Holzinger, A., Keiblinger, K., Holub, P., Zatloukal, K., & Müller, H. (2023). AI for life: Trends in artificial intelligence for biotechnology. *New Biotechnology*, *74*, 16–24.

Hu, W.-S. (2020). Overview of cell culture processes. In Wei Shou Hu (Ed.), *Cell Culture Bioprocess Engineering* (2nd ed., pp. 1–35). CRC Press.

Isoko, K., Cordiner, J. L., Kis, Z., & Moghadam, P. Z. (2024). Bioprocessing 4.0: A pragmatic review and future perspectives. *Digital Discovery*, *3*(9), 1662–1681.

Jain, N. S. (2022). Pioneering the future of technology: Integrating advanced cloud computing with artificial intelligence for scalable, intelligent systems. *World Journal of Advanced Research and Reviews*, *13*, 847–862.

Krishnakumar, A. (2020). *Quantum Computing and Blockchain in Business: Exploring the Applications, Challenges, and Collision of Quantum Computing and Blockchain*. Packt Publishing Ltd.

Kumar, K., Sharma, R., Ranjan, K., & Pandey, D. K. (2025). Artificial intelligence in biotechnology. In *Artificial Intelligence and Biological Sciences* (pp. 192–218). CRC Press.

Liang, W., Tadesse, G. A., Ho, D., Fei-Fei, L., Zaharia, M., Zhang, C., & Zou, J. (2022). Advances, challenges and opportunities in creating data for trustworthy AI. *Nature Machine Intelligence*, *4*(8), 669–677.

Liu, S. (2020). *Bioprocess Engineering: Kinetics, Sustainability, and Reactor Design*. Elsevier.

Luo, L., Song, Y., Liu, Y., & Jiang, X. (2025). The research on intelligent safety risk assessment and early warning mechanisms for power systems based on big data and artificial intelligence technology. *Advances in Resources Research*, *5*(2), 666–688.

Maharjan, R., Lee, J. C., Lee, K., Han, H.-K., Kim, K. H., & Jeong, S. H. (2023). Recent trends and perspectives of artificial intelligence-based machine learning from discovery to manufacturing in biopharmaceutical industry. *Journal of Pharmaceutical Investigation*, *53*(6), 803–826.

Malabadi, R. B., Nethravathi, T., Kolkar, K. P., Chalannavar, R. K., Mudigoudra, B. S., Abdi, G., & Baijnath, H. (2023). Applications of artificial intelligence (AI) in cannabis industries: In vitro plant tissue culture. *International Journal of Research and Innovation in Applied Science*, *8*(VII), 21–40.

Memeti, V. (2025). *Start up and Scale Up Strategies for SME's Faculty of Business and Economics*, South East European University.

Mendes, A. R., Spínola, M. P., Lordelo, M., & Prates, J. A. (2024). Advances in bioprocess engineering for optimising Chlorella vulgaris fermentation: Biotechnological innovations and applications. *Foods*, *13*(24), 4154.

Mondal, P. P., Galodha, A., Verma, V. K., Singh, V., Show, P. L., Awasthi, M. K., Lall, B., Anees, S., Pollmann, K., & Jain, R. (2023). Review on machine learning-based bioprocess optimization, monitoring, and control systems. *Bioresource Technology*, *370*, 128523.

Narayanan, H., Luna, M. F., von Stosch, M., Cruz Bournazou, M. N., Polotti, G., Morbidelli, M., Butté, A., & Sokolov, M. (2020). Bioprocessing in the digital age: The role of process models. *Biotechnology Journal*, *15*(1), 1900172.

Neudeck, S. (2024). *Harmonizing Transformations: Key Considerations for Implementing AI-Driven Sustainability in Pharmaceutical Manufacturing*. Utrecht University.

Ögmundarson, Ó., Sukumara, S., Herrgård, M. J., & Fantke, P. (2020). Combining environmental and economic performance for bioprocess optimization. *Trends in Biotechnology*, *38*(11), 1203–1214.

Ogunjobi, T. T., Ohaeri, P. N., Akintola, O. T., Atanda, D. O., Orji, F. P., Adebayo, J. O., Abdul, S. O., Eji, C. A., Asebebe, A. B., & Shodipe, O. O. (2024). Bioinformatics applications in chronic diseases: A comprehensive review of genomic, transcriptomics, proteomic, metabolomics, and machine learning approaches. *Medinformatics*, *1*(1), 110–127.

Pan, Z. (2023). Machine learning for real-time optimization of bioprocessing parameters: Applications and improvements. *Artificial Intelligence and Machine Learning Review*, *4*(3), 30–42.

Pandey, K., Pandey, M., Kumar, V., Aggarwal, U., & Singhal, B. (2024). Bioprocessing 4.0 in biomanufacturing: Paving the way for sustainable bioeconomy. *Systems Microbiology and Biomanufacturing*, *4*(2), 407–424.

Park, S. Y., Kim, S. J., Park, C. H., Kim, J., & Lee, D. Y. (2023). Data-driven prediction models for forecasting multistep ahead profiles of mammalian cell culture toward bioprocess digital twins. *Biotechnology and Bioengineering*, *120*(9), 2494–2508.

Puranik, A., Dandekar, P., & Jain, R. (2022). Exploring the potential of machine learning for more efficient development and production of biopharmaceuticals. *Biotechnology Progress*, *38*(6), e3291.

Qudus, L. (2025). Leveraging artificial intelligence to enhance process control and improve efficiency in manufacturing industries. *International Journal of Computer Applications Technology and Research*, *14*(2), 18–38.

Rathore, A. S., Mishra, S., Nikita, S., & Priyanka, P. (2021). Bioprocess control: Current progress and future perspectives. *Life*, *11*(6), 557.

Rathore, A. S., Nikita, S., Thakur, G., & Mishra, S. (2023). Artificial intelligence and machine learning applications in biopharmaceutical manufacturing. *Trends in Biotechnology*, *41*(4), 497–510.

Saraf, S., De, A., & Tripathy, B. (2024). Effective use of computational biology and artificial intelligence in the domain of medical oncology. In S. Panda, A. Abraham, V. Gopi, & L. P. Suresh (Eds.), *Computational Intelligence for Oncology and Neurological Disorders* (pp. 228–252). CRC Press.

Sharma, D., & Singh, K. (2025). AI-enhanced bioprocess technologies: Machine learning implementations from upstream to downstream operations. *World Journal of Microbiology and Biotechnology*, *41*(8), 278.

Sharma, V., Mottafegh, A., Joo, J.-U., Kang, J.-H., Wang, L., & Kim, D.-P. (2024). Toward microfluidic continuous-flow and intelligent downstream processing of biopharmaceuticals. *Lab on a Chip*, *24*(11), 2861–2882.

Shekhar, K., Sreeja, R. S., Krishna, M. J., Goud, G. N., & Sri, G. T. (2025). AI-powered dissolved oxygen monitoring and prediction. *Journal of Computational Analysis & Applications*, *34*(4), 359–372.

Shoshi, A., Gündüz, B., & Miehe, R. (2024). Identifying intelligent data utilization in bioprocesses: Overview of current research activities, opportunities and barriers. *Procedia CIRP*, *126*, 869–874.

Singh, A., & Singhal, B. (2023). Role of machine learning in bioprocess engineering: Current perspectives and future directions. In *Design and Applications of Nature Inspired Optimization: Contribution of Women Leaders in the Field* (pp. 39–54).

Stalidzans, E., & Dace, E. (2021). Sustainable metabolic engineering for sustainability optimisation of industrial biotechnology. *Computational and Structural Biotechnology Journal*, *19*, 4770–4776.

Subramanian, G. (2021). *Process Control, Intensification, and Digitalisation in Continuous Biomanufacturing*. John Wiley & Sons.

Vinestock, T., Short, M., Ward, K., & Guo, M. (2024). Computer-aided chemical engineering research advances in precision fermentation. *Current Opinion in Food Science*, *58*, 101196.

Wasalathanthri, D. P., Rehmann, M. S., Song, Y., Gu, Y., Mi, L., Shao, C., Chemmalil, L., Lee, J., Ghose, S., & Borys, M. C. (2020). Technology outlook for real-time quality attribute and process parameter monitoring in biopharmaceutical development—A review. *Biotechnology and Bioengineering*, *117*(10), 3182–3198.

Wlodkowic, D., & Karpiński, T. M. (2021). Live-cell systems in real-time biomonitoring of water pollution: Practical considerations and future perspectives. *Sensors*, *21*(21), 7028.

Yang, C.-T., Kristiani, E., Leong, Y. K., & Chang, J.-S. (2023). Big data and machine learning driven bioprocessing–Recent trends and critical analysis. *Bioresource Technology*, *372*, 128625.

Yee, C. S., Zahia-Azizan, N. A., Abd Rahim, M. H., Mohd Zaini, N. A., Raja-Razali, R. B., Ushidee-Radzi, M. A., Ilham, Z., & Wan-Mohtar, W. A. A. Q. I. (2025). Smart fermentation technologies: Microbial process control in traditional fermented foods. *Fermentation*, *11*(6), 323.

Yetgin, A. (2025). Revolutionizing multi-omics analysis with artificial intelligence and data processing. *Quantitative Biology*, *13*(3), e70002.

Zha, D., Bhat, Z. P., Lai, K.-H., Yang, F., Jiang, Z., Zhong, S., & Hu, X. (2025). Data-centric artificial intelligence: A survey. *ACM Computing Surveys*, *57*(5), 1–42.

Živanović, M. N., & Filipović, N. (2024). System biology modeling for drug optimization. In N. Filipović (Ed.), *In Silico Clinical Trials for Cardiovascular Disease: A Finite Element and Machine Learning Approach* (pp. 105–137). Springer.

6 Real-time Monitoring of the Treatment of Agri-Food Wastes

Mahalakshmi Jeyabalu, Balakrishnan Chinnaiyan, N. Gnanasankaran, Gayathry S. Warrier, and Shivam More

6.1 INTRODUCTION

Food and agriculture sectors are the most crucial industries in global food security and national economic sustenance, but they are equally the greatest producers of waste products that pose serious environmental, economic, and social challenges. According to the Food and Agriculture Organization (FAO), nearly 1.3 billion tons of waste from agri-food is generated globally every year, nearly one-third of all food produced for human consumption (Gustavsson et al., 2011). These wastes include crop residues, fruit and vegetable skins, dairy effluent, by-products of meat processing, and large amounts of food wasted while stored, transported, and processed (Jha et al., 2015). If agri-food wastes are not managed effectively, they could directly contribute to greenhouse gas (GHG) emissions, water pollution, and soil degradation, all of which are contrary to sustainable development goals (SDGs) (Wang et al., 2016).

For instance, India is one such country where agriculture plays a major role in contributing to the gross domestic product (GDP); therefore, large quantities of waste are produced from on-farm and downstream industries. Improper handling, storage, and processing lead to postharvest losses of about 20–25% of fruits and vegetables every year (Jha et al., 2015). Similarly, effluents from the dairy industry often contain high biological oxygen demand (BOD) and chemical oxygen demand (COD) and pose environmental risks when discharged without treatment (Prazeres et al., 2012). The processing of livestock and poultry also produces a huge amount of organic by-products which is required to be treated prior to disposal or reuse. The agri-food wastes thus need to be treated and valorized efficiently to promote environmental sustainability, renewable energy making, and resource optimization (Ren et al., 2018).

Traditional waste treatment methods such as composting, anaerobic digestion, incineration, biochar production, and wastewater treatment have been utilized for a long time and are still used (Ren et al., 2018). Composting is a natural process through which organic matter is recycled and turned into a nutrient-rich soil amendment. It is an aerobic biological process that transforms organic waste into stable

DOI: 10.1201/9781003684374-6

compost. The composting process is vulnerable to variations in feedstock composition, microbial activity, and environmental conditions such as pH, temperature, and moisture (Xu et al., 2021). For instance, very small changes in anaerobic digester temperature can lead to big reductions in methane yield (Ren et al., 2018). Similarly, low oxygen levels in composting can cause the compost to be only partially decomposed and also produce bad smell. Conventional monitoring practices—based on manual sampling and laboratory testing—are slow, labor-intensive, and unable to provide real-time insights into ongoing treatment processes (Kumar et al., 2020).

The system is quite aged and conventional. The outdated method totally lacks the benefits of modern tracking and forecasting. Kumar et al. (2020) noted that the lack of constant oversights often leads to the poor controlling of processes, the unproductive consumption of resources, and the crying over spilled finances. The scenario of modern technologies offering no real-time feedback is similar to that of operators being unable to act fast during uncertain environmental conditions. The tiniest variations in the microbial activity of the biogas plants that go unnoticed may be responsible for up to 30% loss in gas yield (Karthikeyan et al., 2021). Similarly, non-monitored composting systems can produce compost of low quality (Xu et al., 2021). The limitations of the traditional system are pushing the demand for intelligent methods in waste management that support continuous monitoring, early detection of unusual trends, and more informed decision-making (Bouzembrak et al., 2019).

The emergence of artificial intelligence (AI), machine learning (ML), deep learning (DL), and Internet of Things (IoT) integration has set a trend to turn the waste treatment program from a traditional reactive system into a predictive, proactive, and data-driven approach (Bouzembrak et al., 2019). In this system, networked sensors are capable of performing the measurements of parameters such as temperature, moisture, pH, oxygen levels, and gas composition, wherein the information is sent to the servers or edge devices for immediate analysis (Ramanathan et al., 2023). The ML and DL models gather these data streams, detect hidden patterns, and create short-term forecasts, which help the operators in their decision-making process. Supervised algorithms, for instance, Random Forest and XGBoost, based on sensor inputs, can provide an estimate of compost's maturity, whereas time-series models such as RNNs and LSTMs can forecast biogas production trends hours or even days ahead (Xu et al., 2021). As a result of these methods, the waste treatment systems become stable, resource-efficient, and eco-friendly (Ren et al., 2018).

The adoption of advanced technology like AI and ML in waste disposal comes with a multitude of benefits:

1. The technology has led to better fault detection by quickly spotting abnormal measurements or strange behavior in the process.
2. The AI system suggests necessary changes in operations for better efficiency thus helping to make the process more efficient.
3. The technology supports the operators in making decisions, thus minimizing the operational risks and cutting down on unnecessary expenses.
4. The system enables predictive monitoring to estimate the compost readiness, methane yield, and water quality much earlier.

Internationally, research in this particular area is getting popular. Studies have shown that new models using ML predict methane generation in anaerobic digestion with up to 95% accuracy. Meanwhile, DL models recognize waste types and detect issues in treatment plants. It detects anomalies in treatment plants. These are yielding impressive results. Additionally, several pilot projects now use "digital twin" systems. These are virtual copies of waste plants that operate with real counterparts to constantly improve performance using live data worldwide.

The intelligent monitoring system also helps in achieving the SDGs. The SDG goals like SDG goal 2 (Zero Hunger), SDG goal 6 (Clean Water and Sanitation), SDG goal 7 (Affordable and Clean Energy), and SDG goal 13 (Climate Action) are positively affected by the efficient solution of agri-food waste (Kumar et al., 2020; Ren et al., 2018; Wang et al., 2016).

Consider the following instances:

- SDG 2 (Zero Hunger): Agri-food waste treatment aligns directly with the goal of no hunger. Organic waste when turned into compost or biofertilizers increases the soil fertility and consequently the crop yield and the productivity of the land.
- SDG 6 (Clean Water and Sanitation): A massive amount of water that has been used is being polluted. Monitoring systems endorsed by AI allow the treatment plants to have the overall control of their operations. Identifying key factors could be very helpful for industries to lessen the amount of unclean water going into the rivers and lakes while providing cleaner water for the population.
- SDG 7 (Affordable and Clean Energy): The biogas production is one of the uses of food waste. The application of AI and ML models to the anaerobic digestion system is meant to optimize the process so as to get more methane and be more efficient in energy recovery. This will lead to a larger use of renewable energy sources and a lesser reliance on fossil fuels.
- SDG 13 (Climate Action): AI-powered monitoring systems are one of the reasons for the lower greenhouse gas emissions. It detects the first signs of disturbance in the composting and anaerobic processes. On top of that, the application of compost or biofertilizer leads to reduced usage of fertilizers and pesticides, which are one of the main sources of greenhouse gas emissions from agriculture.

Without a doubt, the foremost gain from the use of AI/ML-based monitoring systems will accrue to India and other developing nations. The latter are experiencing rapid urbanization and high agricultural production. However, these countries encounter significant problems. The regions in question hardly ever have a well-defined waste composition, and they also lack treatment plants as well as qualified workers in the area of technology. Intelligent monitoring can indeed solve these problems as it will make the overall system more reliable, lesser need for professionally trained personnel, and the whole process more stable (Kumar et al., 2020). These technologies can deal with such problems as judiciously managed feedstock, widespread and well-trained personnel shortages, and poor and aging treatment facilities by improving consistency and dependability of the process. Moreover, the dropping prices of the IoT sensors, together with the increasing availability of cloud-based analytics tools, advocate the practical use of such technologies even further (Ramanathan et al., 2023).

One of the key benefits that real-time AI support monitoring comes with is the fact that it fits so well with circular economy. It is not only the agriculture that gets to be closed with these systems in terms of resources, but the environment gets relieved of pressure in the same way too, through the conversion of agricultural waste to biogas, nutrient-rich compost, or organic fertilizers (Wang et al., 2016). More effective process control not only means that the costs of operation can be lowered, but also the quality and predictability of what is produced can be enhanced, which is a win-win for both the farmers and the food supply chain, which consists of food processing and marketing industries (Ren et al., 2018).

While the potential gains from the technology may be significant, but there are still several issues that need to be addressed before the full potential is realized. Some of the problems are periodical sensor adjustments, fluctuating data quality due to different operating conditions, and enormous computational power needed to deploy large-scale advanced ML and DL models (Ren et al., 2018). The AI predictions can adversely affect the data received and collected through the sensors, as it could be noisy or incomplete. On the other side, ML model design needs high-quality datasets, but due to the heterogeneous nature of waste materials, there is a scarcity of data in the waste management field. These issues can be addressed through research on data preprocessing techniques—how to remove noise, development of low-cost and durable sensors—development of cheap and durable sensors, and the exploration of hybrid AI approaches that integrate ML with domain knowledge in waste treatment systems (Kumar et al., 2020).

To summarize, agri-food waste treatment provides an emerging technological opportunity. It is lying at the intersection of pressing global challenges. The high demands of environmental protection, resource sustainability, and energy optimization can't be achieved with traditional methods of waste treatment. They are not sufficient for keeping up with the demand. The need for a new system has led to a new paradigm of a real-time waste monitoring system, which is intelligent and can predict. The system is a result of IoT-sensing technologies converging with AI/ML algorithms (Bouzembrak et al., 2019). This section will provide an in-depth knowledge of how ML and DL algorithms are relevant through the selected research works with the use of case applications and the challenges faced by them. It will provide future directions in this rapidly changing field.

6.2 RELATED RESEARCH WORKS AND CASE STUDIES

Many researchers have already looked into real-time monitoring of agri-food waste treatment systems through the combination of AI, ML, and DL. The goal is to optimize composting, anaerobic digestion, and wastewater treatment while achieving better environmental sustainability and energy recovery. This subpart outlines the recent contributions in the field and also the methodologies, results, and drawbacks of existing works.

Karthikeyan et al. (2021) considered the ML algorithms application for methane yield prediction in anaerobic digestion systems. The authors had collected real-time data on volatile fatty acids, temperature, pH, and organic loading rate from a pilot-scale digester. Random Forest and Gradient Boosting helped in achieving about 94% prediction accuracy for daily methane generation. AI models successfully captured nonlinear relationships among digestion parameters, providing an opportunity for an early intervention if performance deterioration occurs. The authors concluded that a reliable framework for predictive control of biogas plants is the integration of ML with IoT-based monitoring systems (Karthikeyan et al., 2021).

Zheng et al. (2020) developed a fault detection model for wastewater treatment plants (WWTPs) based on DL. The authors combined Convolutional Neural Networks (CNNs) with Long Short-Term Memory (LSTM) networks to capture both spatial and temporal variations in key water-quality parameters. The parameters were ammonia, COD, BOD, and total suspended solids (TSS). The framework detected abnormal sensor behavior. It predicted effluent quality several hours in advance. This helps in timely corrective actions by plant operators. The study illustrated the potential of DL to support more proactive and reliable control within wastewater facilities (Zheng et al., 2020).

Similarly, a recent study states an IoT-based composting setup that integrated Support Vector Machines and regression models to estimate compost maturity. The system continuously monitors temperature and humidity levels to enable compost producers to plan better. It saves time and resources compared to conventional laboratory testing methods. Tests showed that the AI-assisted method reduced both the time and costs associated with traditional laboratory-based maturity assessments.

A practical demonstration of AI-supported compost monitoring was carried out at the VTT Technical Research Center of Finland. In this implementation, IoT sensors recorded temperature, humidity, and gas emissions. They used Random Forest and LSTM models to predict compost readiness. These systems helped reduce composting time by approximately 18%. In addition, it improves nutrient recovery. It reinforces how the combination of DL algorithms and IoT infrastructure can enhance composting efficiency in real-world operations (VTT Technical Research Centre of Finland, 2023).

A related contribution by Li and colleagues (2020) investigated the possible use of reinforcement learning (RL) to control the aeration of the wastewater treatment process. Their RL agent learned how to improve the quality of pollutant-laden water leaving the treatment plant be less dirty while minimizing energy consumption. It avoids about 12% energy saving that does not show up in the effluent quality (Chen et al., 2020). The learning agent of the RL came to know the best samples of aeration that are in line with the consumption of energy and removal of pollutants. The results supported the argument that RL could reliably deal with complex biological processes where manual control is often just as difficult. A study in Tsinghua University supported the RL findings, where an RL-based aeration system model enabled the wastewater plants to cut their total energy consumption by 12% and still meet the regulatory limits on discharges (Tsinghua University, 2021).

Digital twins, in recent years, are one concept that has gained the most acceptance in the waste-treatment industry. In one case study in Spain (Gautam et al., 2021), the digital twin of a biogas production plant based on DL was created from sensors data. The operational status of the plant was continuously simulated along with the physical system. It enables real-time feedstock composition and process parameters optimization. The result of implementation was 15% increase in methane yield. The digital twins are an exciting future approach in which AI monitors and simultaneously optimizes real-world waste treatment systems (Spanish Biogas Research Consortium, 2022).

Gautam et al. (2021) focused on the use of CNN for classifying the waste of the food processing industry. CNN models processed high-resolution images of mixed food waste, such as VGG-16 and ResNet50, resulting in classification accuracies of above 95% to distinguish organic, recyclable, and hazardous wastes. The study also points toward the role of computer vision in large-scale food industry waste management efficiency improvement (Gautam et al., 2021).

A recent work undertaken by Singh et al. (2022) focused on developing an integrated system of IoT and ML for monitoring dairy wastewater. The operation of the system employed low-cost pH, turbidity, and dissolved oxygen sensors connected to an Arduino-based IoT platform. The data collected were analyzed using Random Forest and K-Nearest Neighbor (KNN) algorithms in order to classify effluent quality into safe, moderate, and unsafe. The experimental outcomes resulted in an overall classification accuracy of 92%. The authors stated that such intelligent monitoring systems could help small- and medium-scale dairy industries comply with pollution control norms at a small fraction of the cost of traditional laboratory-based testing (Singh et al., 2022). This aligns with the study conducted at the National Dairy Research Institute, India, which also reported over 90% prediction accuracy for wastewater quality parameters using SVM models (National Dairy Research Institute, 2023).

Another study by Chen et al. (2020) refers to LSTM networks for time-series prediction of composting parameters. Temperature, humidity, and microbial activity were historical measurements included in the dataset. During predictions for temperature profiles on separate composting phases (mesophilic, thermophilic, and maturation), LSTM models were shown to outperform standard regression approaches. This work illustrated the application of DL for the prediction and monitoring of composting facilities. Moreover, researchers have also examined the integration of multi-sensor fusion and ML algorithms for improved reliability (Chen et al., 2020).

Park et al. (2021) developed a multi-sensor system that combined near-infrared (NIR) spectroscopy, gas sensors, and temperature probes to monitor anaerobic digestion in real time. The fused data were analyzed using Support Vector Regression (SVR) and XGBoost models, achieving a prediction accuracy of 93% for methane yield. Multi-sensor fusion was shown to reduce the risk of erroneous predictions due to single-sensor failures or noisy data, thereby enhancing robustness in real-time applications (Park et al., 2021).

The advances notwithstanding, many limitations still plague present research. The majority of the studies are either conducted in a laboratory or pilot scale, with little large-scale deployment; the heterogeneity of the agri-food waste feedstocks creates a great challenge for generalizing the results of the model. For example, composting results widely fluctuate depending on the array of mixes of fruit, vegetable, and livestock wastes, while biogas yields are mostly dependent on substrate composition. Another major limitation is the lack of open-access datasets. Many datasets are either proprietary or collected locally, which impedes the development of generalized models using ML and DL. Furthermore, it is challenging to benchmark algorithms because most research relies on these collected datasets (Wang et al., 2016).

In summary, research shows that ML-and DL-based algorithms with IoT technologies offer the most significant advances for real-time monitoring and predictive control of agri-food-waste treatment systems. This being said that scalability, data availability, and cost-effective deployment are significant challenges that future research must address. Further incorporation of digital twins, RL, and hybrid AI models with low-cost IoT sensors will leave the intelligent waste monitoring systems with greater prospects for implementation in the developed and developing world alike (Ramanathan et al., 2023).

6.3 MACHINE LEARNING ALGORITHMS FOR CROP DISEASE PREDICTION

Dynamic parameters such as temperature, solids, dissolved oxygen, pH, carbon-to-nitrogen ratio, volatile and gas emissions like methane and carbon dioxide influence complex biochemical and physical processes involved in treating agri-food wastes (Ren et al., 2018). These parameters are changing continuously, which are often nonlinear and unpredictable during composting, anaerobic digestion, and wastewater treatment. Manual sampling, laboratory testing, and linear regression models are traditional methods that often neglect dynamic and multivariate systems. Old and used methods are time-consuming. Old methods can't predict results and often fail to adapt to changing waste conditions (Kumar et al., 2020). Whereas ML algorithms have advanced significantly. They can be the first choice for real-time monitoring. It can analyze both past and real-time data to find patterns and predict outcomes. This pattern results in predictions and classification, which drives the decision-making for the systems. ML is used to forecast methane production, compost maturity, and wastewater quality (Karthikeyan et al., 2021; Xu et al., 2021). It has been inclined in strategizing for agri-food waste treatments to determine prediction outputs on methane yield. It also helps in sorting and classifying waste to improve input quality for treatment (Gautam et al., 2021).

ML techniques help in improving the system performance and decision-making. ML is now widely used to forecast methane production, compost maturity, and wastewater quality, as well as for the heterogeneous nature of the dataset, as collected through multiple IoT sensors, imaging systems, and biochemical assays, making this highly suitable for real-time monitoring using ML. Waste treatment plants under this kind of technology would be moved into proactive operations when problems are detected at an early stage, interventions suggested before serious disruptions happen, as opposed to reactive operations where problems are only resolved following a failure (Kumar et al., 2020).

Widely adopted in this domain are several ML algorithms. Random Forest is up in the ranks as one of the most favored due to its robustness against noisy data and non-linearity. It has been increasingly used with successes in methane yields and anaerobic digestion with accuracies above 90%. Classification tasks, such as determining the various composting phases or categorizing wastewater effluents as in the categories of safe, moderate, or unsafe, can be performed using SVM (Singh et al., 2022). KNN is among the algorithms that have been applied for clustering and classification of relatively small datasets, such as compost cycle identification (Chen et al., 2020). One of the major considerations that has attracted industries and academia to boosting algorithms like XGBoost and LightGBM is that their competencies greatly emphasize computational efficiency and high accuracy, with a major emphasis lying on predicting COD and BOD levels in wastewater treatment (Ramanathan et al., 2023). Additionally, ANNs, simplistic as they are compared to DL models, still carry the potential to capture nonlinearities in composting and digestion processes. They have been applied to the prediction of temperature profiles, nutrient recovery, and gas production. Naive Bayes classifiers are sometimes used in tasks where speed and efficiency supersede accuracy, such as waste separation. The latest applications of this approach are in RL, where intelligent agents learn to perform process control tasks efficiently, such as aeration in bioreactors. The RL-based systems are used for decreasing the cost of operation. It optimizes energy use and helps in maintaining effluent quality (Gustavsson et al., 2011).

The RL-based systems are used for decreasing the cost of operation. It optimizes energy use and helps in maintaining effluent quality. These sensors measure parameters such as temperature, pH, moisture, dissolved oxygen, and gas composition, or imaging systems may be used for waste segregation (Ramanathan et al., 2023). Collected data undergoes preprocessing to remove noise, normalize values, and deal with missing entries. The next consecutive step would comprise feature extraction and selection, where critical attributes are identified, such as temperature cycles, pH trends, or gas yield rates. The chosen characteristics are subsequently passed to ML methodologies for the purpose of forecasting and classifying. The Random Forest, SVM, or XGBoost algorithm might be the one used (Karthikeyan et al., 2021). The use of clustering techniques, such as K-Means or PCA, has been reported for identifying abnormal treatment patterns (Karthikeyan et al., 2021). To sum up, the ML models' output is merged, thereby forming intelligent systems that support decisions with suggestions, notifications, and even optimization tactics delivered in real time to the users. The outcomes of this pipeline include optimized treatment processes, reduced operational costs, improved energy recovery, and decreased greenhouse gas emissions (Ren et al., 2018).

The process of ML in agri-food waste treatment starts with data collection using IoT sensors. Data includes parameters like measuring pH, moisture, dissolved oxygen, temperature, and gas emissions (Ramanathan et al., 2023). In the next step, the raw data is pre-processed through noise removal, normalization, and missing values handling. Thereafter, feature extraction is performed to identify the features that have the greatest impact on the treatment process. Moreover, the temperature cycles, gas yield patterns, and microbial activity are among the important features extracted (Xu et al., 2021).

When the important features are ready, they are supplied to ML algorithms such as Random Forest, SVM, K-Means, and RL. These algorithms perform tasks such as prediction, classification, and anomaly detection. The model outputs are forwarded to the decision support system. This system interprets these results through dashboards, alerts, or automated recommendations and assists in decision-making (Kumar et al., 2020). Ultimately, the outcomes of this entire workflow result in effective recovery of resources from agri-food waste using ML (Ren et al., 2018) (Table 6.1).

TABLE 6.1
Machine Learning Algorithms Used for the Crop Yield Prediction and Crop Monitoring

S. No.	Title of Study	Algorithm Used	Outcomes	Future Scope	Ref and Year
1	Methane Yield Prediction in Anaerobic Digestion	Random Forest, Gradient Boosting	Achieved 94% accuracy in biogas yield prediction; improved digester stability	Extend to multi-feedstock and industrial-scale digesters	Bouzembrak et al. 2019
2	Compost Maturity Forecasting with IoT Sensors	Support Vector Machine, Multiple Linear Regression	Predicted compost maturity index 3–5 days in advance	Integrate with deep learning for higher accuracy	Chen et al. 2020
3	Wastewater Effluent Monitoring	Random Forest, KNN	Classified effluent into safe/moderate/unsafe categories with 92% accuracy	Apply across multi-industry wastewater plants	Gautam et al. 2021
4	Waste Segregation in Food Industry	Naïve Bayes, ANN	Classified organic vs. inorganic wastes with >90% accuracy	Combine ML–DL for complex mixed waste	Gustavsson et al. 2011
5	COD/BOD Prediction in Wastewater Plants	XGBoost, LightGBM	Provided highaccuracy regression for COD/BOD levels	Cloud-based real-time deployment	Jha et al. 2015
6	Adaptive Bioreactor Aeration Control	Reinforcement Learning	Reduced aeration energy by 12% without affecting effluent quality	Extend to fully autonomous plants	Karthikeyan et al. 2021
7	Prediction of Compost Temperature and Moisture Trends	Artificial Neural Networks (ANN)	Modeled nonlinear compost dynamics with >93% accuracy	Integrate IoT + ANN for on-field smart composting	Kumar et al. 2020
8	Anomaly Detection in Anaerobic Digesters	K-Means, PCA	Identified abnormal digester behavior and sensor failures	Combine with supervised models for early warning systems	Park et al. 2021
9	Nutrient Recovery from Agri-Food Wastes	Decision Trees, Regression Models	Predicted nitrogen and phosphorus recovery efficiency	Apply in circular economy and fertilizer production	Prazeres et al. 2012

6.4 DEEP LEARNING ALGORITHMS FOR WASTE TREATMENT MONITORING

DL, an advanced subset of ML and AI, has emerged as a potent tool for real-time monitoring and optimization of agri-food waste treatment processes. DL algorithms differ a lot from traditional ML algorithms as they do not need any manual feature extraction; they learn the patterns that may be complex by themselves from the raw data, resulting in a more appropriate approach for the highly dynamic and nonlinear nature of composting, anaerobic digestion, and wastewater treatment (Xu et al., 2021). The collections of diverse datasets are produced from these operations. The dataset may have readings from sensors (temperature, pH, dissolved oxygen, gas composition), images (waste texture, microbial growth), and time series of process variables that can be effectively analyzed by DL techniques.

CNNs are increasingly utilized in image monitoring for waste treatment, from detecting the maturity stage in compost to identifying feedstock contamination (Gautam et al., 2021). CNNs are capable of tracking hierarchical features such as color, texture, and architecture that are critical in detecting minor changes over the treatment process. For analysis of time-series data, RNNs in particular, LSTM models can be used to forecast the pH variations, methane generation, or changes in COD/BOD with time (Xu et al., 2021). Autoencoders are used for anomaly detection. It looks for weird and unusual behavior in the data. If found, they consequently indicate instability of the process. It can also indicate the failure of equipment. Additionally, if sensor datasets are imbalanced, generative adversarial networks hold promise in synthesizing new datasets.

This framework indigenously integrates the DL models with IoT-enabled sensors. Real-time predictions and control can be achieved through this integration. For example, CNNs with thermal imaging were used to forecast the maturity of compost. Meanwhile, the LSTM with a gas sensor forecasted methane yield (Ramanathan et al., 2023). This forecasting is hours in advance. These features help reduce greenhouse gas emissions. This reduction in emissions takes place by feeding energy recovery optimization systems with early warnings.

However, the adoption of DL models is limited in small- to medium-scale waste treatment settings. As it faces challenges such as high demand for larger datasets, which are hard to obtain. Additionally, it requires high computational resources, which increases the cost, and lacks interpretability (Kumar et al., 2020).

Future research focuses on lightweight DL models for edge computing. The research can be focused on DL and enhancing trust and transparency associated with it. Hybrid systems that blend DL with ML or physics-based models can be a topic for research. This research will accelerate the development of intelligent waste treatment facilities. The system that aligns with circular economy principles (Ren et al., 2018) (Table 6.2).

TABLE 6.2
Deep Algorithms Used for the Crop Yield Prediction and Crop Monitoring (Karthikeyan et al., 2021; Xu et al., 2021; Gautam et al., 2021; Kumar et al., 2020)

S. No.	Title of Study	Algorithm Used	Outcomes	Future Scope
1	Crop Disease Identification using Leaf Images	Convolutional Neural Networks (CNN)	Achieved >95% accuracy in classifying multiple crop diseases	Extend to multi-crop, real-time field deployment
2	Transfer Learning for Plant Leaf Classification	VGG16, ResNet, Inception	Improved accuracy on small datasets using pretrained models	Apply lightweight models for mobile diagnosis
3	Time-Series Crop Growth Monitoring	Recurrent Neural Networks (RNN), LSTM	Predicted crop stress and disease progression trends	Integrate with IoT sensor networks for precision farming
4	Crop Yield Prediction from Multispectral Images	Deep Neural Networks (DNN)	Yield estimation accuracy above 90% using remote sensing data	Scale to regional yield forecasting systems
5	Early Detection of Crop Anomalies	Autoencoders	Detected early stress and disease symptoms with limited labeled data	Apply for anomaly detection in resource-poor regions
6	Data Augmentation for Disease Prediction	Generative Adversarial Networks (GANs)	Generated synthetic diseased leaf images to balance datasets	Use for rare disease detection and dataset expansion
7	Crop Monitoring with Drone Imaging	CNN + LSTM Hybrid Models	Combined spatial and temporal features for accurate crop health tracking	Integrate UAV data with satellite monitoring
8	Crop Yield Forecasting under Climate Variability	Deep Reinforcement Learning (DRL)	Optimized irrigation and input allocation to maximize yield	Expand to climate resilient smart farming systems

6.5 REAL-TIME MONITORING SYSTEMS FOR AGRI-FOOD WASTE TREATMENT

The emergence of IoT sensors integrated with the AI/ML/DL algorithms has resulted in an advanced monitoring system for waste treatment. It works for the identification of anomalies and leads to operations optimization (Ramanathan et al., 2023).

The real-time monitoring system can have continuous collection of data, with immediate analysis. This results in generating predictive insights and proactive control within time. In contrast, the old and traditional monitoring methods are time-consuming. It performs sampling and testing at a painfully slow pace. The results are delayed. These drawbacks make them incompatible with the new system. In contrast, the real-time monitoring system can have continuous collection of data, with immediate analysis. This results in generating predictive insights and proactive control within time (Kumar et al., 2020).

A real-time monitoring framework mainly consists of three tiers: data accumulation, intelligent analysis, and decision support systems. In the first tier, parameters (like pH, gas composition [CH_4, CO_2, H_2S, NH_3], temperature, dissolved oxygen, and moisture) are measured for which IoT-based sensors are set up. Classification of waste is performed through imaging systems. It also identifies changes in the texture of the compost. Wireless protocols such as LoRa, Zigbee, Wi-Fi, or NB-IoT are used mainly for transmission to centralized cloud platforms or edge computing devices (Ramanathan et al., 2023).

In the second tier, ML and DL techniques are used to help in processing the data. The purpose of data processing is anomaly detection, prediction, and optimization. For example, we can use CNNs for the classification of compost maturity stages on the basis of the images, meanwhile Random Forest and XGBoost can be used for the prediction of methane yield in anaerobic digestion. We can use RL agents for controlling aeration rates in real time. Additionally, we can use hybrid systems that combine supervised and unsupervised learning for robust performance (Karthikeyan et al., 2021; Xu et al., 2021).

The third tier of the framework is the decision support system, which translates the outputs of AI and ML models into practical guidance for operators. Once the system recognizes unforeseen situations like sudden rises in COD or unexpected changes in processing parameters, it can trigger alerts automatically. Operators get advance real-time insights through dashboards, which help them to take corrective decision. The quickly responding system thus reduces operational costs, lowers the impact on the environment, and makes the whole resource retrieving process from waste materials more efficient (Ren et al., 2018).

Presently, the real-time monitoring installations are being piloted in composting, biogas, and wastewater treatment facilities. As trials are encouraging, the major setbacks are some complications, such as the precision of calibration, the cost of installation, and the privacy of the data. The next upcoming technologies are likely to utilize the digital twin models for uninterrupted virtual process like simulation, edge-AI systems for quicker on-site data processing, and blockchain-based solutions to improve transparency and security in waste-tracking networks (Bouzembrak et al., 2019).

6.6 CHALLENGES

Several challenges limit the large-scale implementation of the suggested agri-food waste management (Bouzembrak et al., 2019; Kumar et al., 2020; Ren et al., 2018):

1. Constraints in Data Quality and Dataset Availability: A lot of datasets that are utilized in this field are small in quantity and have been collected in a very specific, local setting. Sensor data is generally noisy and inconsistent, which leads to the demoting of AI predictions' accuracy.
2. High Processing Requirements of Advanced Algorithms: The most sophisticated ML and deep learning frameworks are usually very much reliant on high processing power and memory. It can be quite a tough task to run such models in facilities without access to high-performance computing resources.

3. Multi-Layer System Integration Challenges: The integration of IoT hardware, cloud services, and AI software into a real-time monitoring system is a very complicated process involving multiple integrations. It is not just a matter of making sure that these elements work together; it also involves a lot of engineering and coordination efforts between different parts of the system.
4. Data Security Weaknesses: The data of sensor sent through cloud networks has been flagged with security concerns. The issue of preventing unauthorized access and providing secure data storage still remains a significant challenge.
5. Insufficient Technical Proficiency among System Operators: A lot of the agricultural system operators lack the proper training to not only manage but also interpret and maintain these systems effectively. These deficiencies in proficiency may turn out to be a major hindrance to adoption as well as reducing the overall efficiency of the system.
6. Economic Limitations of Small-Scale Implementers: The prices of sensors and connectivity tools have been coming down little by little; nevertheless, the initial outlay that is necessary for a complete AI-IoT monitoring setup can still be a barrier for small businesses or rural operations.

6.7 FUTURE WORKS

Future research and development should focus on overcoming the barriers to make the adoption of the intelligent waste monitoring system easy (Bouzembrak et al., 2019; Kumar et al., 2020; Ramanathan et al., 2023; Ren et al., 2018) (Table 6.3).

TABLE 6.3
Future Scope of the Intelligent Agri-waste Management System

S. No.	Focus Area	Description
1	Development of Open and Diverse Global Datasets	Creating large, publicly accessible datasets from varied geographical and operational contexts to build AI models that are robust, generalizable, and reliable across different waste-treatment environments.
2	Design of Lightweight Deep Learning Models for Edge Computing	Developing lightweight DL architectures suitable for edge computing to reduce cloud dependence and enable faster, on-site decision-making with low latency.
3	Integration of Hybrid AI Approaches	Combining ML/DL models with physics-based or rule-driven approaches to address sensor inconsistencies and provide more interpretable outputs.
4	Advancement and Deployment of Digital Twin Technologies	Implementing digital twins to support real-time simulation, performance optimization, and predictive maintenance, allowing operators to test changes virtually before applying them physically.
5	Secure and Transparent Data Sharing through Blockchain	Exploring blockchain-based frameworks to enhance data integrity, traceability, and trust across waste-treatment supply chains by preventing tampering or unauthorized access.
6	Innovation in Sustainable and Low-Cost IoT Sensor Design	Focusing on low-cost, biodegradable, and energy-efficient sensors to support sustainable monitoring solutions and expand accessibility.

6.8 DISCUSSIONS AND CONCLUSION

The suggested model for real-time monitoring could lead to considerable reductions in postharvest food waste because of its ability to monitor, through the incorporation of AI and IoT-based tools, the parameters and storage conditions throughout the supply chain. It can very quickly inform producers of the occurrence of spoilage, so they will be able to make quicker and more informed decisions for loss reduction. Unlike the traditional, slow-moving, labor-intensive methods, these smart systems allow for continuous monitoring, predictive analytics, and automated decision-making as well, which in turn leads to more precise and timely control of waste management processes (Ramanathan et al., 2023).

Still, the benefits of the monitoring system, the evidence from existing studies has pointed out quite a few gaps that serve as barriers to the large-scale adoption. Concerns regarding the lack of standardized datasets, the high computational demands of state-of-the-art models, and the shortage of skilled professionals remain unresolved. In the case of small and medium enterprises, the total amount for the purchase of sensors, the establishment of connectivity infrastructure, and the procurement of analytical tools can be a big financial burden. To overcome these hurdles, it is going to take not only the technical breakthroughs but also strong institutional support and policy initiatives.

Looking to the future, the technologies like hybrid AI models, edge computing platforms, and digital twin systems will be the ones making the waste treatment process more adaptable, transparent, and ready for automation. The mentioned technologies will let the operators conduct the tasks in real time, saving resources, and enhancing the reliability of the operations (Bouzembrak et al., 2019). The ongoing effort in creating low-cost, energy-efficient sensors and secure data exchange systems will further enable the communication lines for intelligent waste management to be very strong.

To a certain extent, the system has drawbacks that include high installation and connectivity costs; however, at the end of the day, its role in reducing waste and maximizing profits makes it a sustainability and profitability enhancer.

REFERENCES

Bouzembrak, Y., Klüche, M., Gavai, A., & Marvin, H. J. P. (2019). Artificial intelligence solutions for enabling circular economy in food waste management. *Trends in Food Science & Technology*, 94, 80–89.

Chen, J., Li, X., & Huang, Z. (2020). Long short-term memory networks for prediction of composting parameters under variable conditions. *Environmental Technology & Innovation*, 18, 100–145.

Gautam, A., Kumar, S., & Yadav, A. (2021). Deep learning-based waste classification system for food-processing industries. *Journal of Cleaner Production*, 315, 128–142.

Gustavsson, J., Cederberg, C., Sonesson, U., van Otterdijk, R., & Meybeck, A. (2011). *Global food losses and food waste: Extent, causes and prevention.* Food and Agriculture Organization of the United Nations.

Jha, S. N., Vishwakarma, R. K., Ahmad, T., Rai, A., & Dixit, A. K. (2015). *Assessment of quantitative harvest and post-harvest losses of major crops and commodities in India.* ICAR – All India Coordinated Research Project on Post-Harvest Technology.

Karthikeyan, O. P., Singh, S., & Raghavan, V. (2021). Machine learning approaches for methane prediction in anaerobic digestion systems. *Bioresource Technology*, 333, 125–210.

Kumar, S., Iyengar, S. R. S., & Caytiles, R. D. (2020). Smart agriculture: An overview of the use of AI and IoT. *International Journal of Advanced Science and Technology*, 29(5), 245–256.

National Dairy Research Institute. (2023). IoT and machine learning–based frameworks for low-cost wastewater monitoring. *Environmental Technology & Innovation*, 32, 103208.

Park, J., Kim, S., & Lee, H. (2021). Multi-sensor fusion and machine learning for real-time monitoring of anaerobic digestion processes. *Journal of Environmental Monitoring and Assessment*, 193(4), 1–14.

Prazeres, A. R., Carvalho, F., & Rivas, J. (2012). Cheese wastewater treatment: A review. *Journal of Environmental Management*, 110, 48–61.

Ramanathan, R., Duan, Y., Ajmal, T., Pelc, K., Gillespie, J., Ahmadzadeh, S., Condell, S., Hermens, J., & Ramanathan, U. (2023). Motivations and challenges for food companies in using IoT sensors for reducing food waste: Insights and a roadmap for the future. *Sustainability*, 15, 1665.

Ren, M. A., Ward, M., & Li, J. (2018). Optimization of anaerobic digestion of agricultural waste: A review of artificial intelligence applications. *Renewable Energy*, 120, 120–128.

Singh, R., Verma, P., & Gupta, M. (2022). IoT–ML integrated system for dairy wastewater classification using low-cost sensors. *Journal of Water Process Engineering*, 49, 102348.

Spanish Biogas Research Consortium. (2022). *Deep-learning digital twin model for real-time optimization of biogas production*. Madrid.

Tsinghua University. (2021). Reinforcement learning–based aeration control for energy-efficient wastewater treatment. *Water Research*, 202, 117512.

VTT Technical Research Centre of Finland. (2023). *AI-driven compost monitoring using IoT sensors and machine learning models*. VTT Research Reports, Espoo.

Wang, H., Xu, Z., Fujita, H., & Liu, S. (2016). Towards felicitous decision making: An overview on challenges and trends of big data. *Information Sciences*, 367, 747–765.

Xu, F., Wang, Y., & Wang, H. (2021). Real-time monitoring of composting processes using Internet of Things and machine learning. *Bioresource Technology*, 320, 124378.

Zheng, X., Li, Y., & Chen, Q. (2020). Deep learning–based fault detection in wastewater treatment using CNN-LSTM models. *Journal of Water Process Engineering*, 38, 101–115.

7 Optimization of Precision Fermentation Using AI/ML

S. Padmavathy, R. M. Murugappan, J. Jenifer Annis Christy, S. Karthik, L. Shamini, V. Roja, and A. Rathisha

7.1 INTRODUCTION TO PRECISION FERMENTATION AND ITS INDUSTRIAL SIGNIFICANCE

The escalating global demand for sustainable, high-quality, and protein-rich foods continues to strain traditional agriculture-based production systems. Conventional livestock and crop-based protein sources are limited by land scarcity, resource depletion, and environmental degradation, contributing significantly to greenhouse gas emissions and biodiversity loss. Simultaneously, the Food Tech and biomanufacturing sectors face challenges such as supply chain disruptions, rising production costs, and the urgent need for improved resource efficiency. These constraints have prompted the biotechnological community to develop alternative production systems that are resource-light, scalable, and environmentally resilient.

Precision fermentation has emerged as a transformative extension of traditional fermentation, engineered to address these sustainability challenges through microbial biotechnology. Unlike conventional fermentation, which produces complex product mixtures, precision fermentation employs genetically programmed microorganisms—such as bacteria, yeast, or filamentous fungi—to biosynthesize specific, high-value compounds with remarkable precision and purity (Terefe, 2022; McClements, 2023). This approach enables the production of functional proteins, lipids, vitamins, enzymes, and colorants that can replicate or even enhance the sensory and nutritional qualities of conventional ingredients. By utilising inexpensive carbon substrates and controlled bioreactors, precision fermentation not only minimizes waste and environmental impact but also significantly reduces production costs. It thus represents a key pillar of modern biomanufacturing, powering industries ranging from food and pharmaceuticals to biomaterials and renewable fuels. Figure 7.1 illustrates the precision fermentation process, depicting how genetically engineered microorganisms convert feedstock-based nutrient inputs into high-value biomolecules under controlled bioreactor conditions—from fermentation scale-up to purification and formulation into final food products (Teng et al., 2021; Boukid et al., 2023; Ajayeoba and Ijabadeniyi, 2025).

DOI: 10.1201/9781003684374-7

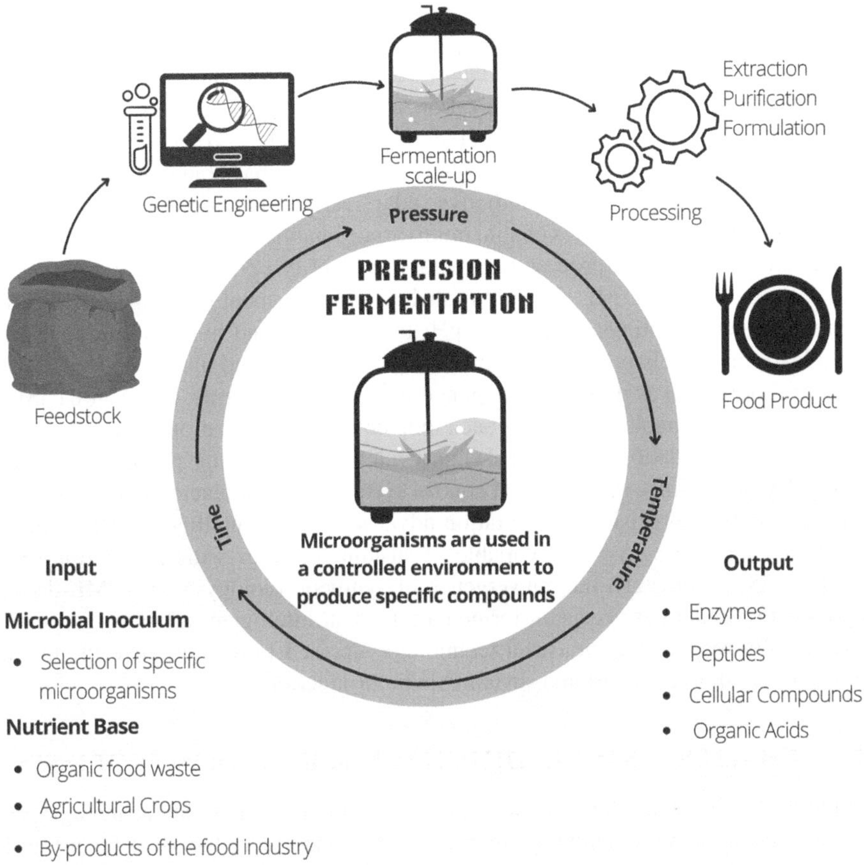

FIGURE 7.1 Shows how the precision fermentation converts from feedstock to food.

The foundation of precision fermentation lies in microbial strain engineering, where genetic modifications are introduced to reprogram organisms toward targeted biosynthesis. This involves integrating or deleting specific genes to enhance metabolic flux toward the desired molecule. Genetically regarded as safe (GRAS) host strains such as *Bacillus subtilis*, *Saccharomyces cerevisiae*, *Trichoderma reesei*, and *Escherichia coli* are frequently employed due to their well-characterized genomes, high productivity, and industrial robustness (Mirsalami and Mirsalami, 2025). Synthetic biology tools, including CRISPR-based systems and modular metabolic pathway design, further enable precise control over gene expression and pathway optimization. These advances allow for the sustainable production of animal-free proteins, enzymes, and metabolites, aligning food innovation with ethical, environmental, and nutritional objectives.

Despite the success of microbial strain engineering, the dynamic and nonlinear nature of fermentation systems presents challenges in achieving consistent yield, quality, and process stability. Biomanufacturing processes are influenced by multiple interdependent factors—pH, temperature, oxygen transfer, substrate concentration, and gene regulatory feedback—that are difficult to optimize manually. Traditional statistical or empirical optimization approaches often fail to capture the complex metabolic–process interactions that determine overall performance. This necessitates the integration of intelligent optimization tools capable of managing large-scale biological and engineering datasets in real time (Yadav et al., 2024).

Artificial intelligence (AI) and machine learning (ML) have radically transformed precision fermentation into a predictive, adaptive, and data-driven bioprocessing discipline. These technologies enable multivariate analysis of "omics" datasets, sensor-derived signals, and process parameters to identify optimal genetic constructs and culture conditions. ML algorithms such as reinforcement learning dynamically adjust bioreactor settings—including pH, temperature, and nutrient feeds—while deep learning models predict gene–promoter interactions and metabolic flux distributions. Tools such as digital twins, generative design platforms, and AutoCRISPR simulations now enable a rapid feedback loop between computational prediction and experimental validation, reducing design cycles and enhancing process reproducibility. By integrating AI/ML-driven control systems with bioreactor automation and edge computing, precision fermentation advances toward intelligent, self-optimizing biomanufacturing systems capable of sustainable large-scale production (Yee et al., 2025). Ultimately, the convergence of synthetic biology with AI/ML-driven optimization redefines precision fermentation as an intelligent, efficient, and environmentally responsible industrial technology—poised to revolutionize the global bioeconomy through smart and sustainable bioproduction.

7.2 CHALLENGES IN TRADITIONAL FERMENTATION PROCESSES

Traditional microbial fermentation processes in industrial biotechnology are constrained by several significant limitations that underscore the need for a transition to data-driven, precision-oriented fermentation approaches. A core challenge in traditional fermentation is the lack of real-time monitoring and control over critical process parameters—including temperature, pH, substrate concentration, and oxygen transfer rates—making it difficult to maintain consistent and optimal bioprocess conditions. This variability often results in batch-to-batch inconsistencies, suboptimal product yields, and extended cycle times, all of which hinder industrial scalability and efficiency. Furthermore, traditional processes rely heavily on empirical optimization strategies, such as trial-and-error experimentation and fixed design-of-experiment protocols, which are inherently limited in their ability to address the underlying complexity and nonlinearity of microbial metabolism. The inability to predict or dynamically adjust for metabolic shifts, product inhibition, or fluctuating feedstock quality often leads to production bottlenecks and costly process interruptions (Koubaa et al., 2019).

Another persistent obstacle is poor scalability—methods that succeed in the laboratory frequently face challenges during scale-up due to differences in mixing, nutrient gradients, and oxygen distribution in larger bioreactors, ultimately affecting product quality and process reproducibility. Additionally, traditional fermentation is slow to adapt to innovations in strain engineering and bioprocess design, limiting its capacity to accommodate novel products or rapidly change production targets (Huang et al., 2023).

Given these technical and operational limitations, there is a clear imperative to adopt advanced, data-driven solutions. The integration of AI and ML into precision fermentation offers the ability to continuously monitor, model, and optimize fermentation dynamics in real time. AI/ML frameworks can unravel complex biological interactions, predict system responses, and autonomously adjust process variables, thus overcoming key barriers of traditional fermentation and unlocking greater consistency, efficiency, and scalability in industrial bioproduction (Ramirez Rojas et al., 2022; Praveen and Brogi, 2025).

7.3 ROLE OF AI/ML IN PRECISION FERMENTATION

Traditional process optimization typically relies on mechanistic modeling and trial-and-error experimentation, which struggle to capture the nonlinear and dynamic behavior of biological systems. In contrast, modern approaches integrate Big Data analytics and AI/ML to enable predictive process control, rapid pathway design, and closed-loop dynamic adjustments based on continuous sensor feedback (Yee et al., 2025).

Integrating AI and ML into precision fermentation is critical for advancing strain performance, achieving real-time automated process control, and facilitating robust scalability in industrial biotechnology. These technologies optimize the use of genetically engineered microbes, streamline bio-compound synthesis, and transform process management from empirical guesswork into predictive, data-driven decision-making.

The integration of AI and ML is pivotal for advancing precision fermentation, as these technologies optimize microbial strain performance, automate process control, and support scalability for industrial bioproduction. In the context of bioprocess engineering, AI is an umbrella term that includes diverse computational techniques—such as ML, deep learning, and neural networks—applied to model the nonlinear, dynamic interactions characteristic of complex fermentation systems (Mannaa et al., 2021; Jiang et al., 2025). Figure 7.2 illustrates the distinct yet complementary roles of ML and AI in precision fermentation, highlighting ML's focus on data-driven

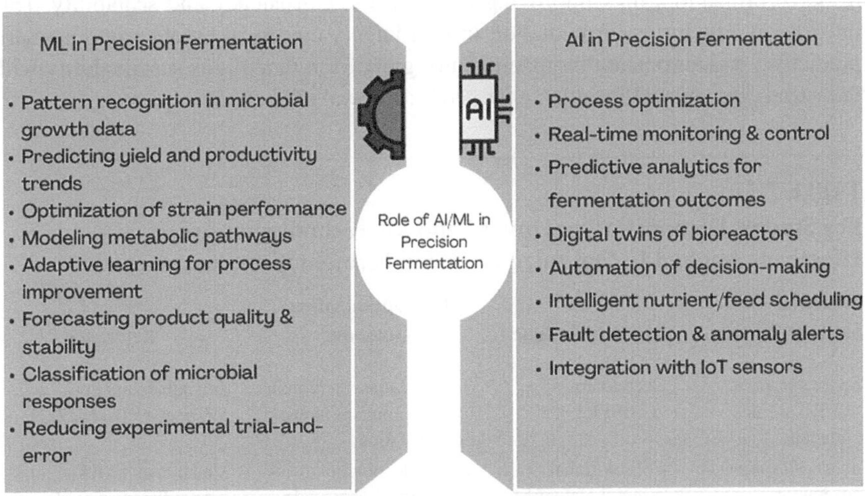

FIGURE 7.2 Depicts role of AI/ML in precision fermentation.

microbial growth analysis and process forecasting, while AI emphasizes real-time process control, digital twin integration, and automation for optimized fermentation outcomes.

Unlike traditional methods that rely mainly on mechanistic ("white-box") models derived from biochemical equations or empirical knowledge, AI/ML approaches enable data-driven ("black-box") modeling, harnessing large datasets to reveal hidden patterns and process relationships without explicit prior assumptions. Between these two extremes, hybrid ("gray-box") models integrate mechanistic knowledge with ML, leveraging the strengths of both approaches for greater prediction accuracy and adaptability in dynamic environments—such as fed-batch fermentation, where both pathway flux and environmental conditions must be tuned in real time (Sharma and Singh, 2025).

Core ML algorithms utilized include Artificial Neural Networks (ANN) for capturing nonlinear process dynamics, Support Vector Machines (SVM), and Random Forests for feature selection and classification, and reinforcement learning for automated control and optimization of bioreactor operations. These modeling frameworks drive applications such as process optimization, accelerated strain engineering, metabolic pathway redesign, real-time anomaly detection, and predictive maintenance—collectively enabling precision fermentation to evolve into a robust, intelligent, and resource-efficient foundation for the next generation of bio-industrial manufacturing (Khanal et al., 2023; Becchi et al., 2025).

7.4 AI-DRIVEN PROCESS OPTIMIZATION AND CONTROL

AI-driven process optimization and control have become indispensable in precision fermentation, enabling real-time monitoring and dynamic adjustment of critical process parameters such as temperature, pH, and agitation (Butean et al., 2025). By leveraging advanced algorithms like reinforcement learning and deep neural networks, these systems enhance metabolite yield, reduce batch failures, and optimize resource utilization, thus improving both process consistency and scalability. This integration transforms fermentation from a largely empirical practice into a smart, predictive, and automated biomanufacturing platform that aligns sustainability with industrial efficiency (Vinestock et al., 2024; Kanwal et al., 2025).

TABLE 7.1
Key Studies Demonstrating How AI and ML Technologies Are Effectively Applied to Optimize Precision Fermentation Processes

Microbial System / Process	AI/ML Technique	Key Optimization Outcome	Reference
Saccharomyces cerevisiae alt - proteins	Deep Learning + CRISPR	300% increase in protein yield; promoter –gene optimization	Priyadharshini et al. (2025); Zrimec et al. (2022)
Aspergillus niger pectinase	Convolutional Neural Networks	70% reduction in CRISPR design cycle	Chuai et al. (2018); Priyadharshini et al. (2025)

(*Continued*)

TABLE 7.1 (*Continued*)
Key Studies Demonstrating How AI and ML Technologies Are Effectively Applied to Optimize Precision Fermentation Processes

Microbial System / Process	AI/ML Technique	Key Optimization Outcome	Reference
Bacillus subtilis plant – based diary	Reinforcement Learning	98% purity with 30% less energy input	Pereira et al. (2025)
Streptomyces violaceoruber enzyme	Artificial Neural Networks	Precise yield prediction and process optimization	Priyadharshini et al. (2025)
Various fermentations	Pattern Recognition ML	Enhanced yield and process efficiency	BioRaptor platform documentation (2024); Hilgendorf et al. (2024)
S.cerevisiae vitamin B2	Reinforcement Learning	Dynamic gene regulation and increased vitamin synthesis	Priyadharshini et al. (2025)
Wine yeast ester pathways	Generative Adversarial Networks	Targeted pathway prediction and flavor profile enhancement	Zrimec et al. (2022); Priyadharshini et al. (2025)
S.cerevisiae cranberry fermentation	Reinforcement Learning + Edge AI	Sub – second pH response; improved anthocyanin stability	Priyadharshini et al. (2025)
Generic precision fermentation	Graph-based Machine Learning	Computational strain and pathway design optimization	Pereira et al. (2025); Hilgendorf et al. (2024)
Various fungal and yeast systems	Deep Learning + Genome Editing	Accelerated strain engineering with off – target prediction	Chuai et al. (2018); Priyadharshini et al. (2025)

7.5 STRAIN ENGINEERING AND GENOME EDITING GUIDED BY AI

Strain engineering and genome editing are foundational to precision fermentation, enabling microbial hosts to be reprogrammed for the efficient biosynthesis of targeted compounds. Major genome editing platforms include zinc finger nucleases (ZFNs), transcription activator-like effector nucleases (TALENs), and, most prominently, CRISPR-Cas systems. While ZFNs and TALENs offer targeted specificity, their design and assembly are complex and less versatile compared to CRISPR-Cas, which provides ease of use, multiplexing capability, and adaptability across species. Traditional genome editing approaches face limitations such as design complexity, unpredictable off-target effects, and the time-consuming nature of iterative optimization—challenges that are magnified when scaling for industrial production (Mirsalami and Mirsalami, 2025).

The integration of AI and ML techniques directly addresses these challenges by enhancing every stage of genome editing and strain engineering. AI-powered predictive models can rapidly screen and prioritize gene targets, simulate the effects of genetic edits, and identify optimal enzyme/protein variants. For example, convolutional neural networks in tools like AutoCRISPR predict potential off-target sites and assess editing efficiency, thereby reducing experimental cycles by up to 70%. Deep learning models, trained on multi-omics datasets, enable the mapping of complex genotype–phenotype relationships, allowing for accurate prediction of metabolic pathway outcomes and informed selection of edits that maximize yield with minimal metabolic burden. Furthermore, AI-assisted platforms such as DeepCRISPR and reinforcement learning-guided dCas9 systems automate the design of guide RNAs and dynamically adjust gene regulation to enhance product synthesis, as seen in boosted alt-protein yields or vitamin biosynthesis pathways (Boukid et al., 2023; Ajayeoba and Ijabadeniyi, 2025).

These AI models operate by ingesting high-dimensional biological data (e.g., genomic sequences, transcriptomic profiles, metabolite fluxes) as input and generating output predictions regarding edit efficiency, likely phenotype, metabolic impact, or product titers. This approach transforms strain improvement from a trial-and-error process into a predictive, data-driven engineering workflow. In industrial contexts, such as alternative protein production or enzyme synthesis, AI-guided genome editing accelerates the generation of high-yield, robust production strains, shortens R&D timelines, and supports precise metabolic pathway optimizations to achieve desired phenotypes. As AI/ML continues to advance, it is redefining strain engineering as a cornerstone of next-generation precision fermentation, enabling manufacturers to achieve superior process efficiency, scalability, and product quality. The Figure 7.3 depicts the integration of AI, ML, and new microorganisms in precision fermentation to develop innovative fermented beverages and dairy products at an industrial scale (Zeng et al., 2022; Dudeja et al., 2025).

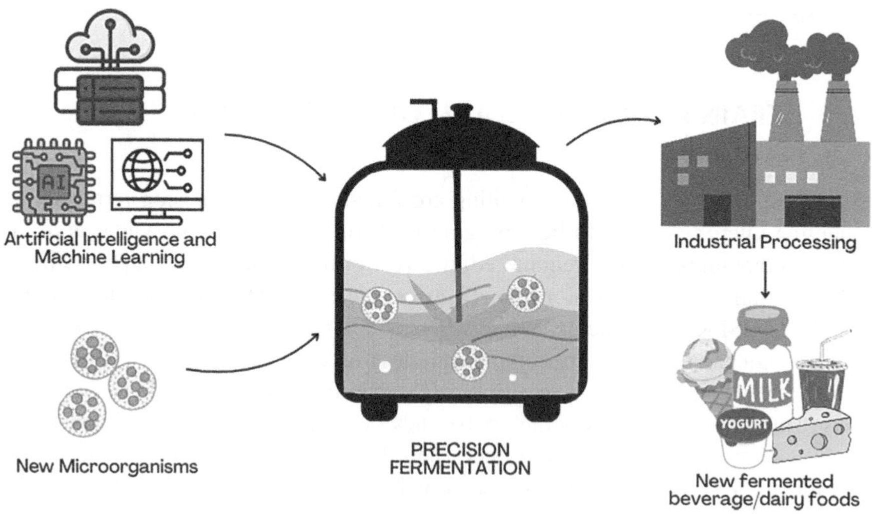

FIGURE 7.3 Depicts AI/ML-driven precision fermentation for next-generation dairy and beverages.

7.6 DIGITAL TWINS AND SURROGATE MODELING FOR FERMENTATION SCALE-UP

Digital Twins (DTs) in bioprocessing have emerged as a transformative technology essential for fermentation process scale-up, providing a comprehensive digital counterpart of physical bioreactors. By integrating multi-scale data from cellular metabolism through process parameters, to plant-wide operations, DTs offer a holistic simulation environment for monitoring, prediction, and control. This approach effectively addresses scale-up challenges like heterogeneous mixing, oxygen transfer inefficiencies, and nutrient gradients that commonly impair microbial productivity and product consistency in larger fermentation systems. The integration of AI and ML algorithms within DT platforms significantly enhances their predictive capability by continuously processing complex, multi-parametric data streams and enabling adaptive optimization and real-time decision-making (Dazzarola et al., 2026).

These ML techniques, including neural networks and reinforcement learning, contribute to the development of surrogate models—reduced-order representations of complex bioprocess simulations—that allow swift scenario analysis and process improvements without the high computational cost of full-scale models. Industry-leading tools such as Siemens Process Digital Twin, AspenTech Hybrid Models, and BioTwin simulators exemplify how AI-assisted DTs are transforming fermentation scale-up workflows, facilitating yield maximization and cost reduction. AI-driven DT applications have demonstrated success in dynamic feed rate control, fault detection, and lifecycle management, improving robustness in industrial fermentation. However, challenges remain, principally in harmonizing diverse data formats, validating model accuracy across scales, and integrating heterogeneous data sources. Ongoing advancements aim to refine AI/ML frameworks within DT ecosystems toward fully autonomous, scalable, and sustainable precision fermentation manufacturing, heralding a new era of bioprocess control and innovation (Zhao et al., 2025).

7.6.1 Steps Involved in the Digital Twin System

7.6.1.1 Define Goal and Experiments

The initial step is to establish the objectives entails the outlining of precise goal setup, such as desired yield, reaction parameters, increasing resource efficiency (Zobel-Roos et al., 2020; Zhao et al., 2025).

7.6.1.2 Data Collection of Real-Time and Historical Data

After establishing the objectives, the system gathers both historical and real-time data from a variety of sensors integrated within the bioprocess. This data encompasses essential parameters such as temperature, pH, oxygen levels, and nutrient concentrations. By merging historical data, which offers baseline insights and trends, with real-time information, the DT achieves a thorough understanding of these parameters (Zobel-Roos et al., 2020; Zhao et al., 2025).

7.6.1.3 Development of the Process Model

This stage entails the creation of a comprehensive process model that incorporates both mechanistic (physics-based) and data-driven elements. Mechanistic models are grounded in scientific principles to depict the bioprocess, whereas data-driven models (for instance, ML algorithms) utilize empirical data to discern patterns and refine predictions (Zobel-Roos et al., 2020; Zhao et al., 2025).

7.6.1.4 DT in Silico Simulation

During this phase, the DT simulates a range of scenarios based on real-time data inputs and the developed process model. This simulation framework permits operators to experiment with modifications, such as altering temperature or nutrient levels, and to observe the anticipated results (Zobel-Roos et al., 2020; Zhao et al., 2025).

7.6.1.5 Application in the Real-Time Process Monitoring and Control

The insights produced by the DT's simulations are utilized in the active bioprocess. Through real-time monitoring and feedback loops, the DT interacts with the physical system to implement accurate adjustments, thereby preserving optimal conditions for yield and quality.

Example: In biogas production through anaerobic digestion of organic waste, such as agricultural residues and food scraps, maintaining optimal parameters like temperature and pH is crucial for maximizing biogas yield and operational efficiency. An augmented DTsystem simulates the anaerobic digestion process, modeling microbial activity alongside environmental conditions and feed rates. This virtual system provides real-time monitoring and predictive insights, enabling the recommendation and implementation of optimized operational settings to enhance biogas production effectively (Zobel-Roos et al., 2020; Zhao et al., 2025).

7.7 PREDICTIVE MAINTENANCE AND ANOMALY DETECTION IN FERMENTATION SYSTEMS

As fermentation processes scale and diversify, the need for predictive maintenance in bioreactors becomes crucial—issues such as microbial contamination, sensor fouling, and unplanned equipment failures can disrupt efficiency and product quality if not detected early. Manual inspection or conventional rule-based monitoring struggles with the high volume and complexity of fermentation data, making proactive intervention both tedious and less effective. This is where AI and ML methods play a vital role: they extract patterns from continuous bioreactor signals and historical logs, enabling timely predictions of faults and system degradations before they escalate (Nguyen et al., 2025).

Autoencoders (AEs) and One-Class Support Vector Machines (OCSVM) are particularly well-suited for fermentation anomaly detection. AEs learn to represent normal operational data—such as time-series sensor readings for pH, temperature, dissolved oxygen, and substrate concentration—and flag anomalies when reconstruction errors spike, indicating departures from established patterns. OCSVM models are trained primarily on data collected during healthy operation, defining a boundary for normal performance and triggering alerts when new sensor inputs fall outside this region, identifying potential issues like microbial shifts, contamination events, or sensor malfunctions. In practice, a fermentation system might feed real-time sensor output to these models, which then classify data points and compute anomaly scores; if scores exceed set thresholds, predictive maintenance routines can be launched—such as targeted cleaning, recalibration, or controlled responses—to prevent loss of product or extended downtime (Abbasi et al., 2021; Goffi et al., 2025).

For example, AI-powered bioreactor designs have demonstrated dynamic regulation of pH, temperature, and nutrient supply—reducing human intervention and increasing reliability by autonomously monitoring triggers for maintenance actions. Integrating these anomaly detection algorithms with DT frameworks amplifies their utility: faults flagged by AEs or OCSVM can be instantly cross-referenced with simulated future system states, thus informing more robust real-time interventions and predictive scheduling (Gofffi et al., 2025).

Advantages of AI-based predictive maintenance include early detection of subtle operational problems, reduced reaction times, lower risk of catastrophic failures, and data-driven insights for long-term process improvement. Limitations can arise from the need for large, clean datasets to train robust models, the risk of false positives/negatives due to extraneous variability, and challenges in integrating heterogeneous sensor systems. Overall, AI-driven anomaly detection and predictive maintenance transform fermentation into a proactive, responsive discipline—complementing other smart bioprocessing tools like DTs for optimized industrial performance.

7.8 SUSTAINABILITY AND ECONOMIC FEASIBILITY THROUGH AI-ENHANCED PROCESS SYSTEMS ENGINEERING

Sustainability and economic feasibility are essential objectives in modern bioprocess engineering, and AI-enhanced process systems engineering provides powerful tools to achieve these goals. Sustainability assessment commonly employs Life Cycle Assessment (LCA), Techno-Economic Analysis (TEA), and process synthesis methodologies, each serving complementary roles in evaluating environmental impact, economic viability, and process design optimization. AI and ML techniques enhance these frameworks by processing diverse data inputs such as process parameters, emissions inventories, and cost databases to generate more accurate, predictive models. For instance, in LCA and TEA, regression models, neural networks, and optimization algorithms reduce computational complexity by approximating complex biological and chemical interactions, enabling rapid scenario analysis and sensitivity studies. In process synthesis, AI drives flowsheet generation and surrogate modeling, facilitating multi-objective optimization that balances productivity, resource use, and emissions reduction. Industrial applications demonstrate these benefits: AI-augmented LCA helped optimize feedstock selection in microbial fermentation, lowering overall carbon footprint; TEA combined with ML optimization enhanced cost-effectiveness in enzyme production; and AI-driven process synthesis redesigned fermentation pathways for improved energy efficiency. By integrating LCA, TEA, and process synthesis within AI frameworks, bioprocess engineers can make more informed, sustainable, and economically sound decisions—accelerating innovation while minimizing environmental impact and operational cost (Vinestock et al., 2024; Ajayeoba and Ijabadeniyi, 2025).

7.8.1 Life Cycle Assessment in Fermentation

LCA gives a "big picture" view of the environmental costs of a product or process. When applied to fermentation, it looks at every stage of production. This includes the sourcing of feedstocks, energy, and water requirements during fermentation, the often energy-intensive downstream processing steps such as separation and purification, and finally, the treatment or reuse of wastes and emissions (Knychala et al., 2024).

By tracing impacts across the full chain, LCA identifies where the biggest environmental burdens occur. For example, some studies have shown that downstream operations such as drying or distillation can contribute more to energy use and emissions than the fermentation itself. This kind of insight helps companies focus their improvement efforts where they will make the greatest difference (Vanapalli et al., 2023; Oraby et al., 2024; Kossmann et al., 2025).

7.8.2 Adding AI to LCA

Adding AI to LCA has increasingly supported the rapid and effective prediction of vital sustainability indicators, especially in earlystage research and process development where comprehensive datasets are often lacking. Among ML methods, regression algorithms (including Random Forest and Extreme Gradient Boosting), Gaussian Processes, and ANNs are particularly suited for forecasting environmental metrics such as energy demand, water usage, and emissions across different process scenarios. Reinforcement learning is valuable for scenario exploration and adaptive process optimization, while surrogate models enable quick simulations and sensitivity analysis in complex systems (Mohammadi Kashka et al., 2023).

AI models fill data gaps by learning from existing industrial databases or experimental results and can impute missing life cycle inventory data using statistical or supervised learning techniques, even when only limited input features—like process parameters, physicochemical properties, and historical emissions inventories—are available. These methods also reduce uncertainty in LCA predictions by leveraging large datasets to better capture underlying patterns, and Gaussian Process models, in particular, offer probabilistic estimates that quantify prediction confidence (Mohammadi Kashka et al., 2023).

Instead of conducting labor- and resource-intensive experiments for each scenario, AI can rapidly simulate thousands of possibilities, efficiently identifying environmentally preferable designs at early development stages. Case studies have shown that integrating ML with LCA models, such as those involving bioethanol or enzyme production, leads to higher prediction accuracy, improved process optimization, and significant reductions in greenhouse gas emissions and resource consumption compared to manual assessment methods. By automating data gap filling, uncertainty quantification, and scenario forecasting, AI-enhanced LCA streamlines sustainability evaluation and strengthens decision-making in bioprocess development.

7.8.3 AI in Process Synthesis

AI-driven process synthesis moves beyond simple prediction to designing and optimizing entire bioprocess routes, delivering substantial benefits over conventional simulation-based approaches like Aspen Plus or MATLAB. Instead of relying solely on deterministic models for stepwise energy and mass balance calculations, AI systems—using MLalgorithms such as neural networks, genetic algorithms, and generative models—analyze numerous process configurations, rapidly exploring design spaces and highlighting routes with higher efficiency. For fermentation, AI can pinpoint energy-intensive stages, recommend opportunities for heat integration or waste recycling, and suggest optimal connections between upstream and downstream units, resulting in reduced resource use and lower environmental impact (Tarafdar et al., 2026).

A key advancement is the integration of AI models as surrogate models within traditional simulators. These surrogate models drastically speed up computation by approximating complex process behaviors, allowing swift exploration of various design scenarios without the heavy computational cost of full-resolution simulations. For instance, generative AI and multi-objective optimization have enabled researchers to tailor metabolic pathways for higher-yield bioproducts, while ML-augmented flowsheet synthesis can rapidly assess recycling, heat recovery, and stream integration options. Compared with classical approaches, AI-driven process synthesis provides faster optimization cycles and greater flexibility in assessing sustainable alternatives. Recent industrial case studies demonstrate that coupling AI surrogate models with AspenTech Hybrid Models or Siemens Process DT enables automated process route selection, dynamic evaluation of trade-offs, and real-time process re-design for large-scale fermentation operations—leading to higher yield, reduced waste, and improved environmental outcomes (Wang et al., 2025).

7.8.4 Linking with Economic Evaluation

A sustainable fermentation process must also be economically feasible, making the combination of AI-based environmental assessment with TEA vital for modern industry. Multi-objective optimization frameworks—most commonly using algorithms such as Genetic Algorithms, Particle Swarm Optimization, and Bayesian Optimization (BO)—enable the simultaneous evaluation and quantification of trade-offs among product yield, carbon emissions, operating costs, and overall profitability. These algorithms help decision-makers explore thousands of scenarios and efficiently navigate complex design spaces. For example, BO applies probabilistic surrogate models (often Gaussian Processes) to predict outcomes and guide iterative experiments, while neural networks trained on process parameter, emissions, and cost data can rapidly evaluate scenario impacts (Hilgendrof et al., 2024).

AI-driven multi-objective optimization assigns quantitative weights or constraints to each objective, generating Pareto fronts that allow stakeholders to visualize and select process conditions where economic and environmental impacts are balanced. Systems may identify strategies—like slightly reduced productivity in exchange for significant energy cost savings—that optimize total sustainability rather than any single metric.

DT platforms and AI-supported decision support systems push this further, continually updating process and economic performance in real-time by integrating streaming production, resource, and financial data. These advanced tools facilitate automated trade-off analysis and adaptive process updates, ensuring that both financial viability and sustainability are dynamically addressed as conditions evolve. Case studies from syngas fermentation, enzyme manufacturing, and bioethanol production demonstrate these frameworks in industrial practice, leading to evidence-based strategies that do not force an exclusive choice between economic and environmental goals but enable simultaneous progress on both fronts.

7.9 ROLE OF AI IN PHARMACEUTICALS, FOOD, AND ALTERNATIVE PROTEIN INDUSTRIES

AI is profoundly transforming the pharmaceutical, food, and alternative protein industries by enabling more precise, efficient, and data-driven processes. In pharmaceuticals, advanced AI tools such as deep learning, generative adversarial networks (GANs), and reinforcement learning analyze complex datasets—including chemical structures, genomic data, and clinical trial records—to accelerate drug discovery, optimize molecule design, and improve patient stratification. For example, Insilico Medicine's AI platform employed GANs and reinforcement learning to develop a novel drug candidate that reached clinical trials much faster than traditional methods, reducing development time and cost by nearly 50%. Similarly, Roche utilizes AI-driven natural language processing and predictive models on electronic health records and pharmacovigilance databases to identify adverse drug reactions in real time, enhancing drug safety and regulatory compliance (Augustin et al., 2024).

In the food and alternative protein sectors, AI algorithms optimize microbial strain engineering and fermentation conditions by integrating multi-omics data and process parameters, thereby maximizing product yield, nutritional content, and sensory qualities while minimizing resource consumption. Platforms leveraging ML also provide real-time monitoring, predictive maintenance, and adaptive control of bioprocesses, as seen in fermentation process DTs, improving scalability and consistency. These AI implementations rely on large-scale data integration, automated feature extraction, and iterative optimization to refine processes continuously. Quantifiable outcomes include significantly shortened R&D timelines, cost savings, improved safety profiles, and enhanced product quality, underscoring AI's transformative impact across these sectors through technically sophisticated, outcome-focused applications (Chai et al., 2022).

7.9.1 AI in the Pharmaceutical Industry

7.9.1.1 Drug Discovery and Development

AI is revolutionizing drug discovery and development by enabling the analysis of complex biological systems and vast multi-omics datasets to identify promising drug targets with unprecedented speed and precision. ML models, including deep learning architectures, predict three-dimensional protein structures, thereby facilitating the rational design of molecules that precisely bind to these targets. AI-powered virtual screening platforms rapidly evaluate large chemical libraries, while generative AI techniques design novel drug candidates optimized for properties such as toxicity, solubility, and bioavailability. Noteworthy examples include IBM Watson's oncology platform, which uses natural language processing and predictive analytics on clinical and genomic data to personalize treatment regimens, and BenevolentAI's AI-driven drug repurposing system, which employs knowledge graphs and reinforcement learning to identify existing drugs for new therapeutic applications. By leveraging AI, BenevolentAI and similar platforms have cut drug discovery timelines by up to 50% and reduced associated costs by approximately 60%, significantly accelerating the path to clinical trials and improving success rates, highlighting AI's transformative potential in making drug development faster, more cost-effective, and more precise (Boukid et al., 2023; Vinestock et al., 2024).

7.9.1.2 Clinical Trials

AI significantly enhances the efficiency and accuracy of clinical trials by leveraging ML models to select ideal patients based on genetic and clinical data, thereby optimizing recruitment and retention. Algorithms optimize trial design parameters, such as dosage and duration, while technologies like natural language processing and computer vision enable real-time monitoring of patient outcomes and early detection of adverse reactions. A prominent example is Pfizer's use of AI during the COVID-19 vaccine trials, where their Smart Data Query (SDQ) tool automated the data cleaning and validation process for nearly 44,000 participants, processing millions of data points.

Traditionally, this data cleaning could take over 30 days, but SDQ reduced this to 22 hours, accelerating the final analysis and regulatory submission. Pfizer also employed AI-powered cloud infrastructure for supply chain optimization and manufacturing scale-up, enabling rapid response to logistical challenges and increasing production efficiency. Notably, AI-driven production scheduling at Pfizer's Freiburg manufacturing site, utilizing hybrid quantum-classical optimization, resulted in a roughly 66% reduction in manufacturing time and a 67% decrease in supply chain cycle time, enabling the production of an additional 20,000 vaccine doses per batch. This integration of AI across clinical and manufacturing operations exemplifies how data-driven automation significantly compresses timelines, reduces costs, and increases output while maintaining high-quality standard (Boukid et al., 2023; Vinestock et al., 2024).

7.9.1.3 Manufacturing and Supply Chain

AI significantly advances personalized medicine through platforms like Tempus and Roche's Foundation Medicine, which integrate comprehensive genomic and clinical data to tailor treatments with remarkable precision. Tempus employs ML algorithms on multi-omics and real-world clinical data to identify actionable genetic variants, optimize biomarker detection, and match patients to targeted therapies or clinical trials, resulting in a 96% clinical trial matching rate for patients with advanced cancers. Their platform incorporates tools such as natural language processing for electronic health records and predictive analytics for therapy response.

Roche's Foundation Medicine complements this by utilizing its Foundation One CDx assay—a comprehensive genomic profiling test—combined with validated bioinformatics pipelines to analyze tumor DNA and RNA, accurately matching genetic alterations with FDA-approved and investigational therapies. Collaborative projects, such as the partnership with France's Institute Gustave Roussy, have enabled onsite access to liquid biopsy testing and accelerated genomic diagnostics, enhancing clinical decision-making speed and accuracy. Evidence from ongoing studies shows that patients receiving AI-guided matched targeted therapy experience higher response rates, longer progression-free survival, and improved overall survival compared to those receiving unmatched treatments. These platforms continually refine treatment recommendations through iterative AI-driven data analysis, enabling health systems to optimize resource use and improve patient outcomes, thereby realizing the promise of precision oncology on a global scale.

7.9.2 AI in the Food and Beverage Industry

7.9.2.1 Quality Control and Safety

AI-powered vision systems combined with deep learning can detect defects, contaminants, and mislabeling in food products within milliseconds. Predictive analytics further enhances safety by analyzing sensor data throughout the supply chain, enabling proactive interventions to prevent potential hazards (Verma et al., 2025).

7.9.2.2 Supply Chain Optimization

AI improves demand forecasting, reduces food waste, and enhances transparency through the integration of IoT devices and blockchain technology. These systems allow food companies to optimize inventory, streamline logistics, and maintain traceability from production to delivery.

7.9.2.3 Personalized Nutrition and Product Innovation

ML algorithms analyze consumer preferences, allergies, and health data to deliver personalized meal recommendations. AI also supports the development of new food products by identifying ingredient combinations that align with emerging trends, while apps and platforms offer tailored dietary guidance for individual users (Verma et al., 2025).

7.9.2.4 Sustainable Agriculture and Waste Reduction

AI applications in agriculture optimize irrigation, fertilization, and pest control, reducing water usage and chemical inputs. Predictive models aid in crop yield forecasting and labor management, while deep reinforcement learning in controlled environments like vertical farms and greenhouses enhances energy efficiency. AI also supports food waste reduction by predicting spoilage and optimizing distribution strategies.

7.9.3 AI in Alternative Protein Production

7.9.3.1 Strain Engineering and Process Optimization

AI and ML algorithms enhance the efficiency of microbial strains used in precision fermentation for alternative protein production. Data from bioreactors and growth conditions are analyzed to optimize yield, product quality, and sustainability. Digital twins—virtual replicas of fermentation processes—allow real-time simulation and adjustment of conditions for maximum efficiency (Kanwal et al., 2025).

7.9.3.2 Product Formulation and Sensory Prediction

AI tools predict taste, texture, and flavor profiles, enabling rapid product development and refinement. These predictive models accelerate the creation of plant-based or cultured protein products that meet consumer expectations in flavor and mouth feel.

7.9.3.3 Environmental Sustainability

AI aids in evaluating and minimizing the environmental footprint of alternative protein production by analyzing resource consumption and emissions. This ensures that sustainable practices are integrated throughout production while maintaining product quality and consistency.

7.10 FUTURE DIRECTIONS AND EMERGING TRENDS

7.10.1 Advances in BO, Physics-Informed AI, and Integration with Synthetic Biology for Next-Generation Fermentation

Fermentation technology is experiencing a major transformation thanks to the integration of computational intelligence and synthetic biology. The adoption of advanced techniques such as BO, Physics-Informed AI (PIAI), and AI-assisted synthetic biology is helping researchers and industries optimize fermentation processes more efficiently, precisely, and sustainably. These developments are paving the way for smarter bioprocess design, improved process monitoring, and accelerated microbial production.

7.10.1.1 BO in Fermentation Processes

BO is a powerful technique for improving complex fermentation systems, particularly when traditional trial-and-error or exhaustive optimization would be too resource-intensive or slow. BO excels at efficiently identifying optimal conditions by strategically choosing experiments that are most likely to enhance performance, reducing the number of costly and time-consuming runs required. In fermentation contexts, BO is commonly used to optimize parameters such as temperature, pH, agitation speed, aeration rate, substrate concentrations, and nutrient media composition. The objective functions often focus on maximizing product yield, biomass concentration, growth rate, productivity, or minimizing resource use and waste emissions (Bolmanis et al., 2025; Choi et al., 2025).

BO stands out against conventional optimization methods like grid search or response surface methodology because it achieves high performance with less data—leveraging probabilistic surrogate models (typically Gaussian Processes) to predict outcomes and guide experimentation. This data-efficient approach allows rapid convergence to optimal solutions even when experimental data is limited, which is a frequent constraint in industrial fermentation (Bolmanis et al., 2025).

A practical example is the batch fermentation optimization of BY-2 plant cells in stirred-tank reactors. Here, researchers applied multi-objective BO to five critical input parameters in the medium (sucrose, ammonium, nitrate, phosphate, and initial biomass), aiming to maximize final fresh mass concentration and growth rate. Using only a handful of iterations and experiments, BO guided the selection of medium compositions that led to a 36% increase in biomass yield compared to conventional media, with optimal resource allocation and reduced unnecessary nutrient use. BO's surrogate models allowed researchers to analyze parameter relevance, efficiently explore the design space, and confirm improvements in follow-up experiments—all with significantly fewer resources than traditional approaches.

In other studies, BO has been paired with computational fluid dynamics (CFD) to optimize bioreactor geometry and operating conditions, and with deep neural networks for continuous online monitoring and control of fermentation batches. These integrations further demonstrate BO's flexibility and real-world impact, enabling both theoretical modeling and applied process enhancement in modern fermentation engineering. Figure 7.4 depicts theintegration of AI, modeling, and CFD in fermentation technology.

7.10.1.2 PIAI for Enhanced Modeling

Physics-Informed Neural Networks (PINNs) represent a transformative advancement in modeling and controlling fermentation processes by embedding mathematical representations of physical and biological laws directly into the neural network's architecture. This embedding is typically achieved by incorporating governing equations—such as mass and energy balances or kinetic rate laws—into the network's loss function. During training, PINNs are penalized if their predictions violate these differential equations, ensuring model outputs respect known process physics even in regions with sparse or missing data. This contrasts with conventional "black-box" neural networks, which require large experimental datasets and do not leverage mechanistic process understanding, often leading to poor extrapolation and higher data dependency (Bangi et al., 2022; Li et al., 2024).

PINNs help alleviate the data bottleneck in fermentation modeling: by learning both from limited experimental measurements and from enforced process laws, they generalize more robustly to untested scenarios and improve interpretability. For example, in recent studies, modeling penicillin fermentation, process mass balances, and reaction kinetics were mathematically embedded within the network. Input data included operating conditions (pH, temperature, substrate feed rates), limited time-series measurements of biomass and product concentrations, and online sensor signals. The PINN learned to predict dynamic profiles of penicillin concentration and biomass growth throughout the batch, achieving a mean squared error reduction of over 40% compared to black-box neural networks while requiring up to 60% less training data (Bangi et al., 2022).

Similarly, a multi-stage physics-informed modeling approach to microbial fermentation used process inputs like substrate concentrations and real-time off-gas measurements, with biochemical ODEs (Ordinary Differential Equations) embedded via loss penalty terms. This allowed accurate system identification and parameter estimation with minimal experimental runs, enabling real-time soft sensing and more precise closed-loop control in lab-scale batch fermentations.

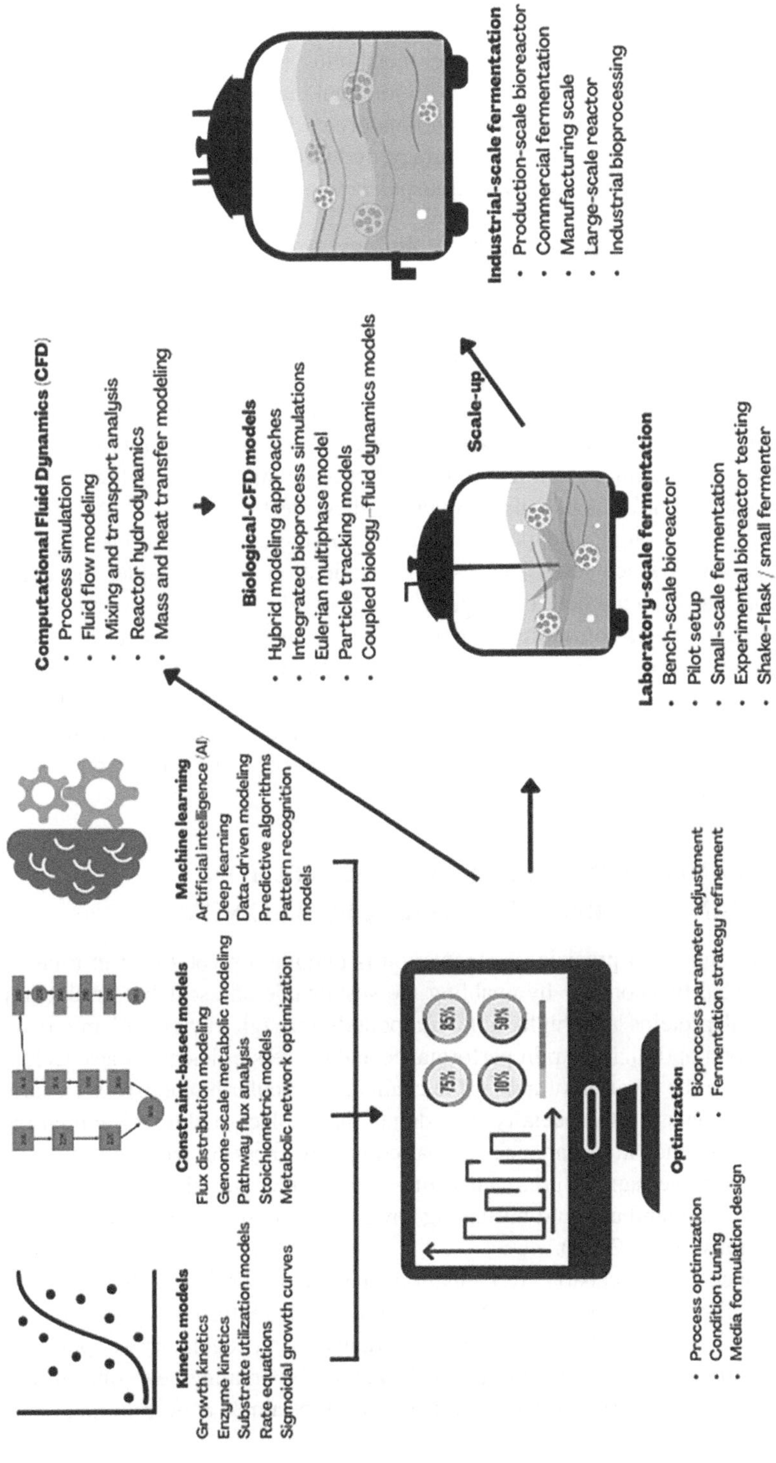

FIGURE 7.4 Depicts integration of AI, modeling, and CFD in fermentation technology.

Development of AI-driven DTs for fermentation further leverages PINNs by providing live, accurate predictions for states that are costly or hard to measure experimentally—such as intermediate metabolite levels—supporting automated monitoring and process optimization with minimal manual intervention. Overall, by uniting first-principles knowledge with data-driven learning, PINNs offer measurable improvements in prediction accuracy, reduced data requirements, improved parameter estimation, and increased transparency in fermentation system modeling.

7.10.1.3 Integration with Synthetic Biology

The integration of AI with synthetic biology is transforming the design of microbial production systems by combining data-driven algorithms with mechanistic biological models. AI-guided pathway design frameworks use ML to analyze omics data and enzyme kinetics, allowing rapid identification and optimization of metabolic routes to enhance product yield and efficiency. A concrete case is the use of embedded PINNs, which incorporate physical and biochemical constraints directly into neural network training, ensuring model predictions comply with biological laws.

Automation platforms like digital microfluidics interface with these AI models by executing high-throughput experiments under AI-suggested conditions, closing the loop between in silico design and wet-lab validation. For example, the ePINNs framework has been applied to synthetic gene networks to model resource competition and gene expression more accurately while maintaining physical consistency. This seamless integration accelerates iterative strain engineering cycles, reduces experimental resource consumption, and fosters interdisciplinary collaboration, thereby driving the development of novel bio-based products ranging from pharmaceuticals to sustainable materials with enhanced precision and reproducibility (Mirsalami and Mirsalami, 2025).

7.11 AI AND ML IN PRECISION FERMENTATION: TRANSFORMING THE GLOBAL BIOECONOMY

AI and ML-powered precision fermentation is playing a pivotal role in transforming the global bioeconomy by enabling the sustainable and scalable production of bioidentical proteins and high-value compounds through engineered microorganisms. Beyond optimizing strain performance and process efficiency, these technologies underpin the creation of new value chains and circular bioindustries by turning renewable biomass into specialty ingredients that replace traditional animal-based sources. This innovation enhances global competitiveness by reducing reliance on resource-intensive agriculture, supporting decarbonization through significantly lower water and land use, and mitigating environmental impacts such as greenhouse gas emissions (Starz, 2024).

Moreover, the precision fermentation market is projected to surge from USD 5.82 billion in 2025 to over USD 150 billion by 2034, fueled by rising consumer demand for ethical, sustainable protein alternatives and regulatory support. These advances stimulate socio-economic transformation by creating new jobs, attracting investments, and fostering public–private partnerships aimed at overcoming scale-up and financial barriers. Industrial-scale applications—from dairy protein substitutes to bioactive compounds in pharmaceuticals—illustrate how AI-driven precision fermentation is the cornerstone of a resilient, sustainable bioeconomy with far-reaching effects on food security, environmental stewardship, and circular economy growth globally (Starz, 2024).

7.11.1 Enhancing Efficiency and Sustainability

AI and ML are pivotal in optimizing fermentation processes. ML models can monitor and control fermentation in real time, improving yield, consistency, and efficiency. One innovative approach is the use of DTs, virtual replicas of physical fermentation systems that simulate and optimize conditions continuously. These models allow for automatic adjustments, early anomaly detection, and proactive measures to prevent process failures, reducing waste, and resource consumption.

Moreover, AI technologies streamline biowaste conversion into valuable resources. Intelligent sorting, predictive analytics, and novel valorization pathways allow biowaste to be transformed into energy or raw materials efficiently, promoting both environmental sustainability and economic growth (Kanwal et al., 2025).

7.11.2 Optimizing Strain Engineering and Bioprocessing

In order to generate particular molecules, precision fermentation uses "cell factories," or modified microbes. ML-guided directed evolution is transforming biological design by predicting which genetic changes will improve microbial performance without requiring complete mechanistic models. This accelerates strain optimization and improves production efficiency.

AI-driven surrogate modeling also speeds up experimentation and process scale-up. It helps address challenges like gas transfer, feedstock variability, and complex low-carbon or waste-based inputs, making industrial-scale precision fermentation more feasible (Boukid et al., 2023; Vinestock et al., 2024).

7.11.3 Advancing the Bioeconomy through Alternative Protein Production

Precision fermentation provides a sustainable solution to climate change and food security challenges. By producing proteins identical to those found in animal products, microbial fermentation drastically reduces greenhouse gas emissions, land use, and water consumption compared to traditional livestock farming. Studies indicate potential reductions of up to 97% in greenhouse gas emissions, 99% in land use, and 94% in water use (Roos et al., 2017). Additionally, precision fermentation improves the efficiency, safety, nutrition, and flavor of fungal-based foods produced from yeasts, molds, and macrofungi. This capability supports the development of sustainable food systems while reducing dependency on conventional animal agriculture (Knychalaet al., 2024; Kossmannet al., 2025).

7.11.4 Addressing Challenges and Future Directions

Despite its transformative potential, precision fermentation faces multiple challenges in scaling from laboratory to commercial production, including technological hurdles, high production costs, and complex regulatory landscapes. AI and ML play a crucial role in addressing these challenges by enabling predictive modeling that accurately simulates scale-up conditions, thereby reducing costly trial-and-error experiments. AI-driven optimization facilitates efficient feedstock utilization by analyzing large datasets to recommend pathway adjustments that maximize yield and minimize waste (Singh and Kumar, 2025).

Furthermore, process automation powered by AI reduces operational costs and enhances consistency by dynamically controlling bioreactor parameters in real time, while AI algorithms support regulatory compliance through automated monitoring and quality control based on robust data analytics. Integrating AI with DTs creates a feedback-rich environment where process performance and production economics are continuously updated, allowing for swift adaptation to evolving conditions and standards. Together, these AI-driven approaches help overcome economic and technical barriers, promoting the scalability, cost-efficiency, and regulatory alignment essential for precision fermentation's full integration into the global bioeconomy (Gumulya et al., 2025).

7.12 CONCLUSION

The integration of AI and ML has strategically transformed precision fermentation from a conventional bioprocessing approach into an intelligent and adaptive biosynthetic platform. Through predictive modeling, self-learning algorithms, and real-time analytics, AI/ML empowers microbial systems to operate with enhanced precision, reproducibility, and sustainability. These data-driven systems optimize nutrient flux, enzyme regulation, and bioreactor conditions dynamically—transforming microbial metabolism into a controllable and evolvable process. As a result, industrial fermentation now achieves higher yield, lower resource consumption, and improved process resilience.

Beyond technical optimization, the AI-augmented precision fermentation framework contributes directly to the global bioeconomy by enabling scalable production of high-value biomolecules across food, pharmaceutical, agricultural, and energy sectors. By improving process efficiency and minimizing dependency on traditional petrochemical inputs, AI-driven fermentation supports circular economy principles and low-carbon manufacturing pathways. This convergence of computational intelligence with biological design not only redefines industrial biotechnology but also catalyzes a shift toward sustainable economic models—where digital innovation and microbial engineering together drive the next generation of bio-based production systems.

REFERENCES

Abbasi, M. A., Khan, A. Q., Mustafa, G., Abid, M., Khan, A. S., & Ullah, N. (2021). Data-driven fault diagnostics for industrial processes: An application to Penicillin fermentation process. *IEEE Access*, *9*, 65977–65987.

Ajayeoba, T. A., & Ijabadeniyi, O. A. (2025). *Transforming Food for the Future: Precision Fermentation as a Key to Sustainability, Nutrition, and Health* (Preprint). SSRN.

Augustin, M. A., Hartley, C. J., Maloney, G., & Tyndall, S. (2024). Innovation in precision fermentation for food ingredients. *Critical Reviews in Food Science and Nutrition*, *64*(18), 6218–6238.

Bangi, M. S. F., Kao, K., & Kwon, J. S. I. (2022). Physics-informed neural networks for hybrid modeling of lab-scale batch fermentation for β-carotene production using Saccharomyces cerevisiae. *Chemical Engineering Research and Design*, *179*, 415–423.

Becchi, P. P., Rocchetti, G., & Lucini, L. (2025). Advancing dairy science through integrated analytical approaches based on multi-omics and machine learning. *Current Opinion in Food Science*, *63*, 101289.

Bolmanis, E., Galvanauskas, V., Grigs, O., Vanags, J., & Kazaks, A. (2025). Leveraging historical process data for recombinant P. pastoris fermentation hybrid deep modeling and model predictive control development. *Fermentation*, *11*(7), 411.

Boukid, F., Ganeshan, S., Wang, Y., Tülbek, M. Ç., & Nickerson, M. T. (2023). Bioengineered enzymes and precision fermentation in the food industry. *International Journal of Molecular Sciences*, *24*(12), 10156.

Butean, A., Cutean, I., Barbero, R., Enriquez, J., & Matei, A. (2025). A review of AI applications for biorefineries and bioprocessing: From data-driven processes to optimization strategies and real-time control. *Processes*, *13*(8), 2544.

Chai, W. Y., Teo, K. T. K., Tan, M. K., & Tham, H. J. (2022). Fermentation process control and optimization. *Chemical Engineering & Technology*, *45*(10), 1731–1747.

Choi, H., Roh, K., & Lee, J. H. (2025). CFD-based determination of optimal design and operating conditions of a fermentation reactor using bayesian optimization. *Biotechnology and Bioengineering*, *122*(4), 870–883.

Chuai, G., Ma, H., Yan, J., Chen, M., Hong, N., Xue, D., Zhou, C., Zhu, C., Chen, K., Duan, B., Gu, F., Qu, S., Huang, D., & Liu, Q. (2018). DeepCRISPR: Optimized CRISPR guide RNA design by deep learning. *Genome Biology*, *19*, 80.

Dazzarola, C., Tighe, R., Pérez-Correa, J. R., & Saa, P. A. (2026). Toward a digital twin for beer quality control: Development of a digital model integrating industrial process data and model-based fermentation descriptors. *Journal of Food Engineering*, *403*, 112726.

Dudeja, C., Mishra, A., Ali, A., Singh, P. P., & Jaiswal, A. K. (2025). Microbial genome editing with CRISPR–Cas9: Recent advances and emerging applications across sectors. *Fermentation*, *11*(7), 410.

Goffi, P.-E., Tremblay, R., & Oakes, B. (2025). Engineering a digital twin for the monitoring and control of beer fermentation sampling. In *Proceedings of the 2025 ACM/IEEE 28th International Conference on Model Driven Engineering Languages and Systems Companion (MODELS-C 2025)* (pp. 167–173).

Gumulya, Y., Peng, H., Ebert, B. E., Johnson, H., van der Pols, J. C., Bansal, N., Marcellin, E., & Turner, M. S. (2025). Advancing Australia's food future: Opportunities and challenges in precision fermentation. *Future Foods*, *11*, 100630.

Hilgendorf, K., Wang, Y., Miller, M. J., & Jin, Y. S. (2024). Precision fermentation for improving the quality, flavor, safety, and sustainability of foods. *Current Opinion in Biotechnology*, *86*, 103084.

Huang, Y., Wang, Y., Shang, N., & Li, P. (2023). Microbial fermentation processes of lactic acid: Challenges, solutions, and future prospects. *Foods*, *12*(12), 2311.

Jiang, L., Lu, Y., & He, Q. (2025). Unlocking the microbial mechanisms in fermented bean products: Advances and future directions in multi-omics and machine learning approaches. *Critical Reviews in Food Science and Nutrition*, *66*(7), 1364–1386.

Kanwal, N., Zhang, M., Khan, S., & Batool, U. (2025). AI-driven precision fermentation: From restaurant food waste to sustainable protein production. *Food and Bioprocess Technology*, *18*(10), 8354–8379.

Khanal, S. K., Tarafdar, A., & You, S. (2023). AI and machine learning for smart bioprocesses. *Bioresource Technology*, *375*, 128826.

Knychala, M. M., Boing, L. A., Ienczak, J. L., Trichez, D., & Stambuk, B. U. (2024). Precision fermentation as an alternative to animal protein, a review. *Fermentation*, *10*(6), 315.

Kossmann, H., Karslioglu, O.O., & Breunig, P. (2025). *From moo to microbes: Pathways for precision fermentation in recombinant protein production. Journal of Agriculture and Food Research*, *22*, 102056.

Koubaa, M., Barba, F. J., Roohinejad, S., Saraiva, J., & Lorenzo, J. M. (2019). New challenges and opportunities of food fermentation processes: Application of conventional and innovative techniques. *Food Research International*, *115*, 552–553.

Li, Q., Zhang, J., Wan, H., Zhao, Z., & Liu, F. (2024). Physics-informed neural networks for multi-stage Koopman modeling of microbial fermentation processes. *Journal of Process Control*, *143*, 103315.

Mannaa, M., Han, G., Seo, Y. S., & Park, I. (2021). Evolution of food fermentation processes and the use of multi-omics in deciphering the roles of the microbiota. *Foods*, *10*(11), 2861.

McClements, D. J. (2023). Biotech meat: Growing meat from cells. In D. J. McClements (Ed.), *Meat Less: The Next Food Revolution* (pp. 149–183). Springer Nature Switzerland.

Mirsalami, S. M., & Mirsalami, M. (2025). Advances in genetically engineered microorganisms: Transforming food production through precision fermentation and synthetic biology. *Future Foods*, *11*, 100601.

Mohammadi Kashka, F., Tahmasebi Sarvestani, Z., Pirdashti, H., Motevali, A., Nadi, M., & Valipour, M. (2023). Sustainable systems engineering using life cycle assessment: Application of AI for predicting agro-environmental footprint. *Sustainability*, *15*(7), 6326.

Nguyen, X. D. J., Liu, Y. A., McDowell, C. C., & Dooley, L. (2025). Methodology for contamination detection and reduction in fermentation processes using machine learning. *Bioprocess and Biosystems Engineering*, *48*(9), 1547–1563.

Oraby, A., Briem, A. K., Bippus, L., Rupp, S., & Zibek, S. (2024). Life cycle assessment as a driver for process optimisation of cellobiose lipids fermentation and purification. *The International Journal of Life Cycle Assessment*, *29*(9), 1652–1671.

Pereira, A. A., Yaverino-Gutierrez, M. A., Monteiro, M. C., Souza, B. A., Bachheti, R. K., & Chandel, A. K. (2025). Precision fermentation in the realm of microbial protein production: State-of-the-art and future insights. *Food Research International*, *200*, 115527. https://doi.org/10.1016/j.foodres.2024.115527.

Praveen, M., & Brogi, S. (2025). Microbial fermentation in food and beverage industries: Innovations, challenges, and opportunities. *Foods*, *14*(1), 114.

Priyadharshini, D., Muthuvel, I., Saraswathy, S., Kavitha, P. S., & Jegadeeswari, V. (2025). Precision to plate: AI-driven innovations in fermentation and hyper-personalized diets. *Frontiers in Nutrition*, *12*, 1659511.

Ramírez Rojas, A. A., Swidah, R., & Schindler, D. (2022). Microbes of traditional fermentation processes as synthetic biology chassis to tackle future food challenges. *Frontiers in Bioengineering and Biotechnology*, *10*, 982975.

Röös, E., Bajželj, B., Smith, P., Patel, M., Little, D., & Garnett, T. (2017). Protein futures for Western Europe: Potential land use and climate impacts in 2050. *Regional Environmental Change*, *17*(2), 367–377.

Sharma, D., & Singh, K. (2025). AI-enhanced bioprocess technologies: Machine learning implementations from upstream to downstream operations. *World Journal of Microbiology and Biotechnology*, *41*(8), 278.

Singh, A., & Kumar, S. (2025). Exploring the functionality of microbes in fermented foods: Technological advancements and future directions. *Fermentation*, *11*(6), 300.

Starz, N. T. (2024). Road to a bioeconomy in the European Union: Mapping drivers of precision fermentation adoption. *Junior Management Science (JUMS)*, *9*(4), 2024–2049.

Tarafdar, A., Das, A., & Singh, B. (2026). Application of AI and machine learning in solid-state fermentation proceses. In A. Pandey, C. Larroche, C. R. Soccol, & R. R. Singhania (Eds.), *Current Advances in Solid-State Fermentation* (2nd ed., pp. 129–144).

Teng, T. S., Chin, Y. L., Chai, K. F., & Chen, W. N. (2021). Fermentation for future food systems: Precision fermentation can complement the scope and applications of traditional fermentation. *EMBO Reports*, *22*(5), e52680.

Terefe, N. S. (2022). Recent developments in fermentation technology: Toward the next revolution in food production. In P. Juliano, K. Knoerzer, J. Sellahewa, M. H. Nguyen, & R. Buckow (Eds.), *Food Engineering Innovations across the Food Supply Chain* (pp. 89–106). Elsevier. https://doi.org/10.1016/B978-0-12-821292-9.00026-1

Vanapalli, K. R., Bhar, R., Maity, S. K., Dubey, B. K., Kumar, S., & Kumar, V. (2023). Life cycle assessment of fermentative production of lactic acid from bread waste based on process modelling using pinch technology. *Science of The Total Environment*, *905*, 167051.

Verma, K., Duhan, P., Pal, D., Verma, P., & Bansal, P. (2025). *Precision fermentation for the next generation of food ingredients: Opportunities and challenges. Future Foods*, *12*, 100750.

Vinestock, T., Short, M., Ward, K., & Guo, M. (2024). Computer-aided chemical engineering research advances in precision fermentation. *Current Opinion in Food Science*, *58*, 101196.

Wang, Y., Wang, Y., & Xu, F. (2025). AI-driven fermentation optimization for α-amylase hyperproduction enabled by Raman monitoring and metabolic network analysis. *Bioresource Technology*, *439*, 133287.

Yadav, H., Singh, S., & Sinha, R. (2024). Fermentation technology for microbial products and their process optimization. In P. Verma (Ed.), *Industrial Microbiology and Biotechnology: A New Horizon of the Microbial World* (pp. 35–64). Springer Nature Singapore.

Yee, C. S., Zahia-Azizan, N. A., Abd Rahim, M. H., Mohd Zaini, N. A., Raja-Razali, R. B., Ushidee-Radzi, M. A., …& Wan-Mohtar, W. A. A. Q. I. (2025). Smart fermentation technologies: Microbial process control in traditional fermented foods. *Fermentation*, *11*(6), 323.

Zeng, Y., Hong, Y., Azi, F., Liu, Y., Chen, Y., Guo, C., …& Xu, P. (2022). Advanced genome-editing technologies enable rapid and large-scale generation of genetic variants for strain engineering and synthetic biology. *Current Opinion in Microbiology*, *69*, 102175.

Zhao, S., Jiao, T., Adade, S. Y. S. S., Wang, Z., Ouyang, Q., & Chen, Q. (2025). Digital twin for predicting and controlling food fermentation: A case study of kombucha fermentation. *Journal of Food Engineering*, *393*, 112467.

Zobel-Roos, S., Schmidt, A., Uhlenbrock, L., Ditz, R., Köster, D., & Strube, J. (2020). Digital twins in biomanufacturing. In C. Herwig, R. Pörtner, & J. Möller (Eds.), *Digital Twins: Tools and Concepts for Smart Biomanufacturing* (pp. 181–262). Springer International Publishing.

Zrimec, J., Buric, F., Muhammad, A. S., Chen, R., Siewers, V., & Nielsen, J. (2022). Controlling gene expression with deep generative design of regulatory DNA. *Nature Communications*, *13*, 5099.

8 Integration of AI/ML in Biofuel Production

C. P. Shirley, Immanuel Johnraja Jebadurai, Getzi Jeba Leelipushpam Paulraj, S. Thanga Helina, and Amir Seyyedabbasi

8.1 INTRODUCTION

Global industry's unstoppable growth and the resulting need for sustainable energy sources have led to an unprecedented emphasis on the production of biofuel as a competitive substitute for combustion fuels derived from fossil fuels. Traditional approaches have established a crucial basis, but their effectiveness is often limited by three stubborn obstacles: the unpredictable nature of feedstock composition, the intricate intricacies of process optimization, and the constant need for economic viability [1]. The combination of artificial intelligence (AI) and machine learning (ML) becomes more than just a supplementary technology in this furnace of necessity; rather, it is a paradigm shift. This computational vanguard proffers a sophisticated arsenal of data-driven prognostication, predictive modeling, and automated control, thereby promising to dismantle these long-standing barriers across the entire production life cycle, from conception to commercial realization.

AIML's scope goes beyond the initial optimization of biochemical processes to include the very functional sinews of production infrastructure. The implementation of sophisticated sensor networks combined with ML analytics enables a previously unheard-of level of real-time monitoring and predictive maintenance, anticipating mechanical failures and optimizing energy use with extraordinary precision. At the same time, these intelligent systems guarantee a level of product quality consistency that was previously unachievable. Additionally, the use of AI-driven life cycle assessment models offers a thorough and sophisticated analysis of environmental effects, directing the creation of truly scalable and sustainable solutions. This fortunate combination of biotechnological innovation and computational intelligence does more than just promise increased operational efficiency; it fundamentally inaugurates novel pathways for research and development, positioning the biofuel industry at the vanguard of the quest for carbon-neutral energy security.

 DOI: 10.1201/9781003684374-8

8.1.1 The Global Energy Landscape and the Rise of Biofuels

A most urgent and paradoxical duality characterizes the current global energy landscape: an unquenchable desire for power to propel economic growth juxtaposed with the escalating demands of resource depletion and climate change. A significant epistemological shift has resulted from this, taking the discussion beyond energy security to include the critical significance of sustainability and environmental stewardship [2]. With the alluring prospect of renewable, liquid energy made from organic matter, biofuels have emerged from a specialized interest to a key component in the strategic planning for a carbon-neutral future. Their potential to decarbonie sectors notoriously resistant to electrification, such as aviation and heavy freight, renders them not merely an alternative, but an indispensable component of a diversified and resilient energy portfolio.

However, the rise of biofuels has not been smooth or unhindered. They have faced numerous technical and financial challenges along the way from laboratory curiosity to commercial viability, which have limited their widespread adoption until recently. Historically, there have been significant obstacles due to the intrinsic variability of biological feedstocks, the complex intricacies of biochemical conversion processes, and the high expenses involved in attaining efficiency and scale. The story takes its most interesting turn right at this point of great promise and enduring difficulty. Advanced computational intelligence integration has the potential to turn these challenges into opportunities and usher in a new era of biofuel production.

8.1.2 Inherent Challenges in Conventional Biofuel Production

Despite their admirable goal of replacing fossil fuels, traditional biofuel production methods face numerous inherent and significant obstacles that have long prevented them from achieving true scalability, economic equality, and environmental effectiveness. The entire endeavor is essentially tied to the extreme erratic nature of chemical and biological systems, which conventional methods, which mainly rely on empirical experimentation and fixed operational parameters, are ill-equipped to handle consistently with accuracy. This limits the potential of biofuels to be a truly disruptive force in the global energy matrix by producing a production life cycle that is frequently inefficient, unreasonably expensive, and frustratingly inconsistent.

The following is a list of the most notable of these obstacles:

8.1.2.1 Feedstock Variability and Availability

A fundamental problem in the production of biofuel is the selection and steady supply of suitable biomass, which has an immediate effect on the operation's overall scalability and economic feasibility. Seasonal variations, geographic location, climate, farming methods, and competition with food production are just a few of the many variables that affect feedstock availability and variability. Standardizing and optimizing conversion processes is made extremely difficult by the inherent inconsistency in biomass composition, such as the different amounts of cellulose, hemicellulose, and lignin in plant matter. Additionally, harvesting, transportation, and storage logistical issues can impact feedstock availability and quality, resulting in bottlenecks that impede ongoing production.

A sophisticated strategy for overcoming these obstacles is provided by the combination of ML and AI [3].

- AI-driven models are able to predict the best feedstock type and harvesting time for particular regions by analyzing large datasets that include soil health, weather patterns, genetic information, and past crop yields.
- Supply chain logistics are improved by ML algorithms, which forecast storage conditions to reduce deterioration and optimize transportation routes.
- AI systems are able to recognize genetically modified and unconventional feedstocks that have a higher potential yield and are more resilient to environmental stresses.
- AI and ML technologies facilitate more dependable, effective, and sustainable biofuel production processes by offering data-driven insights into feedstock selection, availability forecasting, and supply chain management.

8.1.2.2 Process Inefficiencies and High Operational Costs

The economic competitiveness of biofuels in comparison to fossil fuels is undermined by the fact that conventional biofuel production is often hindered by substantial process inefficiencies that directly lead to high operating costs. The intricate, multi-step process of production, which involves pre-treatment, enzymatic hydrolysis, fermentation, and distillation, is the cause of these inefficiencies. Numerous interdependent factors, including temperature, pH, pressure, enzyme concentration, and microbial activity, can affect each stage. In addition to taking a lot of time and resources, optimizing these parameters through conventional trial-and-error experimentation frequently misses the true global optimum for maximizing yield and minimizing energy input. Furthermore, large-scale bioreactors and distillation columns are energy-intensive assets. Suboptimal process control leads to excessive energy consumption, low conversion rates, inconsistent product quality, and unplanned downtime due to equipment failure or microbial contamination, all of which escalate operational expenditures.

The integration of AI and ML presents a transformative solution for addressing these core economic and operational challenges.

- ML algorithms can analyze vast, high-dimensional datasets from historical operations and real-time sensor networks to build precise digital twins and predictive models of the entire production process.
- These models can simulate countless scenarios to identify the most efficient operating conditions for maximizing yield and purity while minimizing energy and chemical consumption.
- AI-driven process control systems can dynamically adjust parameters in real-time to maintain optimal performance despite feedstock variability or changing environmental conditions.
- For asset management, AI-powered predictive maintenance algorithms analyze equipment sensor data to forecast failures before they occur,

scheduling maintenance only when needed to prevent costly downtime and extend machinery lifespan.
- By enabling such a high degree of optimization, automation, and foresight, AI and ML technologies directly target the root causes of high operational costs, paving the way for a more efficient and economically sustainable biofuel industry.

8.1.2.3 Scalability and Environmental Impact Concerns

A critical challenge facing the biofuel industry is the transition from successful laboratory-scale production to consistent, economically viable industrial-scale operations, while simultaneously ensuring a net positive environmental outcome. Scalability is hindered by several factors, including the non-linear behavior of biological processes when amplified, difficulties in maintaining optimal and uniform conditions in large-volume reactors, and complex supply chain logistics for securing vast quantities of consistent feedstock. In addition, the environmental impact of producing biofuel is closely examined, including issues with water use, fertilizer runoff, land-use change, and overall energy return on investment (EROI) [4]. It is challenging to verify and compare the sustainability of various production pathways because traditional life cycle assessment (LCA) methods frequently fail to adequately capture this intricate, multifaceted impact across the whole value chain, from cultivation to combustion.

De-risking scale-up and guaranteeing environmental accountability can be accomplished with the help of AI and ML.

- To create complex digital twins of production facilities, AI algorithms can analyze large, diverse datasets, such as soil data, weather patterns, geospatial data, and real-time process metrics.
- Before money is spent on physical infrastructure, these models can accurately simulate scale-up scenarios, anticipating possible bottlenecks, and optimizing design parameters.
- AI-enhanced LCA models offer a more thorough and dynamic analysis of environmental impact.
- They can incorporate real-time data from IoT sensors in factories and on farms to update a biofuel's environmental impact over time, taking into consideration factors like shifting energy sources or agricultural practices [5].
- Additionally, ML can determine the best areas for feedstock production that minimize land-use change and water stress, as well as create more effective logistics networks to lower transportation-related carbon emissions.

8.1.3 Chapter Scope and Objectives

This chapter is intended to serve as a thorough academic analysis of the paradigm shift brought about by the incorporation of ML and AI in the biofuel production industry. It offers a thorough examination of how data-driven intelligence is being used to address long-standing inefficiencies and unlock new capabilities, and its scope is carefully crafted to span the whole biofuel value chain.

In order to make the topic understandable to a multidisciplinary readership, the discussion will begin by providing a strong conceptual foundation in AI and ML that is especially suited for their use in bioprocessing contexts. It will then proceed to dissect specific applications, from the initial stage of gnomically informed feedstock selection and pre-treatment optimization through to the core conversion processes—including fermentation, transesterification, and thermochemical reactions—and onward to the operational management of production facilities via real-time monitoring, predictive maintenance, and the deployment of digital twins.

A significant portion of the chapter will also be devoted to the critical role of AI in ensuring sustainability, particularly through advanced LCA and the optimization of supply chain logistics for a minimized carbon footprint. The scope is not merely technical but also forward-looking, incorporating a critical analysis of the economic, scalability, and policy dimensions essential for the widespread adoption of these technologies.

The objectives of this exploration are multifaceted and ambitious. Primarily, the chapter aims to synthesize a vast and often disparate body of research into a coherent narrative that clearly demonstrates the transformative impact of AI/ML on enhancing yield, purity, and economic viability while concurrently reducing environmental impact. By offering tangible frameworks and illustrative case studies—like the use of ML for in-silico Design of Experiments (DoE) or AI-driven dynamic LCAs—that offer a roadmap for real-world implementation, it aims to go beyond theoretical postulation.

Providing an honest and comprehensive assessment of the current issues, such as data standardization, model interpretability (the "black box" problem), high initial capital investment, and the notable skill gap in the workforce, is also a crucial goal. Lastly, the chapter is focused on the future; it will develop a rational industry roadmap, examining the state-of-the-art convergence of AI with advanced biorefining and synthetic biology, and talking about the critical role that investment and policy play in promoting this integration. The ultimate goal is to create a reliable and essential resource that not only informs but also motivates scientists, industrial engineers, and legislators to take an active role in developing the sustainable, data-driven biofuel systems of the future.

8.2 FUNDAMENTALS OF AI AND ML RELEVANT TO BIOFUEL PRODUCTION

The integration of AI and ML into biofuel production is predicated on a foundation of core computational principles. At its essence, AI provides the overarching framework for creating machines capable of intelligent behavior, while ML, a subset of AI, empowers systems to learn and improve from data without explicit programming. For biofuel applications, supervised learning algorithms are particularly vital; these models are trained on historical data—such as process parameters (temperature, pH, reaction time) and corresponding outputs (biofuel yield, purity)—to build predictive models that can forecast outcomes under new conditions.

Techniques like regression analysis, support vector machines, and ensemble methods like Random Forest are adept at handling the complex, non-linear relationships inherent in biological and chemical processes. Furthermore, unsupervised learning methods like clustering can identify hidden patterns and groupings within vast datasets, such as categorizing different biomass feedstocks based on their compositional properties to predict optimal processing routes.

The quality and quantity of data are the only factors that can affect how effective these ML models are, so gathering and pre-processing data is an essential first step. Data streams for biofuels come from a variety of sources, including spectroscopic analysis of reaction components, real-time sensor readings in bioreactors, genomic studies of feedstocks, and the outcomes of innumerable laboratory tests. Before being used to power ML algorithms, this raw data must be painstakingly cleaned, normalized, and converted into pertinent features—a procedure called feature engineering [6]. Features might include, for example, the amount of cellulose, hemicellulose, and lignin in biomass, or the concentration of enzymes and rate of agitation in a fermentation broth. By harnessing these fundamentals, researchers can move beyond traditional trial-and-error experimentation, instead using data-driven models to simulate scenarios, optimize complex multi-variable processes, and accelerate the path to efficient and scalable biofuel production.

8.2.1 A Primer on Key Concepts: AI, ML, and Deep Learning

The application of AI and ML to the production of biofuel is not just a fad; rather, it represents a paradigm shift away from empirical, trial-and-error research and toward data-driven, predictive science. A solid understanding of the underlying computational concepts is necessary to completely appreciate the transformative applications. In addition to demythologizing the hierarchy of AI, ML, and deep learning (DL), this topic acts as a fundamental primer by offering a thorough rundown of the particular algorithms and data-handling techniques that are most pertinent to the opportunities and challenges in the production of biofuel [2].

8.2.1.1 The Hierarchical Relationship: AI, ML, and DL

- Understanding the nested relationship between these terms is essential: The broadest term is AI, which includes any method that makes it possible for computers to simulate human intelligence. This covers everything from straightforward rule-based systems ("if pressure > X, then close valve Y") to intricate neural networks. AI aims to create autonomously reasoning, learning, and acting systems.
- **Machine Learning (ML)** is a subset of AI. It provides the statistical foundation for systems to automatically learn and improve from experience (data) without being explicitly programmed. Instead of following static instructions, ML algorithms identify patterns within data to build models that can make predictions or decisions. ML is the primary engine driving most modern AI applications in biofuel research.
- **Deep Learning (DL)** is a further subset of ML, inspired by the structure and function of the brain. DL models are based on artificial neural networks (ANNs) with many layers—hence "deep." These deep networks can automatically learn hierarchical representations of data. For instance, lower layers might identify simple patterns (edges in an image, basic sequences in DNA), while higher layers combine these into more complex features (plant cell structures, gene motifs). DL excels at processing raw, unstructured data like images, spectra, and genetic sequences.

8.2.2 Overview of Relevant Algorithms for Biofuel Research

The choice of algorithm is dictated by the problem type, the nature of the available data, and the desired outcome. The biofuel production lifecycle presents a diverse array of problems, each amenable to specific classes of ML algorithms.

8.2.2.1 Supervised Learning (Regression, Classification)

Supervised learning is the most common ML paradigm in biofuel research. The algorithm is trained on a **labeled dataset**, where each input data point is paired with the correct output. The model learns the mapping function from the inputs to the output so that it can predict the output for new, unseen data. This is analogous to learning with a teacher or a supervisor who provides the answers [7].

8.2.2.1.1 Regression

Regression is used when the target variable to be predicted is a continuous numerical value.

- Common Algorithms:
  - **Linear Regression**: A fundamental algorithm that models the linear relationship between a dependent variable and one or more independent variables. Useful for establishing baseline relationships.
  - **Support Vector Regression (SVR)**: Effective in high-dimensional spaces and for capturing non-linear relationships using various kernel functions. Robust against outliers.
  - **Random Forest Regression**: An ensemble method that builds multiple decision trees and averages their predictions. Highly effective, robust to overfitting, and can handle complex, non-linear interactions between variables without requiring extensive data pre-processing. Arguably one of the most popular algorithms in bioprocess modeling.
  - **Gaussian Process Regression (GPR)**: A probabilistic approach that provides not only a prediction but also a measure of uncertainty (standard deviation) for that prediction. Invaluable for guiding experimental design, as it can identify regions of the parameter space where the model is most uncertain and where new experiments would be most informative.

8.2.2.1.2 Classification

Classification is used when the target variable is a discrete category or label.

- Common Algorithms:
  - **Logistic Regression**: A linear model for binary classification (e.g., good/bad batch). Simple and interpretable.
    - **Support Vector Machines (SVM)**: Finds the optimal hyperplane that best separates data points of different classes in a high-dimensional space. Effective for both linear and non-linear classification.
    - **Random Forest Classifier**: An ensemble of decision trees that votes on the final class label. Highly accurate and provides feature importance scores, indicating which input variables (e.g., temperature, pressure) are most influential in the classification decision.

- **k-Nearest Neighbors (k-NN)**: A simple, instance-based algorithm that classifies a data point based on how its k-nearest neighbors are classified.

8.2.2.2 Unsupervised Learning (Clustering, Dimensionality Reduction)

Unsupervised learning is used when the data has no labels. The goal is to infer the natural structure hidden within the data, find patterns, or simplify the data for further analysis.

8.2.2.2.1 Clustering

Clustering groups similar data points together based on their inherent characteristics.

- Common Algorithms:
  - **k-Means Clustering**: Partition data into k distinct, non-overlapping clusters. Simple and efficient for large datasets.
  - **Hierarchical Clustering**: Builds a tree of clusters, which is useful for understanding data at different levels of granularity (e.g., from species to genus level in biology).
  - **DBSCAN (Density-based Spatial Clustering of Applications with Noise)**: Finds clusters of arbitrary shape and can identify outliers or noise points, which is useful for detecting anomalous process batches.

8.2.2.2.2 Dimensionality Reduction

Dimensionality reduction preserves the most significant patterns and relationships in complex, high-dimensional data while simplifying it. In bioprocessing, where datasets may contain hundreds of measured variables (from sensors or omics technologies, for example), this is crucial.

- Common Algorithms:
  - **Principal Component Analysis (PCA)**: The most common linear technique. It finds the directions (principal components) that maximize the variance in the data.
  - **t-Distributed Stochastic Neighbor Embedding (t-SNE)**: A non-linear technique particularly well-suited for visualizing high-dimensional data in a low-dimensional space.
  - **Autoencoders (a DL technique)**: Neural networks trained to compress data into a lower-dimensional encoding and then reconstruct it. They are powerful for non-linear dimensionality reduction and learning efficient data representations.

8.2.2.3 Reinforcement Learning

Reinforcement Learning (RL) is a paradigm where an **agent** learns to make decisions by performing **actions** in an **environment** to maximize a cumulative **reward**. The agent learns from trial and error, receiving feedback in the form of rewards or penalties.

- **Common Algorithms**: Q-Learning, Deep Q-Networks (DQN), and Policy Gradient methods. RL is computationally intensive and requires careful simulation environment design, making it a more advanced but highly promising area.

8.2.3 The Role of Data: Acquisition, Pre-processing, and Feature Engineering in Bioprocessing

The performance of any ML model is governed by a simple, unwavering rule: Garbage In, Garbage Out (GIGO). The quality, quantity, and relevance of the data are more important than the complexity of the algorithm. The bioprocessing industry generates a wealth of data, but it is often messy, incomplete, and heterogeneous, making data curation the most critical and time-consuming step in any AIML project.

8.2.3.1 Data Acquisition and Sources

- **Biofuel data** is multi-modal, originating from a variety of sources.
- **Process Data**: Time-series data from sensors in reactors and downstream processing units (temperature, pressure, pH, flow rates, dissolved oxygen).
- **Spectroscopic and Chromatographic Data**: NIR, FTIR, Raman spectra, HPLC, and GC-MS chromatograms providing detailed compositional analysis.
- **Omics Data**: Genomic, transcriptomic, proteomic, and metabolomic data from microbial production strains or biomass feedstocks.
- **Image Data**: Microscopy images of cultures, satellite, or drone images for biomass assessment, and visual inspection data.
- **Text Data**: Scientific literature, patents, and maintenance logs that can be mined for insights using Natural Language Processing (NLP).

8.2.3.2 Data Pre-processing

Raw data is never ML-ready. Pre-processing is essential to transform it into a clean, usable dataset [8].

- **Handling Missing Data**: Strategies include deletion of incomplete records or imputation (e.g., filling missing values with the mean, median, or using a more sophisticated ML model to predict the missing values).
- **Noise Reduction and Outlier Detection**: Smoothing sensor data and identifying and removing erroneous readings caused by sensor drift or process upsets.
- **Normalization/Standardization**: Scaling numerical features to a common range (e.g., 0–1) or to have a mean of zero and standard deviation of one. This is crucial for algorithms that are sensitive to the scale of data, like SVMs and neural networks.
- **Temporal Alignment**: Synchronizing data streams sampled at different frequencies (e.g., a pH reading every second with a lab sample taken every hour).

8.2.3.3 Feature Engineering

This is the art of creating new input variables (features) from raw data to improve model performance and interpretability. Domain expertise in biotechnology and chemical engineering is invaluable here.

- **Creating Informative Features**: Instead of using all 1000 data points from a spectrum, an engineer might create features like "peak height at wavelength X" or "area under curve between Y and Z." From time-series sensor data, one might create features like "average temperature over the last 2 hours," "rate of pressure change," or "time since last nutrient addition."

- **Domain-specific Features**: Calculating known derived quantities, such as reaction rates, yields, or metabolic fluxes, which are more directly related to the underlying biology and chemistry than raw sensor readings.
- **Feature Selection**: Identifying the most relevant features and removing redundant or irrelevant ones. This reduces model complexity, minimizes overfitting, and speeds up training. Techniques include statistical tests (correlation analysis), model-based importance (from Random Forest), and regularization methods (Lasso).

The successful integration of AIML in biofuel production is a meticulous process that begins with a deep understanding of these fundamental concepts. Selecting the right algorithm for the problem and investing heavily in the arduous but essential task of data preparation are the prerequisites for unlocking the transformative potential of these technologies, paving the way for the innovative applications discussed in the remainder of this book.

8.3 AI/ML IN FEEDSTOCK SELECTION AND PRE-PROCESSING

The journey to efficient and cost-effective biofuel production begins long before a single liter of fuel is synthesized. It starts with the fundamental choice of feedstock and the processes used to prepare it for conversion. Feedstock costs can constitute over 50% of the total production expense, and their variability in composition, yield, and availability presents a monumental challenge. Traditional methods of feedstock selection and pre-treatment optimization are slow, expensive, and often fail to capture the complex, non-linear interactions between genetic, environmental, and process variables. The integration of AI and ML is revolutionizing this initial stage, transforming it from an artisanal craft into a predictive, data-driven science. This chapter explores how AI/ML algorithms are being deployed to intelligently select, design, and pre-process biomass, thereby establishing a more robust and economical foundation for the entire biofuel value chain.

8.3.1 Predictive Modeling for Biomass Yield and Quality

The agronomic front-end of biofuel production is fraught with uncertainty. Predicting the yield and compositional quality of biomass is critical for planning supply logistics, assessing economic viability, and ensuring consistent quality input for biorefineries. ML-based predictive modeling leverages historical and real-time data to forecast these attributes with remarkable accuracy [9].

8.3.1.1 Data Sources for Modeling

- **Geospatial and Environmental Data**: Satellite and drone imagery (multispectral, hyperspectral), soil quality maps (pH, nitrogen, organic content), historical and forecasted weather data (precipitation, temperature, solar radiation), and topographic data.

- **Genotypic Data**: Genetic marker data (SNPs—Single Nucleotide Polymorphisms) for energy crops like switchgrass, miscanthus, or poplar, which influence traits like drought tolerance, pest resistance, and lignin content.
- **Phenotypic Data**: Field-measured data on plant height, biomass density, canopy cover, and ultimately, harvest yield.
- **Management Practices**: Data on irrigation, fertilizer application, planting density, and harvest cycles.

8.3.1.2 Modeling Approaches

- **Yield Prediction**: Supervised learning regression algorithms are trained on multi-year datasets linking environmental conditions and genotypic information to final yield. **Random Forest Regression** and **Gradient Boosting Machines (e.g., XGBoost)** are particularly effective as they can handle the high dimensionality and complex interactions within this data. They can predict regional yield potential, allowing for optimized supply chain planning.
- **Compositional Quality Prediction**: The same principles apply to predicting critical compositional metrics. For lignocellulosic biomass, models predict the percentages of cellulose, hemicellulose, and lignin, which directly impact the difficulty of pre-treatment and the eventual sugar yield. For oilseed crops or algae, models predict lipid content and fatty acid profiles. Support Vector Regression (SVR) and Gaussian Process Regression (GPR) are often used for these tasks due to their effectiveness with smaller, noisier datasets.

8.3.2 Genomic and Metagenomic Analysis for Enhanced Feedstock Design

Beyond selecting naturally occurring feedstocks, AI/ML is accelerating the *design* of superior feedstocks through advanced genetic analysis. This involves two key approaches: optimizing the genetics of the feedstock itself and understanding the microbial communities that influence its growth and decomposition.

8.3.2.1 Genomic Analysis for Crop Improvement

- **Genomic Selection (GS)**: Compared to conventional marker-assisted selection, this represents a paradigm shift. ML models, such as Bayesian models and Ridge Regression Best Linear Unbiased Prediction, or RR-BLUP, are trained using phenotypic and genomic data from a "training population" of plants, which includes thousands of SNPs. The trained model then eliminates the need for time-consuming and costly field trials by predicting the phenotypic performance (e.g., high biomass, low lignin) of new plants based only on their genotype. This enables quick cycles of breeding and selection for the best energy crops.
- **Gene Discovery and Functional Genomics**: To find promoters, enhancers, and non-coding regions that control genes of interest, DL models—in particular, Convolutional Neural Networks (CNNs) and Recurrent Neural

Networks (RNNs)—are applied to genomic sequences. This aids in comprehending the genetic foundation of intricate characteristics and can direct genetic engineering initiatives (such as the use of CRISPR) to precisely modify genomes for desired characteristics.

8.3.2.2 Metagenomic Analysis of Associated Microbiomes

- The community of microorganisms (fungi, bacteria) in a feedstock's rhizosphere (root zone) has a significant impact on both its yield and health. From these environmental samples, metagenomic sequencing produces massive, intricate datasets of all genetic material.
- Unsupervised learning techniques like clustering and dimensionality reduction (PCA, t-SNE) are used to visualize and identify different microbial community structures associated with high-yielding or stressed plants.
- Supervised learning can then be used to build predictive models that link specific microbial taxa or functions to positive plant outcomes. This knowledge can be used to develop probiotic "consortia" of microbes that, when applied to crops, enhance nutrient uptake, increase drought resistance, and improve overall biomass yield.

8.3.3 Optimization of Pre-treatment Processes Using ML Algorithms

The most expensive and energy-intensive stage in the production of lignocellulosic biofuel is pre-treatment. Its objective is to release cellulose for enzymes by dissolving the stubborn lignin structure. Temperature, pressure, residence time, catalyst concentration, and feedstock particle size are all interacting variables in the process. Navigating this intricate optimization space requires the use of ML.

8.3.3.1 ML-driven Optimization Workflow

- **Design of Experiments (DoE)**: To guarantee that the first set of experiments yields the most information with the fewest runs, ML algorithms can guide optimal DoE strategies (such as D-Optimal design).
- **Model Building**: Surrogate models (Random Forest, SVR, or ANNs) that translate pre-treatment inputs to outputs are trained using data from experiments. For any combination of parameters within the tested range, these models reliably forecast the formation of inhibitors and sugar yield.
- **Global Optimization**: Using these trained surrogate models, optimization algorithms (such as Genetic Algorithms [GA] and Particle Swarm Optimization [PSO]) effectively search the multi-dimensional parameter space for the combination that maximizes sugar yield while minimizing energy input and inhibitor generation.
- **Robustness Analysis**: The models can also be used to find operating conditions that are less sensitive to small fluctuations in feedstock composition, ensuring consistent performance despite natural variability.

8.3.4 Computer Vision and Spectroscopy for Real-time Quality Assessment

To run a consistent and efficient biorefinery, the incoming feedstock must be rapidly characterized. Traditional wet-lab methods are too slow for real-time decision-making. AI-powered vision and spectral analysis provide instant, non-destructive assessment.

8.3.4.1 Hyperspectral and NIR Spectroscopy

- **Technology**: Near-Infrared (NIR) and Hyperspectral sensors shine light on a biomass sample and measure the reflected or transmitted spectra. The spectral signature is uniquely influenced by the chemical bonds (O-H, C-H, C-O) in the material, providing a fingerprint of its composition.
- **AI/ML Role**: The raw spectral data has hundreds of wavelengths. **Convolutional Neural Networks (CNNs)** are exceptionally adept at processing this spectral "image" to identify complex patterns. Alternatively, **Partial Least Squares Regression (PLSR)** models, a classic technique, are still widely used. These models are trained on paired data: spectra of samples and their corresponding lab-measured composition (e.g., lignin content). Once trained, the model can instantaneously predict the composition of a new sample based on its spectrum alone [10].

8.3.4.2 Computer Vision

- **Technology**: High-resolution cameras capture images of biomass (e.g., a conveyor belt of straw, a pond of algae).
- **AI/ML Role: Deep Learning-based CNNs** (e.g., YOLO—You Only Look Once, U-Net) are trained to perform tasks like:
  - **Sorting and Classification**: Differentiating between feedstock types (e.g., corn stover versus wheat straw) or identifying and removing contaminated or rotten material.
  - **Morphological Analysis**: Estimating particle size distribution after grinding, which is critical for pre-treatment efficiency.
  - **Algal Health Monitoring**: Analyzing images from photobioreactors to estimate algal density, detect contamination by other species, or identify nutrient deficiency based on color changes.

8.3.5 Case Study: AI-driven Selection of Algal or Lignocellulosic Feedstocks for Maximum Lipid/Sugar Yield

8.3.5.1 Optimizing a Regional Biorefinery Network for Lignocellulosic Bioethanol Using an Integrated AI-driven Approach

Background: A bioenergy company aims to establish a network of biorefineries in a specific region. The goal is to maximize annual ethanol production while minimizing average cost per liter. The region has a diverse agricultural landscape producing corn stover, wheat straw, and miscanthus, each with variable yield, composition, and availability [11].

The AI/ML Solution: A Multi-model Pipeline

- **Feedstock Yield and Composition Prediction**:
  - **Data**: Historical satellite imagery (NDVI), soil data, and weather data for the past ten years for every field in the region.
  - **Model**: A **Random Forest regression** model was trained to predict the annual yield and (separately) the lignin content for each field for each potential feedstock crop.
- **Pre-treatment and Conversion Simulation**:
  - **Data**: A database of hundreds of lab-scale pre-treatment experiments (varying temperature, acid concentration, time) and subsequent enzymatic hydrolysis experiments for each type of biomass.
  - **Model**: A **Deep Neural Network (DNN)** was trained as a surrogate model. It takes as input: (Feedstock Type, Cellulose %, Hemicellulose %, Lignin %, Pre-treatment T, Pre-treatment Catalyst, Pre-treatment Time) and outputs: (Glucose Yield [g/g biomass], Inhibitor Concentration).
- **Supply Chain and Optimization Model**:
  - **Input**: The predictions from Model 1 (a map of available biomass) and the simulator from Model 2.
  - **Optimization**: A **Genetic Algorithm** was used to solve a massive optimization problem. Its decisions were:
    - Where to locate biorefineries.
    - Which feedstock to procure from which field.
    - How to blend different feedstocks at each plant to achieve a consistent composition.
    - What pre-treatment parameters to use for each unique blend to maximize sugar yield.
- The objective function for the GA was to minimize the total cost per liter of ethanol, which included feedstock transportation cost, pre-treatment energy cost, and capital costs.

Results and Impact

- The AI system identified a previously overlooked optimal biorefinery location that minimized total transportation distance.
- It recommended a dynamic blending strategy: using high-lignin miscanthus blended with low-lignin wheat straw to create a consistent feedstock blend, allowing the pre-treatment process to be finely tuned to that specific blend.
- By using the DNN simulator, the GA found pre-treatment conditions that increased the predicted sugar yield by 17% compared to the standard industry protocol for a single feedstock.
- The integrated model predicted a potential 22% reduction in the final cost of ethanol production by leveraging regional variability and optimized processing, thereby de-risking the multi-million-dollar investment and ensuring long-term economic and operational sustainability [12].

This case study exemplifies the power of integrating AI/ML across the entire feedstock selection and pre-processing value chain, moving from passive assessment to active, intelligent design and optimization.

8.4 OPTIMIZATION OF CONVERSION PROCESSES VIA AI/ML

The transition from fossil-based to bio-based economies hinges on the efficiency and scalability of conversion processes that transform raw biomass into valuable fuels, chemicals, and materials. These processes—whether biochemical (like fermentation) or thermochemical (like pyrolysis)—are inherently complex, non-linear, and multivariate. Traditional optimization methods, often relying on one-factor-at-a-time (OFAT) experimentation and simplified mechanistic models, are time-consuming, expensive, and frequently fail to capture the true complexity of these systems.

AI and ML are revolutionizing this domain. By learning directly from process data, AI/ML models can uncover hidden patterns, build highly accurate predictive models, and identify optimal operating conditions with unprecedented speed and precision. This enables the development of more robust, efficient, and economically viable bioconversion technologies.

8.4.1 Modeling and Optimizing Fermentation Processes

Fermentation is a cornerstone of industrial biotechnology, used to produce a vast array of products, from biofuels (e.g., ethanol, butanol) to pharmaceuticals (e.g., antibiotics, insulin) and specialty chemicals. It involves the controlled growth of microorganisms (bacteria, yeast, fungi) to convert substrates into desired metabolites. The challenge lies in the dynamic and living nature of the system, where microbial growth, substrate consumption, and product formation are influenced by a multitude of interacting parameters.

8.4.1.1 Predicting Microbial Growth and Metabolite Production

Accurately predicting the behavior of a microbial culture is critical for process control and scaling. Mechanistic models, such as Monod kinetics, are useful but often oversimplified.

- **ML Approach**: Time-series of historical fermentation data that include measurements such as these are used to train supervised learning algorithms.
  - **Inputs** include temperature, pH, dissolved oxygen (DO), substrate concentration, and optical density, which serves as a stand-in for cell density.
  - **Results/Aims**: Product titer, yield, productivity, and final cell density.
- **Key Algorithms and Applications**:
  - **Recurrent Neural Networks (RNNs) and Long Short-Term Memory (LSTM) Networks**: These are exceptionally well-suited for time-series data. An LSTM model can "remember" long-term dependencies in the fermentation broth, allowing it to predict the final product titer hours or even days in advance based on early process data. This enables proactive interventions.

- **Gaussian Process Regression (GPR)**: GPR not only provides predictions but also quantifies the uncertainty associated with each prediction. This is invaluable for risk assessment and for guiding experiments in areas where the model is less confident.
- **Random Forests and Gradient Boosting Machines (e.g., XGBoost)**: These powerful ensemble methods can handle complex, non-linear relationships between process parameters and outcomes. They can rank the importance of different factors (e.g., identifying that a specific nutrient spike is the most critical driver for high yield).

8.4.1.2 Optimizing Parameters (pH, Temperature, Nutrient Feed)

The ultimate goal is not just prediction, but active optimization. Parameters like pH, temperature, and nutrient feeding strategy have a profound impact on preventing by-product formation, avoiding substrate inhibition, and maximizing the target metabolite yield.

- **ML Approach**: This involves a closed-loop optimization strategy.
  - **Surrogate Modeling**: First, an ML model—such as a support vector machine or neural network—is trained to function as a "digital twin" of the fermentation process. Compared to conducting a real, time-consuming fermentation, this surrogate model is computationally inexpensive to evaluate.
  - **Optimization Algorithms**: The optimal set of parameters is then determined by applying optimization techniques to this digital twin.
    - **Bayesian Optimization**: This method is especially effective for maximizing costly black-box processes (such as fermentation). It balances exploration (testing unknown regions) and exploitation (improving known good regions) by using the surrogate model's predictions and uncertainty estimates to intelligently choose the subsequent set of conditions to test in the real world. In a relatively small number of experimental iterations, this results in the global optimum.
    - **Genetic Algorithms (GAs)**: In order to replicate natural evolution, GAs generate a "population" of parameter sets, assess their "fitness" using the surrogate model, and combine the best ones to evolve in the direction of the ideal solution.
- **Application Example**: Optimizing a fed-batch process for antibiotic production. The ML system can recommend a dynamic temperature profile and a non-intuitive nutrient feeding strategy that a human operator might never conceive, leading to a significant increase in yield while reducing fermentation time.

8.4.2 AI in Thermochemical Processes: Pyrolysis and Gasification

Solid biomass is converted into liquid bio-oil, syngas, and biochar by thermochemical processes such as gasification (partial oxidation to produce syngas) and pyrolysis (thermal decomposition without oxygen). Extreme conditions and intricate reaction pathways that are challenging to mechanistically model define these processes.

8.4.2.1 Predicting Bio-oil Yield and Composition

The yield and quality of the bio-oil, a complex mixture of hundreds of organic compounds, are critical to the economic feasibility of pyrolysis. Its stability, acidity, and heating value are determined by its composition.

- **ML Approach**: Regression models are trained to forecast results according to process parameters and feedstock characteristics.
  - **Input Features**: Biomass ultimate analysis (C, H, O, N, S content), proximate analysis (moisture, volatile matter, fixed carbon, ash content), particle size, heating rate, final temperature, reactor type.
  - **Target Outputs**: Bio-oil yield, syngas yield, biochar yield; and more advanced targets like the concentration of specific compounds (e.g., phenols, furans) or overall quality indices (e.g., Higher Heating Value—HHV, acidity).
- **Key Algorithms**:
  - **Artificial Neural Networks (ANNs)**: When it comes to capturing the highly non-linear relationships in thermochemical conversion, ANNs are the preferred option. They are able to accurately forecast yields by modeling the intricate relationship between reaction temperature and biomass composition.
  - **Support Vector Regression (SVR)**: Effective in high-dimensional spaces, SVR can be robust when the number of training examples is relatively small.
- **Benefit**: In order to determine whether a new, untested biomass feedstock is a suitable raw material, an operator can input its characteristics, and the ML model will immediately predict the expected yield and quality of bio-oil.

8.4.2.2 Reactor Condition Optimization

The reactor is the heart of the process. Optimizing conditions like temperature, pressure, residence time, and catalyst type is crucial for maximizing desired products and minimizing undesired ones (like tar in gasification) [13].

- **ML Approach**: Similar to fermentation optimization, this combines surrogate modeling with advanced search algorithms.
  - **Digital Twin of the Reactor**: An ML model is built to predict reactor performance (e.g., conversion efficiency, tar yield) based on operating conditions and feedstock input.
  - **Multi-objective Optimization**: The goal is often not a single target but a trade-off. For example, "maximize bio-oil yield" AND "maximize bio-oil HHV" (which often have an inverse relationship). Algorithms like **Multi-Objective Bayesian Optimization** or **NSGA-II (a genetic algorithm)** can find a set of optimal solutions, known as the **Pareto front**. This allows engineers to visually see the trade-offs and select an operating point that best suits their economic and technical goals.
- **Application Example**: In catalytic gasification, an ML model can identify the precise combination of temperature, steam-to-biomass ratio, and catalyst loading that simultaneously maximizes syngas ($CO+H_2$) production while minimizing energy-consuming tar formation.

8.5 AI FOR PROCESS MONITORING, CONTROL, AND MAINTENANCE

The convergence of AI with industrial operations is revolutionizing the manufacturing and process industries. Moving beyond traditional automation, AI introduces a layer of cognitive capability that enables systems to not just execute pre-defined tasks but to perceive, learn, reason, and act autonomously. This paradigm shift is most impactful in the core areas of process monitoring, control, and maintenance [14]. AI turns reactive operations into proactive, predictive, and self-optimizing businesses by utilizing enormous volumes of data. This section explores the main uses of AI that are achieving previously unheard-of levels of effectiveness, dependability, and quality.

8.5.1 Integration of IoT and Advanced Sensor Networks

By establishing a smooth connection between the digital and physical realms, the IoT and sophisticated sensor networks are establishing a new paradigm of intelligent, data-driven systems. In order to achieve this synergy, advanced sensors that can record extremely detailed data—from motion and temperature to vibration and air quality—must be deployed. These sensors will then be connected via IoT platforms that will collect, process, and act upon the data in real time. Whether it is for monitoring vital infrastructure, developing smart cities that dynamically manage energy and traffic, or enabling precision agriculture that conserves water and boosts yield, this potent combination allows for previously unheard-of levels of automation, predictive maintenance, and resource optimization across industries [15]. Ultimately, by embedding sensing and intelligence into the fabric of our environment, this integration is not just generating vast amounts of data but is translating it into actionable insights, driving efficiency, sustainability, and transformative innovation.

8.5.1.1 The Foundation of Data-driven Intelligence

Building a strong infrastructure for data collection is the first and most important step in applying AI in an industrial setting. The IoT and sophisticated sensor networks are integrated to accomplish this.

- **From Legacy to Connected**: Traditional factories operate with siloed data from Programmable Logic Controllers (PLCs) and Supervisory Control and Data Acquisition (SCADA) systems. IoT involves retrofitting or installing new, intelligent sensors on every critical piece of equipment—from motors, pumps, and valves to conveyor belts and storage tanks. These sensors go beyond basic measurements (e.g., temperature, pressure) to capture more nuanced data like vibration spectra, acoustic emissions, motor current signatures, and video feeds.
- **The Role of Edge Computing**: Transmitting all raw sensor data to a central cloud server is often inefficient and introduces latency. **Edge AI** addresses this by performing initial data processing and simple AI inference directly on the device (the "edge") or on a local gateway. For instance, an edge device can analyze vibration data in real-time to detect an anomaly and only send

an alert to the central system, rather than a continuous stream of raw data. This reduces bandwidth usage and enables ultra-fast, localized responses.

- **AI's Role in Sensor Fusion and Data Quality**: AI algorithms, particularly neural networks, excel at **sensor fusion**—the process of combining data from multiple, heterogeneous sources (vibration, temperature, pressure) to create a more accurate and contextualized understanding of the equipment's state than any single sensor could provide. Furthermore, AI can identify and flag sensor drift, calibration errors, or failures, ensuring the integrity of the data foundation upon which all other decisions are made.

In essence, the IoT network acts as the nervous system of the smart factory, and AI is the brain that interprets its signals.

8.5.2 Real-time Analytics for Quality Assurance and Control

Real-time analytics for quality assurance and control represents a transformative shift from reactive problem-solving to proactive perfection, leveraging a continuous stream of data from sensors, vision systems, and production equipment directly on the manufacturing floor. By processing this data instantaneously, these systems can immediately detect deviations, microscopic defects, or subtle process variations that would otherwise go unnoticed until a post-production inspection, if at all.

8.5.2.1 Shifting from End-of-Line Inspection to In-Process Perfection

Traditional quality control often relies on sampling and end-of-line inspection, which is reactive and can lead to significant waste. AI-powered real-time analytics transforms quality assurance into a continuous, proactive process.

- **Computer Vision for Defect Detection**: AI-based computer vision systems, using high-resolution cameras and DL models (like CNNs), can inspect products at high speeds with superhuman accuracy. These systems are trained on thousands of images of both "good" and "defective" products, learning to identify subtle imperfections—scratches, discolorations, misalignments, or microscopic cracks—that are often invisible to the naked eye. This allows for 100% inspection rather than statistical sampling [16].
- **Predictive Quality Analytics**: This is a more sophisticated application where AI correlates real-time process parameters (e.g., temperature, pressure, flow rates, raw material properties) with the final product quality. By analyzing historical data, the AI model learns the precise "recipe" of process conditions that yields a perfect product. In real time, it can then predict the quality outcome *before* the product is even finished. If the process parameters begin to deviate from the ideal "golden batch" profile, the system can alert operators or even automatically adjust control parameters to correct the drift, preventing the production of defective goods altogether.
- **Root Cause Analysis**: When a defect does occur, AI can rapidly analyze terabytes of historical process data to identify the root cause. By identifying the precise parameter or set of parameters that changed and when, it can significantly cut down on the amount of time engineers need to spend troubleshooting.

8.5.3 Predictive Maintenance for Downtime Reduction and Cost Savings

In order to significantly reduce unplanned downtime and produce significant cost savings, predictive maintenance represents a paradigm shift from reactive or scheduled upkeep to a data-driven approach. Using sensors that measure vibration, temperature, and acoustic emissions, this method continuously monitors asset condition and uses ML and advanced analytics to identify early indicators of equipment deterioration. This prevents disastrous malfunctions and production stops by allowing maintenance to be carried out exactly when it's needed, right before a possible failure happens [17]. As a result, there are two significant advantages: it increases asset uptime and operational efficiency while lowering expenses related to emergency repairs, spare part inventory, and needless preventive maintenance. This prolongs the life of equipment and improves the bottom line.

8.5.3.1 From "Fix-It-When-It-Breaks" to "Fix-It-Before-It-Fails"

Predictive Maintenance (PdM) is one of the most financially compelling applications of AI in industry. It goes beyond preventive maintenance, which involves planned repairs regardless of actual need, to maintenance that is determined by the asset's current state.

- **How It Works**: AI models are trained on historical sensor data (vibration, temperature, acoustic, etc.) that captures the normal operating behavior of a machine as well as its behavior leading up to a known failure. The model learns the unique "fingerprint" of impending faults.
- **Key Techniques**:
  - **Anomaly Detection**: The AI establishes a baseline of normal operation. Any significant deviation from this baseline is flagged as an anomaly, indicating potential issues.
  - **Remaining Useful Life (RUL) Prediction**: More advanced models can not only detect a fault but also predict the remaining useful life of a component (e.g., a bearing has 45 days of operational life left). This allows maintenance to be scheduled optimally, minimizing downtime and maximizing parts utilization.
- **Benefits**:
  - **Dramatic Downtime Reduction**: Unplanned downtime is virtually eliminated.
  - **Cost Savings**: Reduces unnecessary maintenance activities and prevents catastrophic failures that cause secondary damage.
  - **Improved Safety**: Prevents dangerous equipment failures.
  - **Spare Parts Inventory Optimization**: Maintenance can be planned, allowing for just-in-time ordering of spare parts.

8.5.4 Digital Twins: Creating Virtual Models of Production Facilities

By producing dynamic, virtual representations of actual production facilities, digital twins mark a revolutionary advancement in industrial management. These models are dynamic digital counterparts that are constantly updated with real-time data from sensors, IoT devices, and factory floor operational systems; they are much more than static 3D drawings. Because of this two-way data flow, engineers and managers can monitor operations, model production scenarios, and anticipate maintenance requirements with previously unheard-of accuracy. The digital twin can replicate the precise state, performance, and health of its physical twin. Digital twins enable companies to increase productivity, reduce downtime, speed up product development, and reduce risk in decision-making by acting as a potent sandbox for analysis and optimization. This, in turn, ushers in a new era of intelligent, agile, and data-driven manufacturing.

8.5.4.1 A Virtual Sandbox for the Entire Plant Life Cycle

A dynamic, virtual representation of a real asset, procedure, or system is called a "digital twin." It creates a living simulation by continuously updating with data from its physical counterpart through IoT sensors.

- **The Components**: A true digital twin consists of three parts:
  - **The Physical Asset**: The real-world machine or production line.
  - **The Virtual Model**: A high-fidelity, physics-based, and data-driven computer model.
  - **The Bi-directional Data Link**: The continuous flow of data from the physical to the virtual, and the flow of insights and commands from the virtual back to the physical.
- **AI-powered Applications**:
  - Design and What-If Analysis: Before putting new products, process modifications, or equipment upgrades into a real-world setting, engineers can use the digital twin to model their effects in a risk-free virtual setting.
  - Operator Training: By using the digital twin, new operators can receive training that enables them to experience and react to infrequent fault scenarios without endangering actual production or safety.
  - Optimization and Control: To determine the ideal setpoints for maximizing throughput, yield, or energy efficiency, the digital twin can execute thousands of simulations in parallel.
  - The physical control system can then be configured with these ideal parameters. Predictive Maintenance Integration: By simulating wear and tear, the digital twin offers an even more reliable platform for anticipating malfunctions and scheduling maintenance.

The digital twin, which acts as the operation's central nervous system, is the best tool for understanding and optimizing processes holistically.

8.5.5 Energy Management and Optimization for Sustainable Plant Operations

A strategic, data-driven approach, Energy Management and Optimization for Sustainable Plant Operations goes beyond basic energy conservation to integrate efficiency into the very fabric of industrial processes. Energy consumption across all systems and equipment is continuously monitored, analyzed, and controlled. Waste, inefficiencies, and areas for improvement are found by utilizing technologies such as advanced analytics and IoT sensors. Plants can drastically lower their energy expenses and carbon footprint by putting in place heat recovery systems, streamlining equipment schedules, enhancing process control, and possibly incorporating renewable energy sources. This methodical search for energy efficiency is not merely a strategy to cut operating costs; it is an essential part of corporate sustainability, increasing competitiveness, guaranteeing regulatory compliance, and building resilience against energy price volatility for a more sustainable and profitable future [18].

8.5.5.1 Driving Efficiency and Meeting Sustainability Goals

One of the biggest operating expenses and a major factor in a plant's carbon footprint is energy consumption. AI offers strong tools for facility-wide energy management and optimization.

- **Energy Consumption Forecasting**: To precisely forecast the facility's energy demand, AI models examine past production data, weather predictions, and energy prices. Better procurement tactics can be implemented as a result, utilizing reduced off-peak prices.
- **System-Level Optimization**: Instead of optimizing individual machines in isolation, AI can optimize the entire system. For example, in a complex HVAC or compressed air system, AI can determine the most efficient combination of chillers, pumps, and fans to deliver the required cooling or pressure at the absolute minimum energy cost, considering the part-load efficiency curves of each machine.
- **Anomaly Detection in Energy Usage**: AI can establish a baseline for "normal" energy consumption for different production schedules and products. A sudden, unexplained spike in energy use can indicate equipment inefficiency (e.g., a fouled heat exchanger), a leak in a compressed air line, or even a malfunctioning meter, allowing for quick corrective action.
- **Sustainability Reporting and Carbon Accounting**: AI can automate the tracking and reporting of energy usage and greenhouse gas emissions, helping companies comply with regulations and meet their Environmental, Social, and Governance (ESG) targets. By identifying the largest sources of energy waste, AI directly guides capital investments toward the most impactful sustainability projects.

The integration of AI into process monitoring, control, and maintenance is not a single technology but a synergistic ecosystem. It begins with the data foundation provided by IoT sensors, which feeds real-time analytics for quality and predictive maintenance models.

8.6 SUSTAINABILITY AND LCA

Sustainability has evolved from a corporate social responsibility initiative to a core strategic imperative. The LCA, a thorough methodology for assessing the environmental effects of a product, process, or service throughout its entire life cycle—from raw material extraction ("cradle") to manufacturing, distribution, use, and final disposal or recycling ("grave")—is at the core of quantifying sustainability. Despite its strength, traditional LCA is frequently a static, data-intensive, and time-consuming procedure [19]. LCA is being transformed from a retrospective analytical tool into a dynamic, predictive, and prescriptive engine for true sustainable innovation by the combination of AI and ML. This section delves into how AI and advanced optimization techniques are enhancing LCA to drive meaningful environmental and economic outcomes.

8.6.1 AI-Powered Analysis of Environmental Impact

When working with complex, interconnected systems, traditional LCA can be restrictive due to its reliance on static databases and manual computations. A more nuanced, predictive, and effective analysis is made possible by AI, especially ML, which brings about a paradigm shift [20].

Key Applications:

- **Predictive Impact Modeling**:
  - **Concept**: Instead of calculating impacts based solely on historical data, ML models can be trained to predict the environmental footprint of new materials, designs, or processes before they are physically created or deployed.
  - **Mechanism**: Large datasets of previous LCA results are used to train models using methods like supervised learning. The model learns the intricate connections between output impact categories (such as water use, eutrophication, and global warming potential, or GWP) and input parameters (such as material composition, energy sources, and manufacturing techniques). After training, the model can analyze its parameters to forecast the effects of a new design.
  - **Benefit**: Drastically reduces the time and cost of LCAs for new product development, enabling rapid iteration and "what-if" scenario analysis.
- **Data Gap Filling and Uncertainty Reduction**:
  - **Concept**: Complete LCA data is often unavailable, especially for novel supply chains or emerging technologies. AI can intelligently fill these data gaps.
  - **Mechanism**: Algorithms like k-nearest neighbors or generative models can estimate missing data points by identifying analogous processes or materials within the database. Furthermore, AI can quantify and model the uncertainty associated with LCA results, providing a more robust and honest assessment rather than a single, potentially misleading, figure.
- **Automated Data Collection and Processing**:
  - **Concept**: AI can automate the most labor-intensive part of an LCA: data collection.
  - **Mechanism**: NLP algorithms can scan through thousands of scientific papers, supplier reports, and environmental product declarations (EPDs) to extract relevant impact data. Computer vision can analyze satellite imagery to monitor land-use changes or emissions from industrial facilities, providing real-time, site-specific data for the LCA model.

- **Anomaly Detection in Environmental Performance**:
  - **Concept**: AI can continuously monitor a product's life cycle data to identify unexpected deviations or inefficiencies.
  - **Mechanism**: Unsupervised learning algorithms can analyze real-time data streams from IoT sensors in factories, logistics fleets, or energy grids. They can detect anomalies—such as a sudden spike in energy consumption during a specific manufacturing step—allowing for immediate corrective action.

8.6.2 Optimizing Supply Chain Logistics for Reduced Carbon Footprint

The supply chain is often responsible for the majority of a product's carbon footprint, particularly in the transportation and distribution phases. AI-driven optimization is critical for minimizing this impact [21].

Key Strategies:

- **Intelligent Route and Load Optimization**:
  - **Concept**: Moving beyond simple shortest-path algorithms, AI systems optimize for carbon efficiency, not just distance or time.
  - **Mechanism**: ML models incorporate real-time and predictive data on:
    - **Traffic Congestion**: Avoiding idling, which wastes fuel.
    - **Weather Conditions**: Routing around storms or leveraging favorable winds.
    - **Vehicle Type and Load**: Optimizing palletization and container space to maximize load capacity per trip, reducing the total number of journeys.
    - **Modal Shifts**: Automatically evaluating and suggesting the most carbon-efficient transport mode (e.g., rail versus truck versus ship) for each leg of the journey.
- **Dynamic and Predictive Logistics Networks**:
  - **Concept**: Instead of a static supply chain, AI enables a dynamic, responsive network that adapts to minimize emissions.
  - **Mechanism**: By analyzing historical sales data, market trends, and even social media signals, AI can forecast demand with high accuracy. This allows for:
    - **Warehouse Placement**: Optimizing the location of distribution centers to be closer to predicted demand hotspots, reducing final delivery distances.
    - **Inventory Optimization**: Preventing overproduction and stockouts, which lead to wasteful expedited shipping or disposal of obsolete goods.
- **Green Supplier Selection and Evaluation**:
  - **Concept**: AI assists in evaluating and selecting suppliers based on their environmental performance, integrating LCA data directly into procurement decisions.

- **Mechanism**: An AI platform can aggregate and analyze sustainability data from potential suppliers (e.g., their energy mix, waste management practices, transportation footprints). It can then score and rank suppliers not just on cost and quality, but on their contribution to the final product's overall carbon footprint, creating a truly sustainable supply chain.

8.6.3 Balancing Economic and Environmental Objectives through Multi-objective Optimization

The classic challenge in business is the perceived trade-off between profitability and sustainability. Multi-objective Optimization (MOO) powered by AI provides a mathematical framework to navigate this complex trade-space and find optimal compromises, or "Pareto fronts" [22].

8.6.3.1 The Pareto Front Concept

A Pareto front represents a set of optimal solutions where improving one objective (e.g., reducing cost) necessarily worsens another (e.g., increasing carbon footprint). There is no single "best" solution, but rather a frontier of non-dominated solutions.

8.6.3.2 AI's Role in Multi-objective Optimization

- **Problem Formulation**:
  - Objectives are defined mathematically. For example:
    - *Objective 1 (Minimize)*: Total Cost = (Material Cost + Manufacturing Cost + Logistics Cost)
    - *Objective 2 (Minimize)*: Carbon Footprint (kg CO_2eq) = (LCA result for GWP)
  - **Constraints are set**, such as minimum product strength, maximum delivery time, or budget limits.
- **Algorithmic Solution with AI**:
  - **Concept**: For complex problems, it is computationally impossible to evaluate every possible combination of variables (suppliers, materials, routes, etc.) in detail. On the Pareto front, AI algorithms are particularly adept at identifying high-quality solutions, especially evolutionary algorithms, a kind of metaheuristic influenced by natural selection.
  - **Mechanism**: Algorithms like NSGA-II (Non-dominated Sorting Genetic Algorithm II) work by:
    - **Initialization**: Creating a random population of potential solutions.
    - **Evaluation**: Scoring each solution against both objectives (cost and carbon).
    - **Selection, Crossover, and Mutation**: "Breeding" new solutions by combining traits of the best-performing ones and introducing small random changes.
    - **Iteration**: Repeating this process over many generations, the population evolves toward the Pareto front, revealing a spectrum of optimal trade-offs.

- **Decision Support**:
  - The output is not a single answer but a clear visualization of the Pareto front. Management can then make an informed strategic decision. For instance, they might choose a solution that increases cost by 5% but reduces the carbon footprint by 30%, depending on their corporate sustainability goals.

8.7 CHALLENGES AND FUTURE DIRECTIONS

The integration of AI and ML into the biofuels sector promises a revolution in efficiency, yield, and sustainability. However, the path from promising pilot studies to widespread industrial adoption is fraught with significant challenges. This section delves into the current hurdles, explores the economic realities, and charts a course for the future, outlining the technological advancements, strategic roadmaps, and policy frameworks necessary to fully realize the potential of this powerful synergy.

8.7.1 Current Technical and Operational Hurdles

Before the industry can reap the full benefits of AI/ML, it must first overcome a series of foundational technical and operational obstacles [23]. These challenges often represent the gap between theoretical models and practical, reliable implementation in complex bioprocessing environments.

8.7.1.1 Data Availability, Quality, and Standardization

Data is the lifeblood of any AI/ML system. The biofuels industry faces a unique "data famine" compared to other digitalized sectors.

- **Data Scarcity**: High-quality, annotated data from biorefineries is limited. Unlike digital transactions, biological experiments (e.g., testing a new strain of algae or a novel enzyme) are time-consuming and expensive to run, resulting in small, often proprietary datasets. This scarcity can lead to overfitted models that perform well on lab data but fail in real-world conditions.
- **Data Silos and Proprietary Formats**: Critical data is often locked within individual companies, research institutions, or even different departments within the same organization. The lack of data-sharing incentives and standardized formats prevents the aggregation of large datasets needed to train robust, generalizable models.
- **Data Quality and Noise**: Bioprocess data from sensors is notoriously noisy, incomplete, and susceptible to drift. Factors like sensor calibration errors, microbial contamination, and fluctuating raw material composition introduce significant variance that can confuse AI models if not properly cleaned and contextualized.
- **Lack of Standardization**: Without industry-wide standards for data labeling (e.g., what exactly constitutes "high lipid content" or "process instability"), it is nearly impossible to combine datasets or compare model performances across different studies or production facilities.

8.7.1.2 Model Interpretability ("Black Box" Problem)

Many of the most powerful ML models, particularly deep neural networks, are often "black boxes." While they can identify complex, non-linear patterns with high accuracy, understanding *why* they made a specific prediction is difficult [24].

- **Implications for Bioprocessing**: In an industry governed by biological and chemical principles, a recommendation from an AI (e.g., "lower the temperature by 1.5°C and increase agitation") that cannot be explained in terms of known biology or thermodynamics is unlikely to be trusted or acted upon by process engineers and microbiologists.
- **Regulatory and Safety Concerns**: For biofuels to be certified and used commercially, processes must be validated and safe. Regulatory bodies may be hesitant to approve processes controlled by an inscrutable AI system, especially if it leads to an unexpected failure or safety incident. Interpretability is crucial for debugging, optimization, and regulatory acceptance.

8.7.1.3 High Initial Investment

The transition to an AI-driven operation requires substantial capital expenditure (CapEx), which can be a major barrier, especially for small- and medium-sized enterprises (SMEs).

- **Infrastructure Costs**: This includes investment in high-performance computing (HPC) resources for model training, cloud computing subscriptions, and the installation of a dense network of advanced sensors (IoT devices) for real-time data acquisition.
- **Software and Licensing**: Acquiring and maintaining licenses for AI/ML platforms, data analytics suites, and digital twin software represents a significant ongoing operational expense (OpEx).
- **Integration and Deployment Costs**: The cost of customizing AI solutions to fit specific biorefinery layouts and processes, and the associated project management, is often underestimated.

8.7.2 Economic and Scalability Considerations

Beyond the initial technical challenges, long-term success depends critically on the scalability and economic feasibility of AI/ML solutions.

- **Return on Investment (ROI) Uncertainty**: Theoretically, the advantages of AI—like higher yield, less downtime, and lower energy consumption—are obvious, but it can be challenging to calculate the return on investment for a particular installation. For investors looking for quick profits, biofuel processes are difficult to sell because of their lengthy development and validation cycles, which can result in a payback period of years.
- **Scalability from Lab to Plant**: An AI model that optimizes a 10-liter laboratory fermenter may not scale linearly to a 100,000-liter industrial

fermenter. Factors like mass transfer limitations, heat dissipation, and population heterogeneity in large-scale systems introduce complexities not present at smaller scales. AI models must be trained and validated on pilot-scale data before they can be trusted for full-scale deployment.

- **Business Model Evolution**: AI could enable new business models, such as "Biofuel-as-a-Service" or highly decentralized, small-scale biorefineries optimized by a central AI platform. The industry must explore these models to determine their economic feasibility.

8.7.3 The Future of AI/ML in Biofuels: Key Technological Frontiers

The future of AI in biofuels lies not just in overcoming current challenges but in leveraging emerging technological paradigms that will redefine what is possible [25].

8.7.3.1 The Convergence of AI with IoT, Digital Twins, and Robotics

AI will not operate in isolation. Its true power will be unlocked through convergence with other Industry 4.0 technologies.

- **AI-powered Digital Twins**: These are virtual, dynamic replicas of a physical biorefinery. Fed by real-time data from IoT sensors, a digital twin, driven by AI, can simulate and predict the outcome of process adjustments, perform "what-if" analysis without risking the actual plant, and enable proactive maintenance.
- **Robotics and Automation**: AI will guide robotic systems for automated sampling, strain screening, and even reactor maintenance, drastically accelerating R&D and ensuring consistent operational quality.

8.7.3.2 Explainable AI (XAI)

To solve the "black box" problem, the field of Explainable AI is rapidly advancing. XAI techniques aim to make model decisions transparent and interpretable to human experts.

- **Application**: An XAI system wouldn't just predict a drop in yield; it would highlight that the prediction was primarily driven by a specific change in the concentration of a key metabolite, linking the AI's finding back to known metabolic pathways. This builds trust and facilitates collaboration between data scientists and domain experts.

8.7.3.3 Transfer Learning

This technique involves taking a pre-trained model (e.g., one trained on a vast dataset of protein structures or microbial genomes) and fine-tuning it for a specific, data-scarce biofuel task (e.g., predicting the efficiency of a novel enzyme).

- **Impact**: Transfer learning can dramatically reduce the amount of expensive, proprietary data needed to develop effective models, lowering the barrier to entry for smaller players and accelerating innovation.

8.7.3.4 AI-driven Synthetic Biology

This is perhaps the most transformative frontier. AI is moving from optimizing processes to designing the biological agents themselves.

- **Generative AI for Molecule and Pathway Design**: AI models can be used to *design* new enzymes, metabolic pathways, or even synthetic microorganisms from scratch that are optimized for high biofuel production, resilience, and the ability to consume low-cost, non-food feedstocks.
- **Accelerating the Design-Build-Test-Learn Cycle**: AI can rapidly analyze results from high-throughput "Build" and "Test" cycles, and then "Learn" to propose improved designs for the next iteration, compressing R&D timelines from years to months.

8.7.3.5 Advanced Biorefining: The AI-optimized Circular Economy

Future biorefineries will be integrated systems that produce not just fuel, but a portfolio of high-value chemicals, materials, and energy. AI will be essential for managing this complexity.

- **Multi-objective Optimization**: AI systems will be tasked with simultaneously optimizing for multiple, often competing, objectives: maximizing fuel yield, minimizing energy and water consumption, reducing waste, and maximizing the value of co-products, all within economic and environmental constraints.

8.8 CONCLUSION

The journey through the complexities of the biofuel sector reveals a landscape of immense promise, significant challenges, and a critical inflection point. This report has systematically deconstructed the technological, economic, and environmental dimensions of biofuels, from first-generation feedstocks to the nascent potential of third and fourth generations. The analysis underscores that biofuels are not a monolithic solution but a diverse portfolio of technologies whose viability is contingent upon a sophisticated interplay of factors. As we stand at the nexus of climate urgency, energy security concerns, and a technological revolution driven by data, the path forward for biofuels must be strategic, integrated, and relentlessly innovative. This concluding chapter synthesizes the key insights garnered, outlines a visionary framework for a data-driven and sustainable biofuel ecosystem, and offers final reflections on the concerted actions required to navigate the path forward.

8.8.1 Final Remarks on the Path Forward

The transition to a sustainable energy future is not a sprint but a marathon, and biofuels are a vital part of the relay team, alongside electrification, hydrogen, and energy efficiency. The analysis presented herein leads to a set of unequivocal calls to action for various stakeholders.

For Policymakers

- **Provide Long-Term, Technology-Neutral Signals**: Move beyond short-term subsidies to implement stable, long-term carbon pricing and policies that reward verified carbon intensity reductions, allowing the most efficient technologies to compete and flourish.
- **Fund Foundational R&D**: Increase public investment in fundamental research on advanced feedstocks, conversion biology, and carbon capture and utilization (CCU) technologies that underpin next-generation biofuels.
- **Mandate Robust and Transparent LCA**: Implement regulations requiring full life cycle accounting, including ILUC, for any biofuel to qualify for incentives, ensuring that public support truly accelerates climate progress.

For Industry and Investors

- **Embrace Collaboration**: Break down silos. Foster partnerships across the agricultural, technology, energy, and logistics sectors to create integrated value chains. Investment should focus on scalable, integrated systems, not isolated technologies.
- **Prioritize Data Infrastructure**: Invest in the digital tools and talent required to harness the power of data. The first-movers in building intelligent, efficient supply chains will gain a significant competitive advantage.
- **Adopt a Circular Mindset**: Design for circularity from the outset. View waste as a resource and seek opportunities to create multiple revenue streams from a single feedstock.

For the Research Community

- **Focus on Cross-Disciplinary Solutions**: The biggest breakthroughs will occur at the intersections of biology, data science, chemical engineering, and environmental science. Collaborative, mission-oriented research is essential.
- **Bridge the "Valley of Death"**: Actively engage with industry and policymakers to ensure that promising laboratory discoveries are translated into pilot and demonstration projects, accelerating the path to commercialization.

In conclusion, the biofuel narrative is at a turning point. The simplistic optimism of the early 2000s has been replaced by a more nuanced and sophisticated understanding of the challenges. This is a sign of maturity. By learning from past missteps, embracing a data-driven and circular approach, and fostering unwavering collaboration among all stakeholders, we can pave a viable path forward. The goal is not merely to create an alternative fuel, but to cultivate a resilient, equitable, and sustainable bio-economy that powers our world while healing our planet. The imperative is clear, the tools are within our grasp, and the time for decisive action is now.

REFERENCES

1. Biofuel Production from Conventional Feedstocks: Challenges and Alternatives. (2022). *Clean Energy Production Technologies*, 1–15. https://doi.org/10.1007/978-981-19-3582-4_1
2. Okolie, J. A. (2024). Introduction of machine learning and artificial intelligence in biofuel technology. *Current Opinion in Green and Sustainable Chemistry*. https://doi.org/10.1016/j.cogsc.2024.100928
3. Ali, M., Shabbir, K., Ali, S., Mohsin, M., Kumar, A., Aziz, M., Zubair, M., & Sultan, H. M. (2024). A new era of discovery: How artificial intelligence has revolutionized the biotechnology. *Nepal Journal of Biotechnology*. https://doi.org/10.54796/njb.v12i1.312
4. Collier, Z. A., Connelly, E. B., Polmateer, T. L., & Lambert, J. H. (2017). Value chain for next-generation biofuels: Resilience and sustainability of the product life cycle. *Environment Systems and Decisions*, *37*(1), 22–33. https://doi.org/10.1007/S10669-016-9618-1
5. Shi, Z., Ferrari, G., Ai, P., Marinello, F., & Pezzuolo, A. (2023). Artificial intelligence for biomass detection, production and energy usage in rural areas: A review of technologies and applications. *Sustainable Energy Technologies and Assessments*, *60*, 103548. https://doi.org/10.1016/j.seta.2023.103548
6. Singh, B., Srivastava, A. K., & Prakash, O. (2024). *Review on Machine Learning as a Key Technology Enabler for Sustainable Biodiesel Production* (pp. 219–238). IGI Global.
7. Bain & Company. (2017). The future of the global energy landscape. https://www.bain.com/insights/energy-trends-2030-infographic/
8. Habu, J. (2023). *Data Collection, Wrangling, and Pre-processing for AI Assurance* (pp. 321–338). Elsevier eBooks. https://doi.org/10.1016/b978-0-32-391919-7.00022-6
9. Agrawal, P., Gnanaprakash, R., & Dhawane, S. H. (2024). Prediction of biodiesel yield employing machine learning: interpretability analysis via shapley additive explanations. *Fuel*. https://doi.org/10.1016/j.fuel.2023.130516
10. Liu, W., Zhao, R., Li, H., & Huang, J. (2018). *Portable Substance Analysis based on Computer Vision, Spectroscopy, and Artificial Intelligence*. https://patents.google.com/patent/US10664716B2/en
11. Naira, V. R., Mahesh, R., Panda, S. K., & Maiti, S. K. (2020). *Biorefinery Approaches for the Production of Fuels and Chemicals from Lignocellulosic and Algal Feedstocks* (pp. 141–170). Springer. https://doi.org/10.1007/978-981-15-1804-1_7
12. Motadayen, M., Nehru, D., & Agarwala, S. (2024). Advancing sustainability: Biodegradable electronics and materials discovery through artificial intelligence. *International Journal of AI for Materials and Design*, *1*, 3173. https://doi.org/10.36922/ijamd.3173
13. Ali, S. O., Hashmi, Z., Usman, T. M. M., Muhammad, A., Idriss, I. M., Abbas, S. H., & Hassan, M. (2024). Exploring biomass conversion technologies: From raw materials to valuable products. *Amplitudo*, *3*(2), 87–96.
14. Lawal, Q. (2025). Leveraging artificial intelligence to enhance process control and improve efficiency in manufacturing industries. *International Journal of Computer Applications Technology and Research*. https://doi.org/10.7753/ijcatr1402.1002
15. Badrinarayanan, A. (2024). AI in manufacturing: Driving operational excellence. *International Journal of Scientific Research in Computer Science, Engineering and Information Technology*, *10*(6), 585–594. https://doi.org/10.32628/cseit24106199
16. Плясов, С. В., & Клопов, І. О. (2024). Transforming industries with artificial intelligence: Practical aspects. *Pìdpriêmnictvo Ta Ìnnovacìï*. https://doi.org/10.32782/2415-3583/31.8

17. Nadaf, S. (2024). AI for predictive maintenance in industries. *International Journal For Science Technology And Engineering*, *12*(6), 2013–2017. https://doi.org/10.22214/ijraset.2024.63442
18. Patil, D. T. (2025). *Artificial Intelligence-Driven Predictive Maintenance In Manufacturing: Enhancing Operational Efficiency, Minimizing Downtime, And Optimizing Resource Utilization*. https://doi.org/10.2139/ssrn.5057406
19. Aviso, K. B. (2024). *Overview of LCA—History, Concept, and Methodology*. Elsevier BV. https://doi.org/10.1016/b978-0-323-90386-8.00113-3
20. Muralikrishna, I. V., & Manickam, V. (2017). *Life Cycle Assessment* (pp. 57–75). https://doi.org/10.1016/B978-0-12-811989-1.00005-1
21. Bassey, K. E., Juliet, A. R., & Stephen, A. O. (2024). AI-Enhanced lifecycle assessment of renewable energy systems. *Engineering Science & Technology Journal*, *5*(7), 2082–2099. https://doi.org/10.51594/estj.v5i7.1254
22. Santibañez-Aguilar, J. E., González-Campos, J. B., Ponce-Ortega, J. M., & Serna-González, M. (2011). *Optimal Biorefinery Planning Considering Simultaneously Economic and Environmental Objectives* (Vol. 29, pp. 1653–1657). Elsevier. https://doi.org/10.1016/B978-0-444-54298-4.50109-4
23. Xia, J., Liu, J., & Zhuang, Y. (2022). Opportunities and challenges for fermentation optimization and scale-up technology in the artificial intelligence era. *Chinese Journal of Biotechnology*, *38*(11), 4180–4199. https://doi.org/10.13345/j.cjb.220606
24. Baskar, G., Sivakumar, R., Kadry, S., Dong, C., Singhania, R. R., Praveenkumar, R., & Sathendra, E. R. (2024). Comparative studies on modeling and optimization of fermentation process conditions for fungal asparaginase production using artificial intelligence and machine learning techniques. *Preparative Biochemistry & Biotechnology*, 1–7. https://doi.org/10.1080/10826068.2024.2367692
25. Khaleghi, M. K., Pour Savizi, I. S., Lewis, N. E., &Shojaosadati, S. A. (2021). Synergisms of machine learning and constraint-based modeling of metabolism for analysis and optimization of fermentation parameters. *Biotechnology Journal*, *16*(11), 2100212. https://doi.org/10.1002/BIOT.202100212

9 AI/ML-powered Biogas and Methane Generation

Leena Ajay Sawaikar, Swati Mayur Keny, Celina Nazareth, Kumarappan Chidambaram, Gnanaprakash Kalimuthu, and Panneerselvam Theivendren

9.1 INTRODUCTION

The need to have sources of renewable energy has brought about great enthusiasm in the production of biogas, especially the production of methane, by using organic waste through an anaerobic digestion process. Such a process provides a variety of sustainable energy source besides solving the environmental issue by reducing greenhouse gas emissions due to organic waste (Ahmad et al., 2021; Ali et al., 2023; Cinar et al., 2021). Although the traditional approaches to the biogas production have provided significant potential, further progress in artificial intelligence (AI) and machine learning (ML) has provided new opportunities to streamline the process of biogas and methane production (Figure 9.1).

AI and ML have been useful in revolutionising different industrial processes and their application in the production of biogas is not an exception. These technologies allow gaining new information about how to optimise the composition of feedstock, microbial activities, and system parameters by analysing huge amounts of data obtained in the course of the anaerobic digestion process. Algorithms of ML can now forecast the yield of biogas, optimise the energy, and increase the efficiency of methane generation, which minimises operational costs and augments system stability to a large degree (Czekała, 2022; Lisowyj & Wright, 2020; Obileke et al., 2021).

The application of AI/ML in the biogas production begins with the data collection. The sensors and monitoring systems are used to give real-time data of temperature, pH levels, rate of organic loading, and other specific parameters that are vital in the process of fermentation. This information is subsequently utilised to train predictive models that fail to predict the behaviour of the anaerobic digester under different conditions. These models are able to maximise the contribution of organic wastes, able to control microbial communities participating in the digestion process, and

 DOI: 10.1201/9781003684374-9

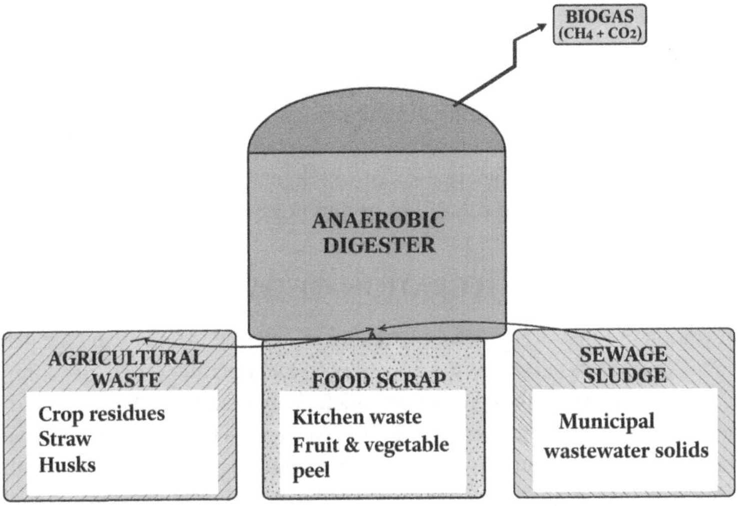

FIGURE 9.1 The process of biogas production from various organic waste materials. Agricultural waste, food scraps, and sewage sludge are fed into the digester, producing biogas as an energy source.

adjust to changing environmental conditions (e.g., the changing temperature or the humidity level) (Kunatsa, 2025; Takase et al., 2021).

Also, the AI/ML-based models may forecast possible failures or inefficiencies in the system to conduct proactive maintenance and management. These models are real time in nature, and thus it is possible to make updates immediately and avoid unnecessary downtimes or energy inefficiencies. These AI systems can also reduce the overall environmental footprint of the process by optimising the production of methane, as well as using less energy by changing the parameters of operation (Akash et al., 2025; Fajobi et al., 2022; Swami et al., 2025).

A second promising use of AI/ML is in optimisation of co-digestion whereby organic waste of various types are used in order to enhance the overall amount of biogas. The ML algorithms could be used to analyse the different feedstock blends and determine the most efficient blend to produce methane considering the nutrient content of the feedstock, its biodegradability, and compatibility with other wastes. This causes a higher production of the methane using minimal resource use which makes biogas production even more sustainable.

Essentially, the technologies of AI/ML are transforming biogas and methane production through the optimisation of the anaerobic digestion process, system optimisation, and real-time modifications on the basis of predictive data. This solution will probably become a part of large-scale biogas with the further development of technology, which will help in shifting to more efficient and cleaner renewable energy systems faster (Farzin & Moghaddam, 2025; Saju et al., 2025; Sharma et al., 2022).

This chapter is aimed at discussing the possibilities of AI and ML technologies to maximise the process of producing biogas and methane. It is hoped that through the combination of predictive model, real-time data analysis, and system optimisation methods, the efficiency, sustainability, and cost efficiency of the anaerobic digestion process would be improved. Also, this study aims at determining the most significant factors that affect the production of methane and to come up with the algorithms which can maximise the biogas production and reduce the environmental effect. The final objective is to help in the creation of intelligent, automated systems capable of enhancing the scalability and reliability of the biogas production.

9.2 SENSORS AND INTEGRATION OF DATA

The integration of sensors and data collection is a key role in enhancing the production of biogas. Constant monitoring of parameters like temperature, pH, gas composition, and organic loading rate among other parameters are continuously monitored by sensors and have an effect on anaerobic digestion process. These sensors can give data in real time and, thus, they can be adjusted immediately to provide the best conditions in generating methane. With this data incorporated into the AI/ML systems, a system will be able to learn and adjust to changes in the environment and enhance efficiency and consistency of biogas generation. The predictive models were based on the accurate and sustained data collection, which increases the reliability and sustainability of biogas plants (Ahmad et al., 2021; Akash et al., 2025; Rutland et al., 2023; Zhang et al., 2024).

9.2.1 Sensors Applied in Biogas Plants

Biomass sensors are also used to detect the anaerobic digestion in biogas plants whereby proper conditions are maintained to produce the maximum amount of methane (Figure 9.2). Common types include (Khatri & Khatri, 2022; Song et al., 2024):

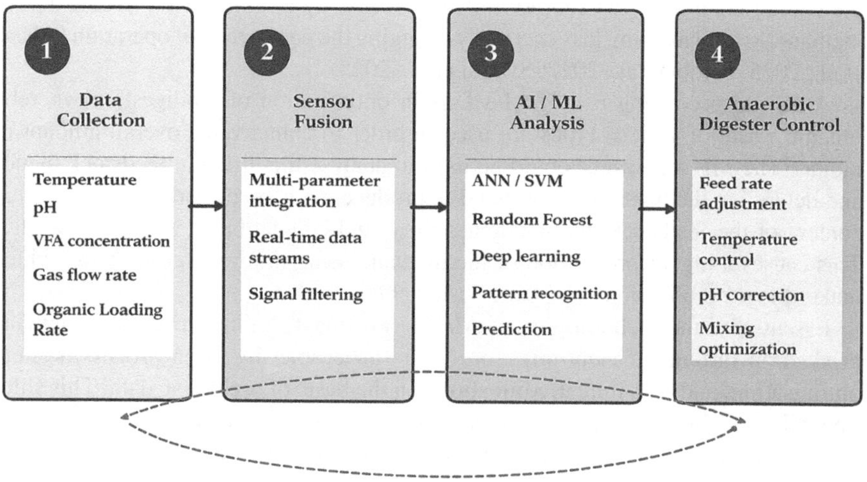

FIGURE 9.2 The process of AI/ML optimisation in anaerobic digestion. Sensors collect data (e.g., pH, temperature), which is analysed by AI/ML algorithms to enhance the performance of the anaerobic digester.

Temperature Sensors: These are necessary in order to keep the optimum temperature of the microbial activity between 35 and 40°C of mesophilic digestion or 50–60°C of thermophilic digestion. They make sure that the digester is running effectively to avoid the fluctuation of temperatures that may affect the production of methane.

pH Sensors: pH value is an important parameter in the health of microbes when subjected to anaerobic digestion. The optimal pH should be between 6.8 and 7.4 and a pH sensor will ensure that the pH remains within the correct range and, therefore, when it deviates, the sensor will activate some form of adjustment.

Gas Composition Sensors: This type of sensor detects the levels of various gases such as methane (CH_4), carbon dioxide (CO_2) and hydrogen sulphide (H_2S) and can be used to understand the level of efficiency of the digestion process. Particularly methane sensors are essential in measuring quality and quantity of biogas generated.

Biological Oxygen Demand (BOD) Sensors: These are used to determine the amount of organic content in the waste which is then used to adjust the amount of feedstock added and the production of biogas becomes more effective.

All these sensors lead to a solid data collection framework that is used to make real-time decisions to optimise the production of methane within biogas plants.

9.2.1.1 Monitoring and Data Collection Systems in Real Time

Contemporary biogas plants have real-time data collection and monitoring systems. These systems collect information on different sensors, such as temperature sensors, pH sensors, and gas composition sensor and sent it to a control unit located centrally. Constant monitoring allows operators to notice anomalies, modify parameters as well as forecast future performance. The AI-based systems will be capable of processing this data to deliver actionable information, improve the anaerobic process of digestion, and eliminate system failures. Moreover, the possibilities of immediate adjustments depending on the real-time data allow to sustain the constant level of methane production, increase its efficiency, and ensure the sustainability of biogas production processes in the long run.

9.2.2 Process Optimisation AI/ML Algorithms

AI and ML algorithms are critical to optimisation of biogas production processes through the analysis of complex data on different sensors. These algorithms are used to detect the best conditions in which the generation of methane can occur. Through constant learning about real-time information, they are able to ensure they forecast the most effective operational parameters, such as the feedstock composition, temperature, and pH levels. These parameters are automated using AI systems so that microbial activity and energy production are optimised. AI/ML application and optimisation of the processes not only increase the biogas yield, but also increase the stability of the system and reduce the operational expenses, which facilitate sustainability (Chiu et al., 2022).

9.3 FEEDSTOCK ANALYSIS ML MODELS

ML systems can be of great help in the analysis of the feedstock used to create biogas, to maximise its mix in terms of the output of methane. Feedstock which is usually made of organic waste materials like food scraps, agricultural residues, and sewage sludge is highly variable in terms of its nutrient content and biodegradability. The ML algorithms may be used to study historical and real-time data on these feedstocks to determine the main features, e.g., moisture content, organic carbon, and nitrogen that affect the efficiency of biogas production (Kothale & Sadgir, 2025).

Regression analysis or classification algorithms are supervised learning models that are utilised to forecast biodegradability of various feedstocks and their effects on biogas yield. These models can forecast the combination of feedstock that is most likely to optimise the production of methane based on the data about previous feedstocks. It is also possible to employ the unsupervised learning methods, such as clustering, to determine patterns in the diverse feedstock mixtures, which can then be used to establish which combination of various types of organic waste can yield the most successful co-digestion, or the combination of several types of organic waste.

As sensors of feedstock properties continue to obtain data, ML models have the potential to fully optimise the process of converting feedstock to methane, which is dynamically prepared to manage the input of waste based on real-time data. This increases the utilisation of resources and enhances the production of methane, which increases the economic and environmental sustainability of biogas plants.

9.3.1 Methane Yield Forecasting Predictive Modelling

Methane yield forecasting involves predictive modelling that depends on past and current records to predict the amount of methane that will be produced using biogas plants. ML algorithms predict the future methane production in different conditions by integrating data to characterise the composition of the feedstock, the conditions of the digester, and the activity of microbes. These types are based on analysing trends and correlations in the data using the regression analysis or neural networks. Predictive models can improve the efficiency of biogas plants and minimise waste by making adjustments in advance amid proper forecasting of methane production that would allow the operators to control the parameters of different biogas plants. This forecasting potential will contribute to the sound decision-making, whereby the biogas generation will be constant and reliable all year round (Olatunji et al., 2023).

9.3.2 Optimisation of the Anaerobic Digestion

The parameter of anaerobic digestion must be optimised in order to achieve maximum biogas and stability of the systems. The parameters that have a direct influence on the microbial activity and production of methane include temperature, pH, organic loading rate, and hydraulic retention time. The systems of AI and ML are essential to ensure the parameters are continuously monitored and changed to achieve optimal conditions. With real-time data gathering and analysis, AI-based models anticipate the optimal environment in every digester, and proper digestion of microbes and

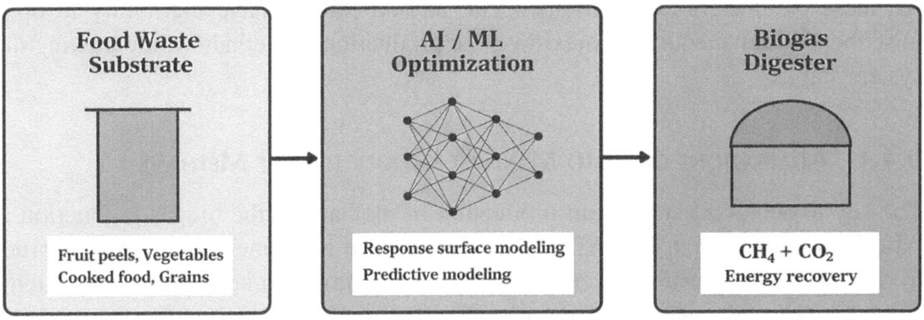

FIGURE 9.3 The integration of AI/ML algorithms with biogas production. The laptop represents data analysis, while the vegetables in the container symbolise organic waste for anaerobic digestion in the bioreactor.

reliable methane production. The optimisation process results in the achievement of not only better methane yield but also the overall efficiency of biogas production, and it improves the sustainability (Figure 9.3).

9.4 EFFECTS OF TEMPERATURE, PH, AND ORGANIC-LOADING RATES

The rate of anaerobic digestion with regard to temperature, pH, and organic loading is important parameters that directly influence the productivity of biogas production. These parameters have an effect on the activity of microorganisms that break down organic matter and produce methane.

Temperature: Temperature is a major aspect that influences the microbial metabolism. Advantages of mesophilic digestion (between 35 and 40°C) and thermophilic digestion (50–60°C) are both unique. Stability is provided in mesophilic conditions and rate of biogas production is increased in thermophilic conditions. AI-based systems observe temperature changes and modify heating systems to provide the optimal thermal conditions in which microbial activity can occur (Yildirim & Ozkaya, 2023).

pH: The digester should have a pH range of 6.8–7.4, which would be the best to have all the microbes. Any variation below this range may slow down or even stop the digestion process. Constantly changing the AI models based on the pH information, it helps to keep the balance between the two to avoid instability in the system.

Organic Loading Rates (OLR): OLR is used to refer to the level of organic material that is added to the digester. High OLR may also cause overloading and hence decrease the efficiency of the system. Through the use of data related to the composition of the feedstock and the performance of the digesting machine, ML systems can be used to optimise OLR in such a way as to avoid disruption of the process, and instead achieve a greater biogas yield.

All these parameters are interdependent, and AI models have the ability to optimise them simultaneously to maximise the production of methane while making the system stable.

9.4.1 Adaptations of AI to Maximise Production of Methane

The key to achieving maximum production of methane in the biogas production is AI-driven adjustments. The AI models can achieve real-time adjustments in order to maximise the digestion by continuously monitoring such key parameters as temperature, pH, and organic loading rate. As an illustration, AI systems can be used to turn on heating or cooling systems in case of temperature anomaly to ensure that the temperature remains at optimum levels. Equally, when the pH levels either fall or exceed the optimal levels, AI will be able to run acid or alkaline dozers to maintain the equilibrium. These computerised adaptations make sure that the anaerobic digestion process is running at its best resulting in the yielding of more methane, an increase in the efficiency of the system, and an increase in sustainability of the resources (Egbuna et al., 2025).

9.4.2 Optimisation to Co-digest with AI

Co-digestion, which is the act of mixing various kinds of organic material, improves the amount of methane by making the microbial environment more favourable and the degradation of the feedstock is improved. AI is also useful in the optimisation of co-digestion, whereby information of different feedstocks is analysed to determine the most appropriate combinations that can result in maximum methane production. ML systems provide constant cheques on such variables as organic content, moisture, and nutrient balance and adjust the feedstock mixture. The optimisation of these variables dynamically guarantees that the co-digestion process is efficient and sustainable so that the production of biogas is maximised and the operation costs are minimised.

9.5 ADVANTAGES OF CO-DIGESTION IN THE PRODUCTION OF METHANE

The concept of co-digestion, which involves the simultaneous treatment of several classes of organic waste, has a number of advantages in the production of biogas, especially in the maximisation of methane production. Among the main benefits, there is the improved yield of methane. By mixing into various feedstocks, e.g., agricultural waste with food scrap or sewage sludge, a more balanced nutrient profile is formed, and a variety of microbial communities is formed that can more effectively break down organic matter. This leads to an increased biogas generation as compared to the single-feedstock digestion (Auyelbekov et al., 2025).

Another major advantage is better stability of the processes. The rate of degradation of various feedstocks is different and co-digestion assists in balancing these rates and ensures that the rate is not overloaded and the system produces methane.

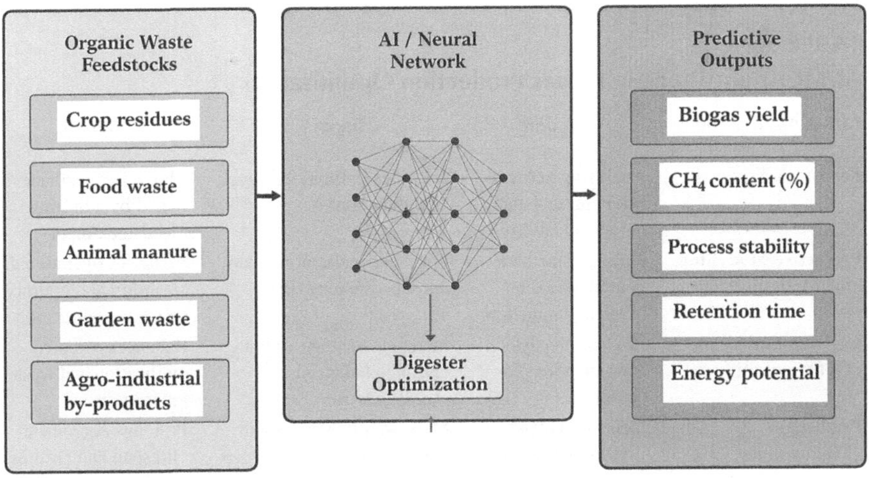

FIGURE 9.4 The process of AI/ML-driven optimisation for biogas production. The organic waste pile represents feedstock, while the laptop shows data analysis and graphs used to optimise anaerobic digestion.

With the presence of different organic materials, the digester becomes less sensitive to changes in pH or the volatile OLR, resulting in the stabilisation of the digestion process (Figure 9.4).

Also, co-digestion is a way of improving the benefits of waste management. This approach has the potential to process a wider variety of organic wastes, thus eliminating external disposal and creating a circular economy, by using a broader range of sources of waste. Finally, the co-digestion will also contribute to the decreased general cost of feedstock purchase since it will be possible to utilise less expensive or waste products that would not be used otherwise. All these combines to make co-digestion a very effective and environmentally friendly process in the generation of methane.

9.5.1 AI/ML Techniques of Feedstock Mixture Optimisation

The AI and ML techniques are important in maximising the feedstock mixes to be co-digested. AI algorithms can be used to determine the most efficient combinations of different types of organic waste to produce methane by analysing historical and real-time data. The ML approaches take into account the nutrient content, biodegradability, and the amount of moisture to forecast the interactions between various feedstocks on a digester anaerobic environment. These models also allow maximising the percentage ratio of each feedstock so that the microbial community grows, the maximum amount of methane is produced, and acid formation or nutrient imbalance is avoided. Optimisation using AI will result in the production of more efficient and cost-effective biogas (Table 9.1).

TABLE 9.1
AI/ML Applications in Biogas Production Optimisation

AI/ML Application	Function	Impact	Benefit
Feedstock Analysis	Analysing nutrient content, moisture, biodegradability	Identifies optimal feedstock combinations	Maximises methane yield by adjusting feedstock input
Predictive Modelling for Methane Yield	Forecasts methane output based on system parameters	Ensures consistent methane production over time	Enhances operational stability and optimises methane generation
Co-Digestion Optimisation	Analyses feedstock combinations for co-digestion	Increases methane output through balanced feedstock use	Enhances resource utilisation and waste management
Predictive Maintenance	Identifies potential system failures or inefficiencies	Minimises downtime and reduces maintenance costs	Prolongs equipment lifespan and ensures smooth operation

9.5.2 Proactive Maintenance and Prevention of Failure

Biogas production systems require predictive maintenance and failure prevention to help them become highly efficient in the long term. AI models utilise real-time sensor data to draw patterns that can be used to predict possible failures or inefficiencies. These models can identify the onset of wear and tear in equipment or inefficient operation by tracking such aspects as heat, pressure, gas, and so on. Predictive maintenance allows the operator to resolve the problem before it leads to expensive system failure or breakdowns. This proactive strategy minimises downtime, the cost of repairs, and enhances the general level of reliability of the biogas production plants.

9.6 UNCOVERING THE SYSTEM INEFFICIENCIES USING AI

The importance of AI lies in its ability to identify inefficiencies in biogas production systems and implement more specific and timely solutions. AI systems can detect small changes in the system parameters (i.e., gas flow rates, pH, temperature, and OLR) to indicate inefficiencies (Figure 9.5).

Algorithms of ML are used to analyse vast amounts of data to identify trends or anomalies that hairy operators might miss, such as minor changes in gas content or digester malfunctions (Mukasine, 2024).

As an example, AI models will be able to notice situations when the production of methane is lower than expected, even with a stable operation. This may be a sign of microbial imbalances, faeces problems, or mechanical problems. AI may also forecast when energy usage may skyrocket or equipment is not functioning efficiently, and this can be used to identify areas where efficiency could be enhanced. These systems allow operators to change parameters on the fly to optimise resource utilisation and get as much as possible out of the system by constantly measuring data in order to maximise the methane output.

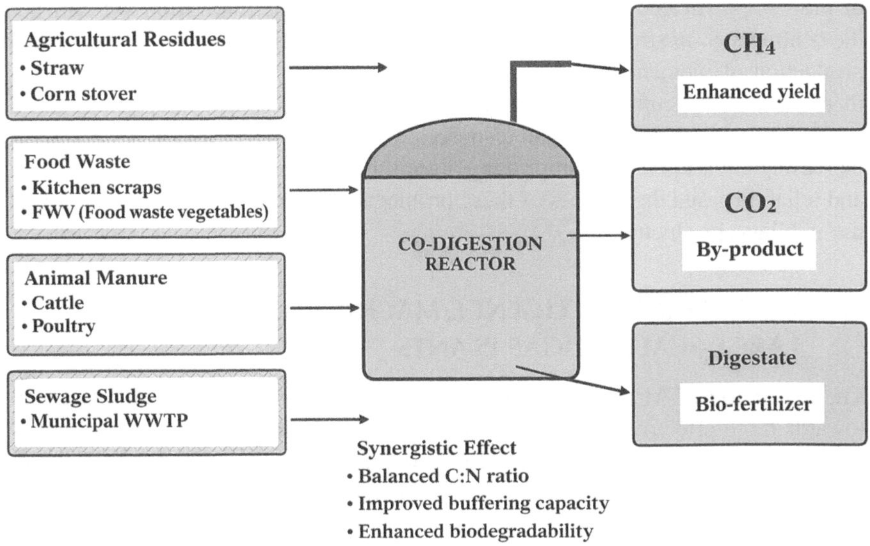

FIGURE 9.5 The co-digestion in anaerobic digestion. Various organic waste materials, such as food scraps and agricultural residues, are processed in the digester to produce methane (CH4) for energy.

Detection of inefficiencies based on AI is also useful in the discovery of process bottlenecks. Indicatively, when biogas production is delayed after feeding, the AI models can be used to trace the source of the delay; the cause could be a failure of the digester, a feeding stock ratio, or even microbial action, and corrective measures can be taken immediately. This ensures maximum performance by the system and reduction of waste and unwarranted expenditures.

9.6.1 Predictive-based Proactive Maintenance Policies

Predictive models can be used to implement proactive maintenance strategies that help biogas plant operators to minimise unexpected downtimes and decrease maintenance expenses. The predictive maintenance systems of AI are based on sensor data that is analysed to predict when the equipment, e.g., pumps, mixers, or valves, might fail or demand a service. By forecasting these events, prior to their occurrence, AI enables operators to schedule maintenance when the demand is low, so that the production of biogas will not be disrupted. This will not only increase the life cycle of equipment but also ensure that the system is running smoothly and productively. Through proactive maintenance, it leads to a reduced number of breakdowns, reduced cost of repair, and an overall performance of the plant (Xu et al., 2025).

9.6.2 Potential Trends and Theses

The further development of AI/ML in biogas generation is favourable, and the main trends are the further automation, processing of real-time data and the introduction of more sophisticated predictive algorithms that fuel the innovation. AI will be at the centre stage in streamlining massive biogas production facilities, enhancing the production of methane, and providing long-term energy production. Nevertheless, there are still difficulties of data quality, complexity of system integration, and the requirement of competent human resources. With the development of AI models, the increasing attention paid to improving algorithms in accordance with their accuracy and reliability, and the success of these problems will be the key to their widespread use in biogas production.

9.7 ARTIFICIAL INTELLIGENCE/MACHINE LEARNING IN LARGE-SCALE BIOGAS PLANTS

The realisation of AI and ML in massive biogas facilities is a revolutionary move towards more effective and sustainable generation of energy. Artificially intelligent models are also being used to streamline different components of biogas production, including feedstock management to the maintenance of the system. On a large-scale operation, where the variables to consider include multiple digesters, differences in feedstock types, and changes in the environment, AI/ML is able to handle and optimise the variables in real time (Table 9.2).

AI models allow monitoring the performance of the system in real time, changing the values of temperature, pH, and OLR to achieve the maximum production of methane. Sensors provide ML algorithms with the ability to analyse huge volumes of data, anticipate the risk of system failures and schedule maintenance even before a system falls into the wrong. In addition, AI may be used to optimise the co-digestion process and assist in defining the most promising feedstock combinations to get a maximum yield of methane.

TABLE 9.2
Types of Sensors Used in Biogas Plants

Sensor Type	Parameter Monitored	Purpose	Ideal Range	Impact on Biogas Production
Temperature Sensor	Temperature	To maintain the ideal temperature for microbial activity	35–40°C (mesophilic), 50–60°C (thermophilic)	Ensures optimal microbial metabolism for biogas generation
pH Sensor	pH Levels	To monitor and maintain stable pH for microbial health	6.8–7.4	Prevents system instability and maintains microbial activity
Gas Composition Sensor	Methane, CO_2, H_2S	To measure the composition of biogas produced	Methane: >50% for efficient production	Helps assess efficiency of methane generation
BOD Sensor	Biological Oxygen Demand (BOD)	To monitor organic content in feedstock	Variable depending on feedstock	Optimises feedstock input for improved biogas yield

The operational cost in large-scale biogas plants is also reduced through the integration of AI/ML, which predicts the energy consumption, reduction in waste, and overall enhancement of the operational efficiency of the plant. The capability to increment the AI solutions within these plants can guarantee that biogas generation is more efficient, affordable, and sustainable. Nevertheless, the integration needs a robust data infrastructure, human resources, and deep knowledge of AI technology as well as biogas processes to achieve its potential (Ahmad et al., 2024; Kabeyi & Olanrewaju, 2022; Onu et al., 2023).

9.7.1 Possible Obstacles and Recommended Interventions

Although the effectiveness of the use of AI/ML in biogas plants is rather considerable, there are a number of obstacles to its implementation. The major issue associated with AI infrastructure and sensor systems is the scalability of this technology to small operators due to the high cost of initial investment. The quality and consistency of data also become a problem because the quality of AI models depends on the high quality and correct data collected by the sensors. In addition, insufficient human resources who are knowledgeable in biogas procedures as well as AI/ML solutions may restrain adoption. The possible solutions are that the government can offer subsidies or grants to facilitate adoption, invest in employee training to upgrade skills, and invest in the standardisation of the collection of data to enhance the accuracy of the models.

9.8 DISCUSSION

The implementation of AI and ML technologies in the biogas generation is an important step towards the optimisation of the process of anaerobic digestion. This transformation is centred on real-time data collection and sensor integration. The sensors are used to monitor important parameters such as temperature, pH, and organic loading rate which are critical in creating the best conditions to support the activity of the microbes that produce methane. The assimilation of this data into AI/MLs allows the operators to constantly modify parameters, which promotes the stability of the system and maximises the amount of methane. Moreover, the predictive modelling on forecasting the yield of methane allows operators to predict the output of biogas in different circumstances so that they can make proactive changes. The systems based on AI/ML can be used to improve the efficiency of the processing of biogas as well as decrease the negative environmental effect of biogas production by means of the optimisation of energy use and the costs of operations.

Another area, in which AI/ML has immense potential, is that of co-digestion optimisation. The mixture of the various organic waste leads to enhanced performance of the microbes and higher levels of production of methane. ML methods may process different feedstocks and identify the most suitable combination to co-digest considering such factors as biodegradability, moisture level, and nutrient balance. The result of this process is a more efficient and sustainable model of producing biogas by utilising as much as possible of the resources and minimal wastage. With the development of AI technologies, even more advanced achievements in predictability and reliability of models will lead to the further introduction of AI in large-scale biogas processes.

9.9 CONCLUSION

The application of AI and ML technologies to biogas manufacture is of immense importance to the efficiency, sustainability, and cost-effectiveness of the process of producing methane. The AI systems can offer real-time data, which enables the constant optimisation of the key parameters that include temperature, pH, and organic loading rate to enhance the stability and performance of anaerobic digesters. Predictive maintenance, optimisation using co-digestion, predictive modelling in terms of methane yield forecasting, helps in the overall efficiency in the operations of the biogas plants, by minimising working downtime and minimising the use of energy. Such developments are in line with the global transition towards cleaner and renewable sources of energy, providing a sustainable energy solution to meet the increasing energy demand. Yet, there are several barriers to the popularisation of AI in the production of biogas, such as the high initial expenses, the problem of the quality of data, and the necessity to have qualified specialists. Government aid, training, and standardisation of data collection will be important in overcoming these barriers so as to fully realise the potential of AI in biogas plants. Since AI-based models are still on the way of development, it is possible that in the future, they will become a part of the optimisation of large-scale biogas plants, which will result in a transition to more sustainable energy supply worldwide.

REFERENCES

Ahmad, A., Yadav, A. K., Singh, A., & Singh, D. K. (2024). A comprehensive machine learning-coupled response surface methodology approach for predictive modeling and optimization of biogas potential in anaerobic Co-digestion of organic waste. *Biomass and Bioenergy*, *180*, 106995.

Ahmad, M. S., Ali, M. S., & Abd Rahim, N. (2021). Hydrogen energy vision 2060: Hydrogen as energy Carrier in Malaysian primary energy mix–Developing P2G case. *Energy Strategy Reviews*, *35*, 100632.

Akash, M. A. H., Hossain, M. S., Islam, M. M., Islam, K. S., Islam, M. N., & Sony, S. Y. (2025). Optimization of methane production in anaerobic digestion: A hybrid approach using experimental analysis, process simulation and machine learning. *Results in Engineering*, 28, 107464.

Ali, S., Yan, Q., Razzaq, A., Khan, I., & Irfan, M. (2023). Modeling factors of biogas technology adoption: A roadmap towards environmental sustainability and green revolution. *Environmental Science and Pollution Research*, *30*(5), 11838–11860.

Auyelbekov, O., Kozbakova, A., Yessentaev, K., Merembayev, T., & Igibayev, K. (2025). Integration of mathematical modeling and artificial intelligence to predict thermal conditions and optimize the operation of biogas plants. *Engineered Science*, *36*, 1603.

Chiu, M.-C., Wen, C.-Y., Hsu, H.-W., & Wang, W.-C. (2022). Key wastes selection and prediction improvement for biogas production through hybrid machine learning methods. *Sustainable Energy Technologies and Assessments*, *52*, 102223.

Cinar, S., Cinar, S. O., Wieczorek, N., Sohoo, I., & Kuchta, K. (2021). Integration of artificial intelligence into biogas plant operation. *Processes*, *9*(1), 85.

Czekała, W. (2022). Biogas as a sustainable and renewable energy source. In J. M. Sánchez Hervás, J. A. Jurado Alameda, & M. D. Pérez Camacho (Eds.), *Clean fuels for mobility* (pp. 201–214). Springer.

Egbuna, I. K., Dalhatu, A., Nwafor, C. A., Ezeifegbu, C., Nasir, F., & Iheakanwa, F. (2025). Application of artificial intelligence in bioenergy supply chain management from feedstock collection to power generation. *World Journal of Advanced Engineering Technology and Sciences, 16*, 141–153.

Fajobi, M. O., Lasode, O. A., Adeleke, A. A., Ikubanni, P. P., & Balogun, A. O. (2022). Effect of biomass co-digestion and application of artificial intelligence in biogas production: A review. *Energy Sources, Part A: Recovery, Utilization, and Environmental Effects, 44*(2), 5314–5339.

Farzin, F., & Moghaddam, S. S. (2025). Enhancing energy circularity and achieving carbon neutrality in water resource recovery facilities through an AI-driven optimization framework for biogas production. *Journal of Environmental Management, 393*, 127076.

Kabeyi, M. J. B., & Olanrewaju, O. A. (2022). Biogas production and applications in the sustainable energy transition. *Journal of Energy, 2022*(1), 8750221.

Khatri, N., & Khatri, K. K. (2022). Artificial intelligence for modeling and optimization of the biogas production. *Artificial Intelligence for Renewable Energy Systems*, 93–113.

Kothale, A., & Sadgir, P. (2025). Application of artificial intelligence and machine learning with international guidelines for greenhouse gas reduction in wastewater treatment. *International Journal of Environmental Science and Technology*, 1–19.

Kunatsa, T. (2025). The role of artificial intelligence in greening biogas operations. In T. Chikozho & T. Chikozho (Eds.), *Innovations in the global biogas industry* (pp. 361–397). Elsevier.

Lisowyj, M., & Wright, M. M. (2020). A review of biogas and an assessment of its economic impact and future role as a renewable energy source. *Reviews in Chemical Engineering, 36*(3), 401–421.

Mukasine, A. (2024). *Machine learning based-predictive modeling with control mechanisms for enhanced biogas yield through internet of things-enabled sensor networks.* University of Rwanda.

Obileke, K., Nwokolo, N., Makaka, G., Mukumba, P., & Onyeaka, H. (2021). Anaerobic digestion: Technology for biogas production as a source of renewable energy—A review. *Energy & Environment, 32*(2), 191–225.

Olatunji, K. O., Madyira, D. M., Ahmed, N. A., Adeleke, O., & Ogunkunle, O. (2023). Modeling the biogas and methane yield from anaerobic digestion of Arachis hypogea shells with combined pretreatment techniques using machine learning approaches. *Waste and Biomass Valorization, 14*(4), 1123–1141.

Onu, P., Mbohwa, C., & Pradhan, A. (2023). Artificial intelligence-based IoT-enabled biogas production. *2023 International Conference on Control, Automation and Diagnosis (ICCAD).*

Rutland, H., You, J., Liu, H., Bull, L., & Reynolds, D. (2023). A systematic review of machine-learning solutions in anaerobic digestion. *Bioengineering, 10*(12), 1410.

Saju, L., Selvaraj, D., & Vairaperumal, T. (2025). Artificial intelligence and machine intelligence: Modeling and optimization of bioenergy production. In A. K. Bairwa, P. K. Guptaa, & S. Tanwar (Eds.), *Computer vision and machine intelligence for renewable energy systems* (pp. 163–176). Elsevier.

Sharma, A. K., Ghodke, P. K., Goyal, N., Nethaji, S., & Chen, W.-H. (2022). Machine learning technology in biohydrogen production from agriculture waste: Recent advances and future perspectives. *Bioresource Technology, 364*, 128076.

Song, C., Cai, F., Yang, S., Wang, L., Liu, G., & Chen, C. (2024). Machine learning-based prediction of methane production from lignocellulosic wastes. *Bioresource Technology, 393*, 129953.

Swami, S., Suthar, S., Singh, R., Thakur, A. K., Gupta, L. R., & Sikarwar, V. S. (2025). Integration of anaerobic digestion with artificial intelligence to optimise biogas plant operation. *Environment, Development and Sustainability, 27*(5), 9773–9803.

Takase, M., Kipkoech, R., & Essandoh, P. K. (2021). A comprehensive review of energy scenario and sustainable energy in Kenya. *Fuel Communications*, *7*, 100015.

Xu, Y., Yazdinejad, A., Wang, H., & Kong, J. D. (2025). *From detection to decision: A systematic literature review of AI and machine learning evolution in methane modelling* (SSRN Scholarly Paper No. 5218753). Social Science Research Network. https://doi.org/10.2139/ssrn.5218753

Yildirim, O., & Ozkaya, B. (2023). Prediction of biogas production of industrial scale anaerobic digestion plant by machine learning algorithms. *Chemosphere*, *335*, 138976.

Zhang, P., Zhang, T., Zhang, J., Liu, H., Chicaiza-Ortiz, C., Lee, J. T., He, Y., Dai, Y., & Tong, Y. W. (2024). A machine learning assisted prediction of potential biochar and its applications in anaerobic digestion for valuable chemicals and energy recovery from organic waste. *Carbon Neutrality*, *3*(1), 2.

10 Enhancing Biohydrogen Production Yield through Precision Fermentation and AI/ML

K. Arul, Somasundaram Prasath Alias Surendhar, Alagarsamy Sivasamy, S. Kannan, K. Selvakumar, J. Vinothkumar, and S. Rajkumar

10.1 INTRODUCTION

As the demand for clean energy rises, biological gas emerges as a promising alternative to conventional energy sources. Traditional methods for hydrogen production are often inefficient; however, controlled bioprocessing enables the optimization of hydrogen-producing microbial systems, enhancing yield and process stability (Wang & Yin, 2019; Abbasi et al., 2025).

Data-driven analysis of bioprocesses reveals crucial trends and predictive insights, unlocking new opportunities for innovation and optimization. By continuously monitoring and analysing process variables, it becomes possible to optimize fermentation parameters in real time, ensuring that conditions remain ideal for maximum hydrogen production. Predictive modelling enables the early detection and mitigation of potential issues, enhancing process stability and consistency.

Furthermore, artificial intelligence (AI)-driven insights facilitate the design of more efficient microbial strains through targeted genetic modifications, improving productivity and system resilience. In addition, data analytics aids in identifying novel substrates and innovative process configurations that enhance biohydrogen yield, thereby advancing the overall efficiency and sustainability of bioenergy systems.

The synergistic combination of precision fermentation and AI/ML technologies holds great promise for revolutionizing biohydrogen production. Through combining these approaches, scientists aim to overcome existing challenges and produce microbially generated clean energy, addressing global energy demands and advancing a more sustainable, low-carbon future (Manuel et al., 2024).

DOI: 10.1201/9781003684374-10

10.1.1 Biocatalytic Pathways for Alternative Fuel Production

Biological molecular hydrogen (H_2) generation offers an eco-friendly substitute for traditional fossil-based production routes, harnessing microbial and phototrophic systems that generate hydrogen through natural metabolic processes. In anoxic bioenergy conversion, specific microorganisms metabolize organic substrates under oxygen-deprived conditions, releasing hydrogen alongside secondary by-products such as volatile fatty acids and other acidic metabolites. Conversely, light-driven bioconversion employs phototrophic bacteria that capture and utilize light energy to metabolize organic compounds, leading to efficient hydrogen gas production (Kothari, 2017).

Figure 10.1 depicts the oxygen-free and light-driven microbial hydrogen production pathways from diverse organic feedstock, highlighting adaptability for industrial-scale applications. Microorganism-driven clean fuel production offers several advantages that foster renewable energy adoption and efficient resource utilization. This carbon-balanced approach leverages biomass-derived substrates for biohydrogen generation, effectively minimizing overall carbon emissions. By utilizing organic materials, such processes convert agricultural and industrial residues into sustainable energy, thereby reducing environmental pollution while promoting circular resource use. Moreover, the optimal operating conditions of these biochemical pathways typically occurring under near-ambient temperature and pressure significantly lower external energy requirements, enhancing both the efficiency and sustainability of hydrogen production. Renewable clean fuel produced from bioresources serves as an adaptable energy carrier, enabling carbon-free operation and climate-positive benefits. It can be stored, carried, and integrated with existing infrastructure for vehicles and manufacturing processes, while ongoing research on bioreactor optimization and genome editing drives a clean energy system. However, several obstacles hinder the large-scale production of hydrogen from renewable biomass. Low-hydrogen yield remains a key issue due to suboptimal biomass conversion efficiencies and competition between hydrogen-producing and hydrogen-consuming

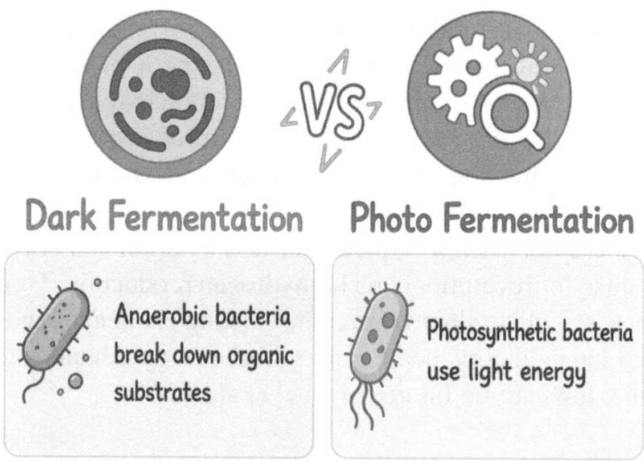

FIGURE 10.1 Process of dark and photo fermentation.

microorganisms. Substrate limitations also pose challenges, as the use of pure substrates is costly, while complex organic wastes exhibit variable composition and often require pretreatment, especially in the case of lignocellulosic biomass. Process instability further complicates operations, with microbial systems being highly sensitive to environmental parameters such as pH, temperature, and inhibitory compounds, resulting in fluctuating hydrogen outputs. At the industrial level, scale-up issues emerge, including difficulties in maintaining microbial performance, high energy requirements for mixing and gas separation, and the need for reactor designs that enhance mass transfer. Additionally, feedstock availability and sustainability concerns such as competition with other biofuel industries, seasonal biomass fluctuations, and resource constraints related to land and water use affect long-term feasibility. From an economic perspective, high production costs, limited market demand, and inadequate hydrogen infrastructure restrict widespread adoption. Finally, genetic engineering limitations, including regulatory barriers and the instability of genetic modifications in mixed microbial cultures, continue to impede progress towards efficient, large-scale biohydrogen production. Integrating AI and machine learning (ML) in biohydrogen production can significantly enhance efficiency, stability, and scalability. Process optimization uses ML to predict ideal conditions and enable real-time control. Microbial community analysis employs deep learning for metagenomic interpretation and interaction prediction. Metabolic engineering benefits from AI-assisted pathway design and genetic modification screening. Feedstock selection is improved through ML models predicting biomass composition and optimizing pretreatment. Reactor design integrates AI with CFD for scale-up and performance modelling, while techno-economic analysis uses AI for cost prediction and supply chain optimization. Advanced sensor systems apply ML for accurate monitoring and fault detection, and hybrid energy systems leverage AI for efficient biohydrogen integration with renewables. Collectively, these AI/ML applications accelerate innovation and enhance the sustainability of biohydrogen production.

10.1.2 Optimized Bioconversion in Microbial Hydrogen Synthesis

Targeted fermentation leverages advanced biotechnology and engineered microorganisms to optimize microbial hydrogen synthesis and enhance yield and efficiency in biological hydrogen production. Strain optimization involves selecting or engineering microbial strains with improved hydrogen-producing capabilities. Metabolic engineering redirects carbon flux towards hydrogen production. Process control involves precise control of fermentation parameters. Substrate utilization optimizes the use of various feedstocks, including waste materials, for cost-effective hydrogen production. Bioreactor design develops specialized bioreactors to enhance microbial growth and hydrogen evolution (Kothari, 2017).

10.2 FUNDAMENTALS OF BIOHYDROGEN PRODUCTION

Biohydrogen production harnesses microorganisms to generate hydrogen gas (H_2) from renewable organic materials such as biomass and waste, offering a sustainable and carbon-neutral energy source. The main biological routes include direct and indirect biophotolysis, photo fermentation, and dark fermentation, each driven by different microbial and environmental conditions. Key factors like pH, temperature, substrate type, and reactor design greatly influence yield and stability. With progress in biotechnology and AI-based optimization, biohydrogen production is emerging as a promising pathway towards clean and renewable energy.

10.2.1 Microbial Hydrogen Production Pathways

Figure 10.2 explains the process of microbial hydrogen production pathways. Microbial hydrogen production involves various metabolic pathways, including dark fermentation, photo fermentation, direct biophotolysis, indirect biophotolysis, and microbial electrolysis. Factors affecting hydrogen production include substrate type, concentration, pH, temperature, hydrogen partial pressure, nutrient availability, and microbial community composition. Challenges include improving hydrogen yield, substrate utilization efficiency, developing robust cultures, scaling up production

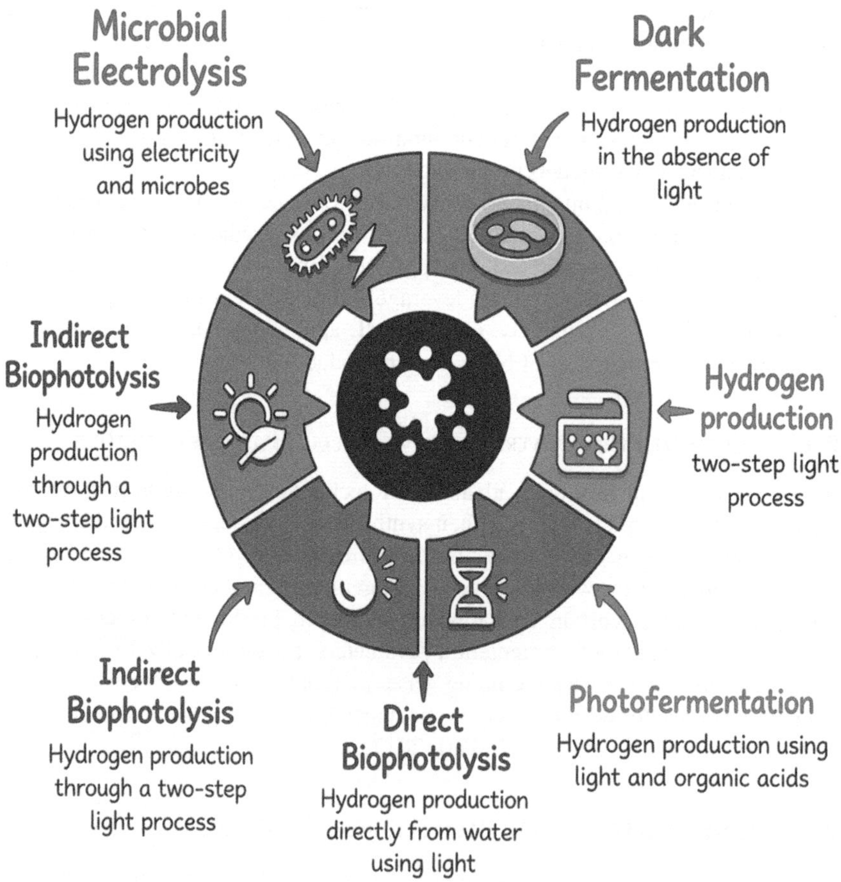

FIGURE 10.2 Process of microbial hydrogen production pathways.

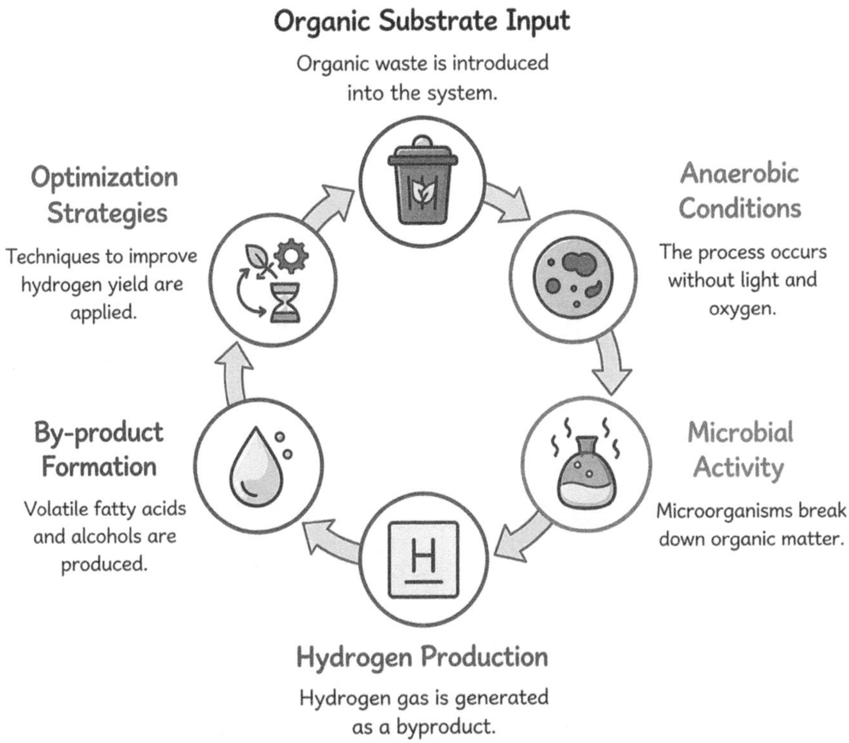

FIGURE 10.3 Process of dark fermentation process.

processes, and integrating hydrogen production with other bioprocesses for sustainability. Research focuses on understanding metabolic pathways, genetic engineering, and optimizing process conditions to enhance hydrogen yields and economic viability (Gupta et al., 2024).

Figure 10.3 shows the process of dark fermentation. Dark fermentation is a biological process that converts organic substrates into hydrogen gas and other by-products without light and oxygen. It involves the metabolic activities of anaerobic microorganisms, primarily bacteria, and operates in an anaerobic environment with a temperature range of 30–80°C and a pH range of 5.0–7.0. The process breaks down complex organic compounds into simpler molecules, with hydrogen production occurring during the acidogenesis phase. By-products include volatile fatty acids and alcohols. Dark fermentation offers renewable energy production, waste treatment, and no light requirement. Challenges include low hydrogen yield, metabolic by-product inhibition, and competition between hydrogen-producing and hydrogen-consuming microorganisms. Optimization strategies include genetic engineering, two-stage fermentation processes, mixed microbial cultures, and integration with other bioprocesses (Gupta et al., 2024).

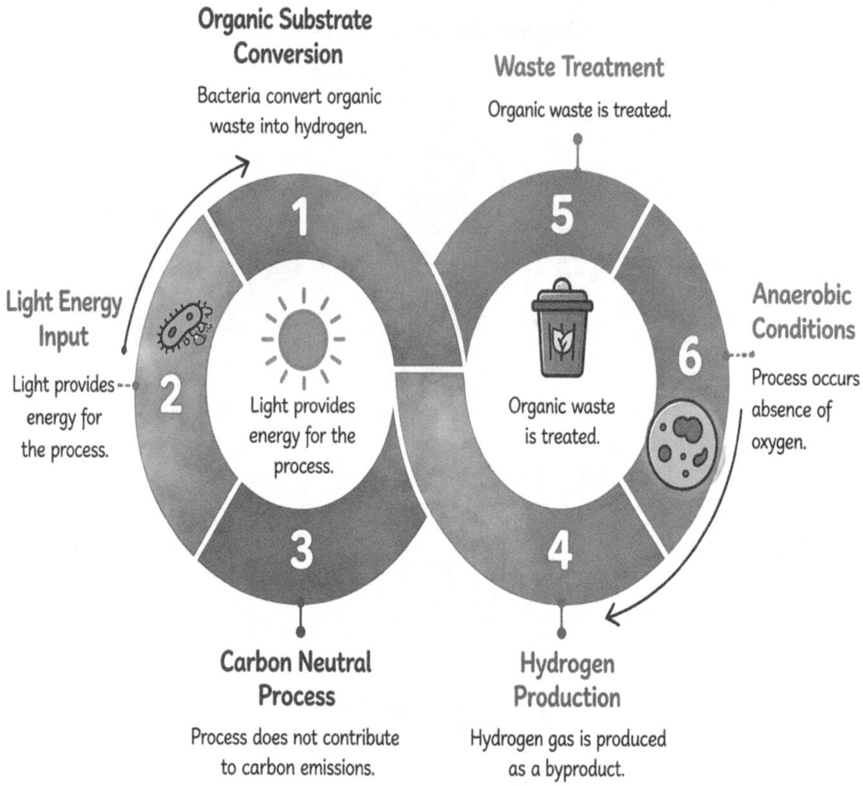

FIGURE 10.4 Process of photo fermentation.

Figure 10.4 shows the process of photo fermentation. Photo fermentation is a biological process that uses photosynthetic bacteria to produce hydrogen gas from organic substrates. It requires light energy to convert organic compounds into hydrogen, occurs in anaerobic conditions, and can utilize various organic compounds, including agricultural and industrial waste. The process is efficient, with a theoretical maximum hydrogen yield of 12 moles of H2 per mole of glucose. It offers sustainable hydrogen production, waste treatment potential, and a carbon-neutral process. Challenges include low light conversion efficiency, optimal reactor design, maintaining anaerobic conditions, and scaling up for industrial applications. Research focuses on genetic engineering, optimizing reactor designs, developing hybrid systems, and exploring new organic waste streams (Wang & Yin, 2019).

Figure 10.5 depicts the process of integrated dark and photo fermentation. Integrated dark and photo fermentation systems combine two fermentation processes to increase hydrogen production from organic waste materials. Dark fermentation, which occurs without light and uses anaerobic bacteria, produces hydrogen, carbon dioxide, and organic acids. It operates faster than photo fermentation but has thermodynamic constraints. Light-powered bacteria convert biogenic precursors into hydrogen via light-driven bioconversion, offering high efficiency, continuous operation, and supporting sustainable energy and circular biotechnology.

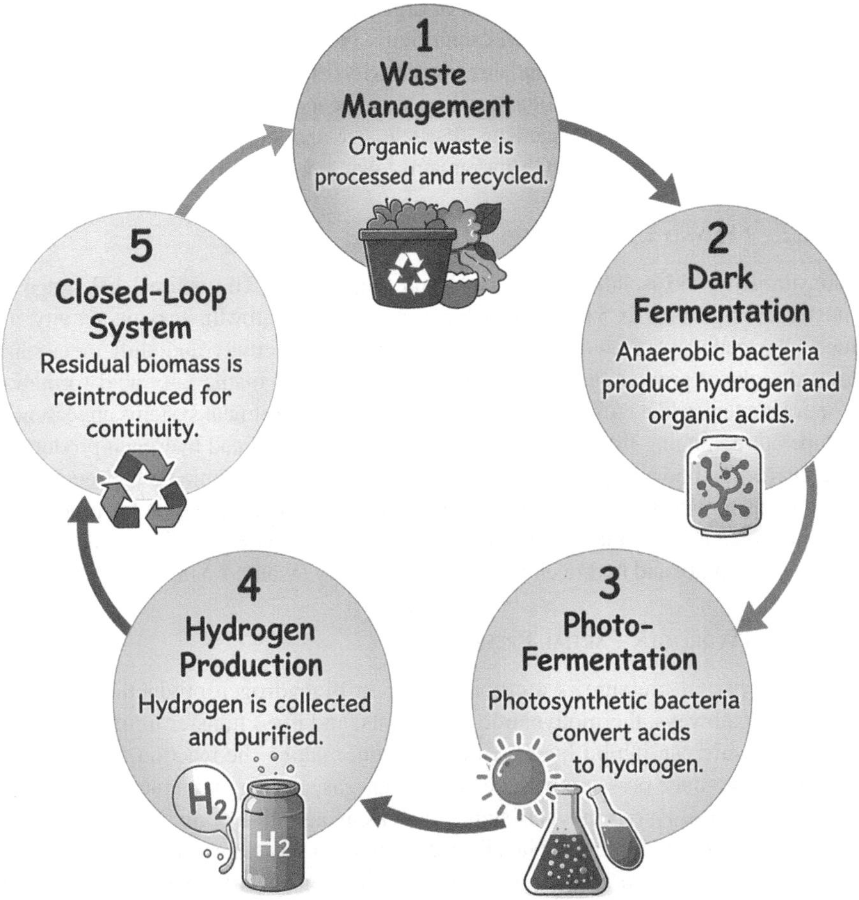

FIGURE 10.5 Process of integrated dark and photo fermentation.

10.3 OPERATIONAL AND BIOCHEMICAL FACTORS IMPACTING HYDROGEN PRODUCTION

Hydrogen production is influenced by operational factors such as temperature, pH, HRT, substrate, reactor design, and hydrogen removal, as well as biochemical factors including microbe type, hydrogenase activity, metabolic pathways, redox potential, cofactors, and inhibitors.

10.3.1 Biomass Composition and Supply Rate

Organic source characteristics and their concentration play a crucial role in determining biohydrogen output. Several parameters influence biohydrogen production, including carbon and nitrogen sources, micronutrients, substrate load and pretreatment strategies, presence of inhibitors, acidity/alkalinity buffering, particle size, uniformity of substrate mixing, and co-substrate supplementation. Carbon sources can include simple sugars, complex carbohydrates, or organic wastes, while nitrogen sources may be biological or mineral-based. Micronutrients include trace elements and essential minerals for microbial growth and enzyme activity. Pretreatment methods enhance the accessibility of fermentable sugars. pH buffering capacity maintains optimal pH for hydrogen-producing microorganisms. Particle size and substrate mixing ensure even distribution of nutrients and microorganisms (Wang & Yin, 2019).

10.3.2 pH and Temperature

Biohydrogen yield is influenced by pH and temperature. The optimal pH range for most microorganisms is 5.5–6.5, which affects microbial growth, enzyme activity, and metabolic pathways. Low pH can inhibit hydrogenase activity and shift metabolism towards solvent production, while high pH may lead to volatile fatty acid formation. Control strategies like buffer addition or automatic pH adjustment systems are essential. Temperature, ranging from 30 to 40°C, affects reaction rates and hydrogen production. The interaction between pH and temperature is crucial for maximizing hydrogen yield. Continuous monitoring and control of pH and temperature are essential for process stability. Gradual acclimation of microbial communities to desired conditions can improve long-term stability and hydrogen production efficiency (Wang & Yin, 2019).

10.3.3 Hydrogen Partial Pressure

Hydrogen partial pressure is a crucial factor in biohydrogen production, affecting metabolic pathways, thermodynamic constraints, and mass transfer limitations. High partial pressure can inhibit hydrogen production, making the reaction unfavourable. Strategies to reduce pressure include continuous gas sparging with inert gases, membrane-based in situ gas removal systems, increasing agitation or mixing, and reducing total pressure in the bioreactor headspace. Low partial pressure can promote hydrogen-producing enzyme activity and shift metabolism towards hydrogen production pathways. Hydrogen partial pressure effects may vary with temperature and pH, and substrate concentration and type can influence sensitivity. Continuous monitoring and closed-loop control are essential for sustaining ideal pressure, while bioreactor configuration must fine-tune the gas phase volume, regulate gas circulation, and ensure effective gas separation and collection (Buitron et al., 2019).

10.3.4 Microbial Community Composition

The composition of microbial communities significantly influences biohydrogen yield. Factors such as the diversity of hydrogen-producing bacteria, the presence of hydrogen-consuming microorganisms, syntrophic interactions, metabolic pathways, substrate utilization efficiency, resistance to environmental stressors, biofilm formation, competition for nutrients, quorum sensing, and adaptation and evolution can all contribute to the production of biohydrogen. A diverse community of hydrogen-producing bacteria can enhance yield by utilizing a wide range of substrates and adapting to varying environmental conditions. Conversely, hydrogen-consuming microorganisms can reduce yield by suppressing them through heat treatment, pH control, or selective inhibitors. Syntrophic interactions between microbial species can provide substrates for hydrogen-producing bacteria, potentially increasing yield.

Additionally, diverse substrate utilization capabilities can efficiently convert complex organic matter into hydrogen. Understanding and manipulating these factors can optimize microbial community composition for improved biohydrogen yield in various production systems (Wang & Yin, 2019).

10.3.5 Historical Methods for Optimizing Hydrogen Fermentation

Early hydrogen fermentation optimization focused on selecting suitable substrates, controlling pH and temperature, and isolating high-yield microbial strains. Advances in reactor design, such as continuous and UASB systems, along with hydrogen removal techniques, improved efficiency. Supplementing nutrients and trace elements, using mixed microbial cultures, and applying adaptive pre-treatments further enhanced hydrogen production and process stability.

10.3.6 Pretreatment of Substrates

Figure 10.6 shows the traditional methods for improving biohydrogen yield. Effective preprocessing of biomass plays a vital role in increasing biohydrogen production, facilitating the breakdown of complex organic compounds for better

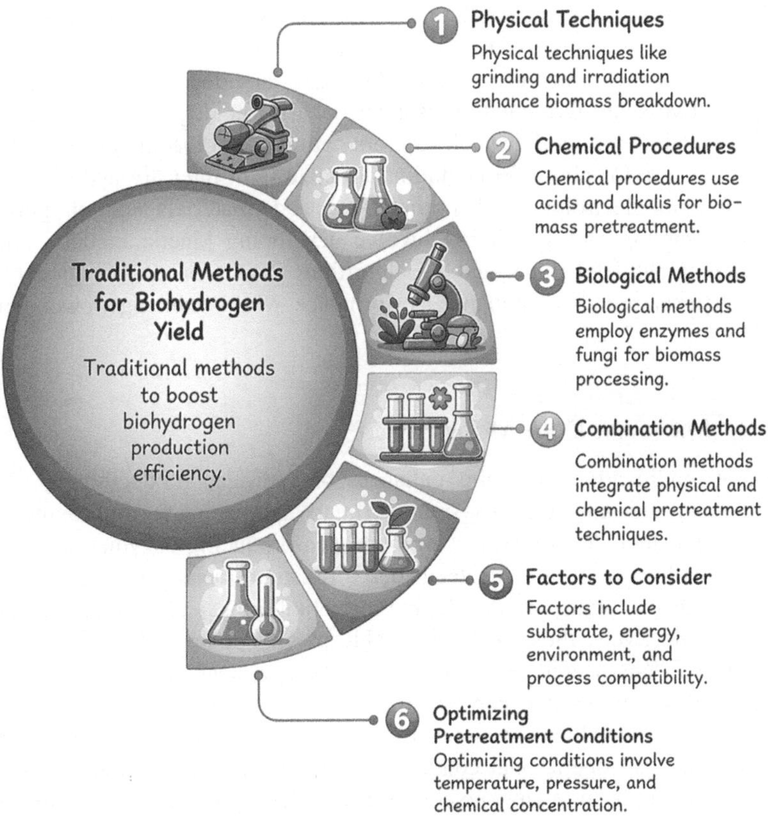

FIGURE 10.6 Standard methods in biohydrogen production enhancement.

microbial utilization. Mechanical grinding, ultrasonication, and microwave irradiation are examples of physical techniques. Acid, alkaline, and oxidative treatments are examples of chemical procedures. Biological methods involve enzymatic hydrolysis and fungal pretreatment. Combination methods combine physical and chemical methods. Factors to consider when selecting a pretreatment method include substrate composition, energy requirements, environmental impact, and compatibility with downstream processes. Optimizing pretreatment conditions includes temperature, pressure, treatment duration, chemical agent concentration, and enzyme loading and reaction conditions (Hitam & Jalil, 2020).

10.3.7 Reactor Design Optimization

Biohydrogen yield can be improved through reactor design optimization methods, which focus on enhancing the efficiency and productivity of biohydrogen production processes. Key approaches include continuous stirred-tank reactor (CSTR) optimization, which involves adjusting hydraulic retention time, organic loading rate, pH control systems, temperature, and mixing conditions. Anaerobic fluidized bed reactor (AFBR) design involves selecting appropriate carrier materials for microbial immobilization, optimizing fluidization velocity and bed expansion, and implementing efficient gas-liquid separation systems. Membrane bioreactor (MBR) integration involves selecting suitable membrane materials and pore sizes, optimizing transmembrane pressure and flux rates, and developing fouling control strategies. Two-stage fermentation systems separate acidogenesis and hydrogenesis phases, while immobilized cell reactors optimize cell loading and distribution. Photobioreactors for photo fermentation optimize light intensity and distribution, while microbial electrolysis cells optimize electrode materials and configurations. Process modelling and simulation involve developing kinetic models, utilizing computational fluid dynamics, and ML algorithms for process optimization. On-site output retrieval processes enhance performance while reducing power demand (Kelif Ibro et al., 2024).

10.3.8 Microbial DNA Modification

Microbial modification of strains enhances biohydrogen production by altering biochemical routes and control systems. Primary approaches for improving biohydrogen output include overexpressing hydrogen-producing enzymes, redesigning metabolic pathways, optimizing feedstock utilization, increasing stress tolerance, enhancing electron transport, developing microbial films, and applying engineered systems. Collectively, these methods enable precise control of hydrogen energy metabolism in engineered microorganisms (Ziegler et al., 2024). They broaden the range of substrates that can be efficiently utilized for hydrogen production. Enhanced resilience allows microbes to maintain high productivity under variable environmental and operational conditions. Additionally, the creation of artificial pathways and engineered enzyme complexes further optimizes energy conversion, resulting in high-performance biohydrogen synthesis.

10.4 AI/ML APPLICATIONS IN PRECISION FERMENTATION FOR BIOHYDROGEN PRODUCTION

ML algorithms are being used for predictive modelling, process optimization, strain design, data analysis, bioreactor monitoring, feedstock evaluation, scale-up prediction, metabolic flux analysis, hybrid modelling, and automated experimentation. These methods facilitate efficient bio-system forecasting, dynamic control, and smart-enabled operation enhancement (Xia et al., 2025). Intelligent computing and predictive modelling significantly advance biohydrogen production by integrating multiple strategies for process optimization and control. Predictive analysis allows forecasting of microbial responses and key bioprocess variables, enabling proactive adjustments to maximize hydrogen yield. Dynamic regulation of process parameters ensures optimal conditions for microbial growth and hydrogen evolution, while microbial strain optimization refines genetic pathways to enhance bioenergetic efficiency. Data-driven assessment leverages real-time process data to identify bottlenecks and improve overall microbial performance. Continuous bioprocess monitoring using sensors and visualization tools allows timely detection of variations, ensuring stable and efficient bioconversion. Material characterization of feedstocks ensures substrates are optimized for hydrogen production, and upscaling projections translate lab-scale results to industrial applications. ML-based reaction pathway monitoring further enables the analysis and optimization of microbial bioenergetics. Integrating these approaches into a unified modelling platform combines mechanistic and intelligent models for comprehensive process understanding and control. Finally, automated analysis and intelligent control systems facilitate rapid optimization and process stability, collectively accelerating efficient, sustainable, and scalable biohydrogen production while supporting the advancement of clean energy technologies.

10.4.1 Overview and Critical Elements of Engineered Biomanufacturing

Precision biomanufacturing is a bioengineering approach that utilizes engineered microbes to efficiently and selectively produce desired compounds. It integrates tools from synthetic biology, metabolic engineering, and systems biology to precisely control cellular pathways. This allows microbes to convert renewable feed stocks into valuable products such as biofuels, pharmaceuticals, and fine chemicals. The process emphasizes accuracy, scalability, and sustainability, ensuring minimal waste and energy use. Overall, precision biomanufacturing represents a key step towards developing smart, eco-friendly bioproduction systems for the future.

Figure 10.7 shows the precision bioconversion process. Engineered microbes with specific genes drive substrate-specific reactions in controlled bioprocessing reactors, with real-time monitoring enhancing efficiency and product quality. Precision biotransformation sustainably generates biologically active molecules, flavours, and biologically active compounds for nutritional and medicinal applications (Ziegler et al., 2024). Advanced biotechnological bioconversion significantly enhances sustainable hydrogen production by integrating genetic engineering, precision control, and optimized process regulation. Through enhanced microbial strain control, genetically optimized microbes improve hydrogen yield and metabolic efficiency, while refined substrate control enables the utilization of a wide range of feedstocks, including polysaccharides and organic residues. Precision regulation of metabolic pathways minimizes by-product formation and maximizes energy conversion efficiency. Controlled bioprocessing of engineered strains leads to refined hydrogen production with faster kinetics, higher yield, and reduced production costs through improved feedstock utilization. Moreover, robust and adaptable production performance ensures stable, scalable, and efficient hydrogen generation under variable industrial conditions. The system's ability to process diverse substrates enhances flexibility, while reduced environmental impact results from efficient substrate conversion and minimal waste formation. Continuous cultivation of optimized microbes supports steady, long-term hydrogen generation, and advanced sensing and control frameworks allow real-time adjustments for consistent, high-efficiency biohydrogen production.

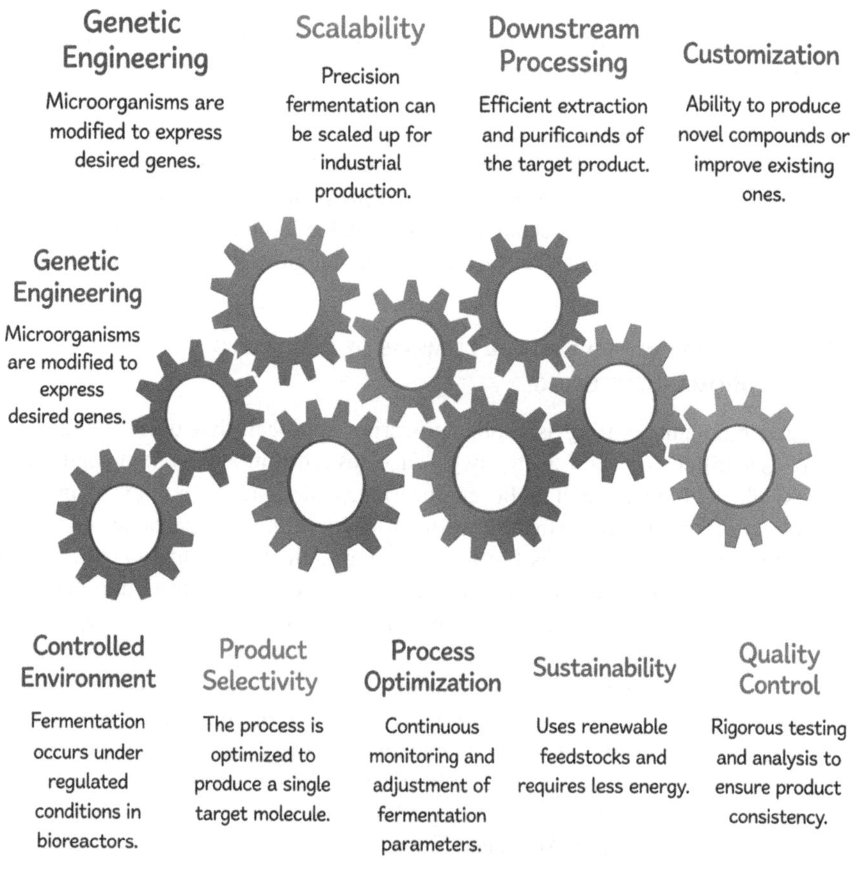

FIGURE 10.7 Precision fermentation process.

10.4.2 Core Elements in Controlled Cell Culture

Optimizing fermentation for high-yield bioprocesses involves precise control over multiple parameters that influence microbial performance and product formation. Strain selection is crucial, as choosing the right microorganism ensures maximum yield and desired product quality. A carefully designed media composition provides the nutrients essential for growth, while maintaining optimal temperature and pH supports stable metabolic activity. Regulation of dissolved oxygen concentration ensures the proper balance between aerobic and anaerobic metabolism, and agitation speed is adjusted to maintain uniform nutrient and microbial distribution. A well-planned substrate feeding strategy helps maximize productivity while minimizing by-product accumulation. In recombinant processes, precise induction timing enhances target gene expression. Continuous process monitoring using advanced sensors enables real-time tracking of metabolites, biomass, and product concentrations, ensuring process stability. During scale-up, maintaining consistent environmental and kinetic parameters is key to preserving performance from lab to industrial scale. Strict contamination control safeguards culture integrity, and efficient harvesting and downstream processing techniques ensure high product recovery and purity, completing an optimized and sustainable biomanufacturing cycle.

10.4.3 Leveraging Machine Intelligence to Improve Biohydrogen Yield

Computational intelligence and data-driven learning algorithms play a vital role in advancing biohydrogen production by integrating intelligent analysis, predictive control, and real-time optimization. Machine intelligence and cognitive computing enhance microbial hydrogen generation by fine-tuning operational parameters, guiding strain engineering, and optimizing bioreactor performance. Through continuous monitoring, biochemical profiling, and biomass assessment, these systems improve hydrogen yield, process stability, and environmental efficiency. Predictive modelling enables precise adjustment of bioprocess variables, enzyme activities, and metabolic pathways while allowing early detection of performance deviations. Intelligent algorithms further enhance mixing, mass transfer, and hydrogen recovery efficiency, while techno-economic integration ensures cost-effective scale-up and alignment with renewable energy frameworks. By combining data-driven intelligence with automated reasoning, biohydrogen production becomes more efficient, scalable, resilient, and sustainable, accelerating progress towards industrial application and clean energy transition.

10.4.4 Opportunities for Intelligent Algorithms to Improve Microbial Cultivation Efficiency

Machine intelligence and cognitive computing can transform bioprocessing by enabling real-time monitoring, predictive modelling, and dynamic process optimization. These approaches enhance microbial yield, product quality, and operational sustainability while improving resource efficiency and reducing costs. Adaptive control supports consistent batch-to-batch performance, predictive maintenance, and scalable operations. Data-driven insights allow optimization of substrates, nutrient utilization, and energy consumption. By integrating historical and real-time process data, intelligent algorithms can monitor microbial cultivation or bioconversion processes continuously. These algorithms detect deviations and predict trends, allowing rapid adjustments to maintain optimal conditions. Fine-tuning of operational parameters enhances process efficiency and resource utilization. The adaptive nature of these systems supports environmentally sustainable production by minimizing waste and energy consumption. Collectively, such AI-driven control strategies enable flexible and scalable bioprocesses suitable for industrial applications.

10.4.5 Data-driven Methods for Biogenic Hydrogen Generation

Computational intelligence approaches are progressively employed to enhance the efficiency and yield of biohydrogen generation. These techniques can be broadly categorized into supervised learning, unsupervised learning, and reinforcement learning approaches. Figure 10.8 show the types of AI and machine ML techniques.

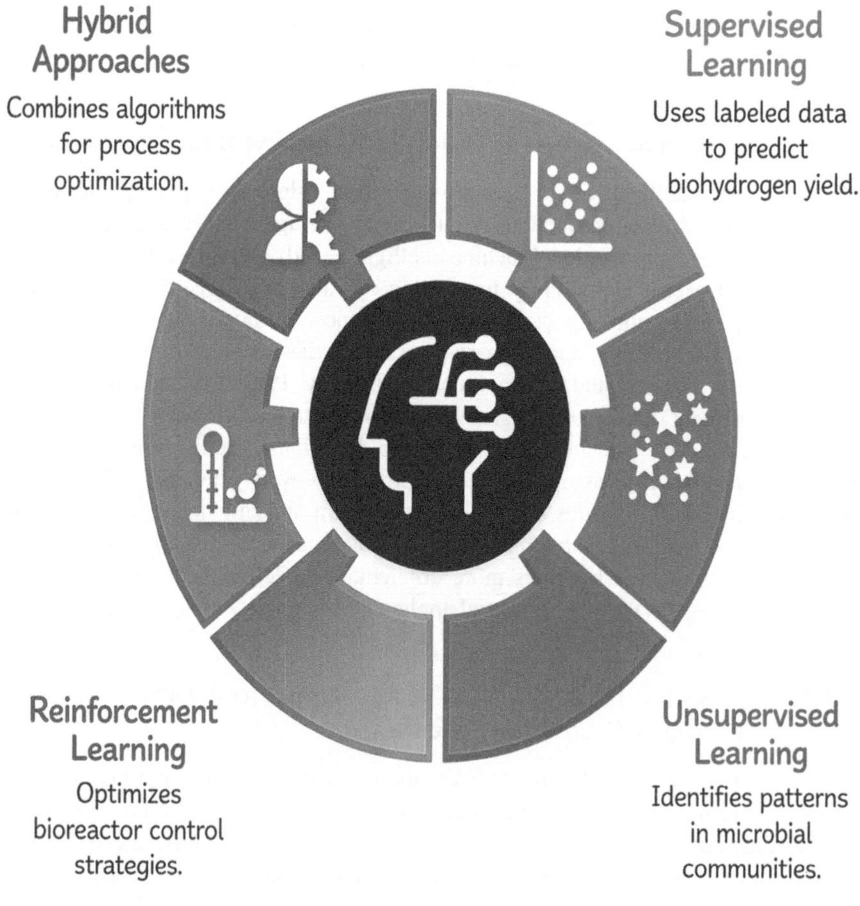

FIGURE 10.8 Types of AI and ML techniques.

10.4.6 Supervised Learning

Neural Networks (NNs) are employed to simulate and understand the complex, non-linear interactions between various process variables and biohydrogen output, enabling accurate prediction and optimization of production conditions. Support Vector Machines (SVMs) help determine the ideal parameters for biogenic hydrogen synthesis by identifying optimal operational ranges that maximize yield. Meanwhile, Random Forests are valuable for feature selection and ranking the importance of process parameters, allowing researchers to pinpoint the most influential factors affecting hydrogen production efficiency. Together, these ML approaches provide a robust framework for modelling, optimizing, and improving biohydrogen generation systems.

10.4.7 Unsupervised Learning

Clustering algorithms such as K-means are used to identify patterns and group similar microbial communities involved in biohydrogen production, providing insights into microbial behaviour and functional relationships within bioreactors. Principal Component Analysis (PCA) further supports data interpretation by reducing the dimensionality of large, complex bioprocess datasets, allowing clearer visualization of variable interactions and key performance factors. Together, these analytical tools enhance understanding, monitoring, and optimization of microbial and process dynamics in biohydrogen generation.

10.4.8 Reinforcement Learning

Q-learning is a reinforcement learning technique that can be applied to optimize bioreactor control strategies for maximizing biohydrogen yield. By continuously interacting with the bioprocess environment, the algorithm learns the best actions, such as adjusting temperature, pH, or substrate feed rate to achieve optimal hydrogen production. Over time, Q-learning models improve decision-making by evaluating rewards from different process states, enabling adaptive, autonomous control of bioreactors. This leads to enhanced process efficiency, stability, and hydrogen yield without requiring manual intervention (Noonan et al., 2025).

10.4.9 Hybrid Approaches

Genetic Algorithms (GAs) combined with Artificial Neural Networks (ANNs) are powerful hybrid computational tools used to optimize process parameters and reactor designs for enhanced biohydrogen production. These techniques enable precise process optimization by predicting ideal conditions for temperature, pH, and substrate concentration to maximize hydrogen yield. They also support microbial community analysis, helping identify key microorganisms and their synergistic interactions within hydrogen-producing consortia. In bioreactor design and control, GAs and ANNs assist in optimizing reactor configurations and developing adaptive, self-regulating systems for stable operation. Additionally, they contribute to feedstock selection by analysing and predicting the most suitable substrates for efficient conversion, while in metabolic pathway engineering, they help identify genetic targets and modifications that can further improve biohydrogen generation. Together, these intelligent techniques provide an integrated framework for achieving efficient, scalable, and sustainable hydrogen production (Alanazi & Aljuhani, 2022).

10.4.10 Metabolic Flux Analysis

Fluxomics is an essential approach for optimizing biohydrogen production Figure 10.9. It quantifies intracellular carbon and energy flows using constraint-based and genome-scale modelling, isotope tracing, dynamic flux analysis, omics integration, and in silico strain design. These approaches reveal biochemical networks, cofactor balance, and feedstock utilization, enabling efficient, continuous, and scalable hydrogen production (Banerjee et al., 2016). This framework also enables in silico strain design, allowing prediction of the effects of gene

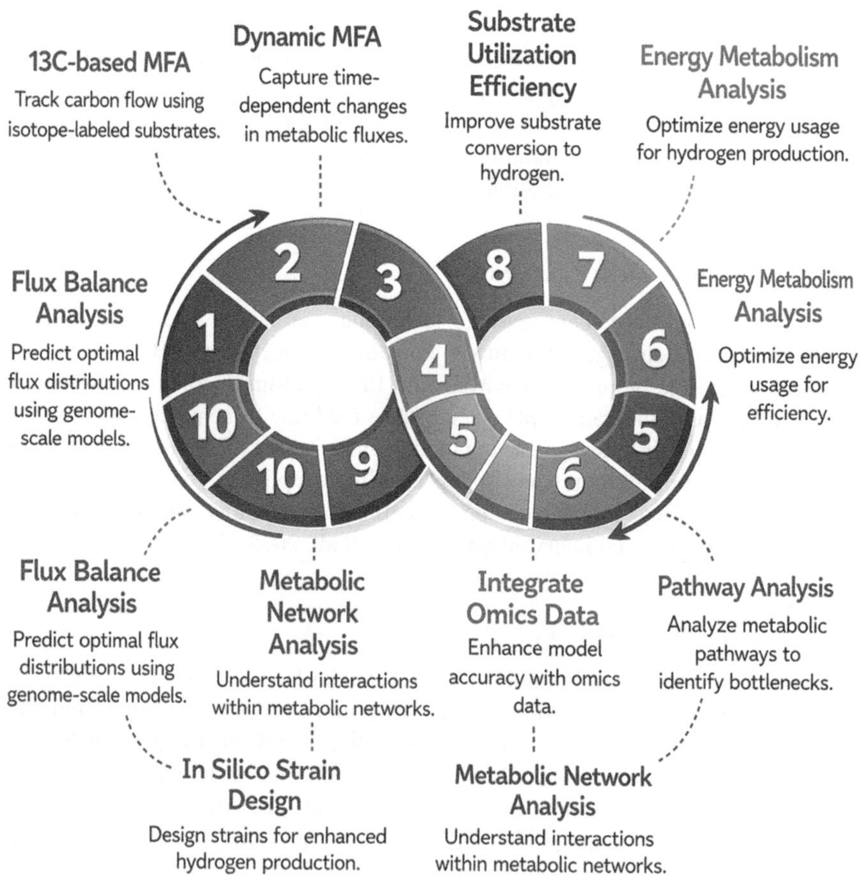

FIGURE 10.9 Workflow of quantitative flow analysis.

knockouts, overexpression, and pathway rerouting on hydrogen metabolism prior to experimental implementation. Coupling fluxomic data with transcriptomic, proteomic, and metabolomic information further elucidates how environmental factors, nutrient availability, and cellular stress impact hydrogen productivity. Advanced fluxomic models identify metabolic bottlenecks, cofactor imbalances, and energy dissipation pathways that limit hydrogen yield, guiding rational network redesign for improved electron transfer, reduced by-product formation, and enhanced process robustness. Integration of these insights into bioreactor operation and process simulation supports real-time optimization of feedstock conversion and energy efficiency. Collectively, fluxomics establishes a quantitative and predictive foundation for continuous, scalable, and high-performance biohydrogen production (Adewumi et al., 2024; Aghbashlo et al., 2024).

10.4.10.1 Industrial Applications and Future Prospects

Biohydrogen production has various industrial applications, including fuel cells, the chemical industry, refining, the food industry, and metallurgy. It can be used in fuel cells for electricity generation, in the chemical industry for ammonia production and hydrogenation reactions, in petroleum refining for sulphur removal, in the food industry for hydrogenation of oils and fats, and in metal processing. Future prospects for biohydrogen production are highly promising, with multiple avenues for technological and industrial advancement. Improved production efficiency remains a primary goal, driven by advances in microbial engineering, process optimization, and AI-assisted control strategies. Integration with renewable energy systems, such as solar and wind power, can further enhance sustainability and reduce reliance on fossil fuels. Waste-to-energy applications offer the potential to convert organic residues into valuable hydrogen, promoting circular bioeconomy practices. Scaling up production while maintaining process stability and cost-effectiveness is a key challenge for industrial deployment. Advancements in hydrogen storage, transportation, and distribution infrastructure will facilitate broader adoption across energy sectors. Supportive government policies, financial incentives, and integration into biorefinery platforms can accelerate commercialization and market penetration. Additionally, coupling biohydrogen production with carbon capture and utilization strategies can contribute to a more sustainable, low-carbon, and carbon-neutral energy system, aligning with global decarbonization goals.

10.4.10.2 Current Industrial Adoption of AI/ML in Biohydrogen Production

Key applications include process optimization, strain selection and engineering, feedstock evaluation, process monitoring and control, techno-economic analysis, scale-up and process integration, and quality control and product purification. AI algorithms analyse real-time data from bioreactors to optimize operating conditions, predict optimal fermentation parameters, and identify promising microbial strains for enhanced hydrogen production. ML models also predict biohydrogen yields from different feedstock combinations, optimizing resource utilization. AI-powered sensors and computer vision systems monitor biohydrogen production in real time, detecting anomalies and potential issues. Techno-economic analysis evaluates the economic feasibility of biohydrogen production processes, optimizing supply chain logistics and resource allocation.

AI simulations help predict challenges and optimize conditions during the scale-up of biohydrogen production, and ML algorithms optimize purification processes. Despite persistent barriers such as insufficient observational records, integration challenges, skill shortages, and policy frameworks, the maturation of algorithm-driven modelling techniques is anticipated to accelerate industrial deployment in biohydrogen production, enhancing efficiency, sustainability, and economic performance.

10.4.10.3 Challenges in Industrial Deployment

The capacity of intelligent algorithms and predictive modelling approaches to advance large-scale bio-based hydrogen generation is substantial. Major domains for enhancement through these methods encompass microbial process optimization, metabolic strain engineering, substrate and feedstock selection, bioreactor design, process integration, predictive maintenance, and supply chain optimization. These multifaceted methodologies involve prognostic modelling of bioprocess performance, continuous reactor condition monitoring, and optimization of substrate composition, microbial strain enhancement, biomass analysis, pretreatment techniques, and computational simulation of scale-up impacts on hydrogen yield. Furthermore, these approaches focus on evaluating economic viability, commercial prospects, ecological consequences, and environmental resilience. Fault detection and predictive maintenance are also essential, allowing for early detection of process anomalies and equipment failures, minimizing downtime, and optimizing plant operations and resource allocation. Finally, supply chain optimization involves predicting feedstock availability, quality, logistics, transportation networks, demand forecasting, and inventory management (Alex et al., 2023; De Medeiros et al., 2023).

10.4.10.4 Challenges for Large-scale Implementation

Critical considerations in applying AI and ML to bio-based hydrogen production encompass several interconnected factors. Data availability and quality are foundational, as AI/ML models require large, diverse, and accurate datasets to learn patterns, make reliable predictions, and avoid biases that could compromise process efficiency. Ensuring standardized data collection, preprocessing, and storage protocols is therefore essential. Model interpretability is another key aspect, since explainable AI frameworks not only facilitate regulatory compliance but also allow process engineers to understand and trust model decisions, supporting transparent optimization of bioprocess parameters. Integration with existing infrastructure presents both technical and operational challenges, requiring the seamless deployment of AI/ML tools alongside current bioreactors, sensors, control systems, and automation platforms without disrupting ongoing production. Cybersecurity is a critical component in AI-enabled biohydrogen production, as it safeguards sensitive operational data, intellectual property, and process control networks from malicious attacks, data breaches, or system failures. Compromised systems could result in significant economic losses, operational disruptions, or safety hazards. To mitigate these risks, robust encryption protocols, secure data storage, and continuous network monitoring must be implemented throughout the bioprocessing infrastructure. Additionally, access controls and user authentication are essential to limit system vulnerabilities. Alongside cybersecurity, workforce development and expertise play a vital role in ensuring the effective deployment of AI and ML tools. Personnel must possess interdisciplinary skills that combine bioprocess engineering, microbiology, and computational modelling. Targeted training programmes and educational initiatives are necessary to bridge knowledge gaps and foster proficiency in AI-assisted operations. Encouraging collaboration between biotechnologists, data scientists, and process engineers strengthens decision-making and innovation. Continuous professional development ensures that staff remain adept at integrating emerging AI technologies and methodologies. Together, these measures support the safe, sustainable, and efficient adoption of intelligent biohydrogen production systems (El-Sofany et al., 2024; Malathi & begum, 2024).

10.4.10.5 Roadmap for Future Research and Development

To advance biohydrogen production, future research and development efforts should focus on:

Figure 10.10 discusses various strategies for enhancing hydrogen production, including genetic engineering, improving feedstock utilization, reactor design and process optimization, hydrogen separation and purification, integrating biohydrogen production with other processes, addressing techno-economic challenges, life cycle

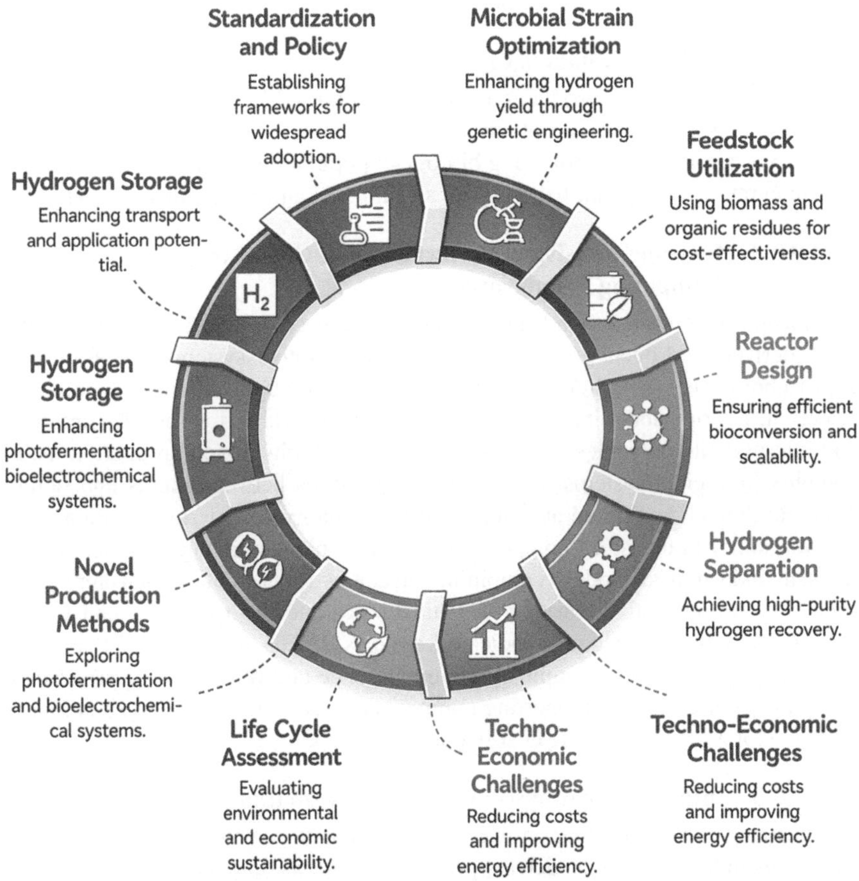

FIGURE 10.10 Roadmap for future research and development.

assessment and sustainability, exploring novel production methods, improving hydrogen storage and distribution, and standardization and policy development. Genetic engineering can be used to develop robust strains tolerant to harsh conditions and explore novel microorganisms with high hydrogen yields. Efficient pretreatment methods for lignocellulosic biomass and alternative feedstocks can also be developed. Process vessel design and process coordination can be improved by implementing production systems at production-level scales and adjusting process conditions (Pascualone et al., 2019).

10.4.10.6 Challenges and Future Directions

Precision biomanufacturing faces challenges such as metabolic complexity, low scalability, and high production costs, which limit its industrial application. The stability of engineered strains and effective integration of multi-omics data also remain key obstacles. In the future, combining AI, ML, and digital twin technologies will enable smarter design and real-time optimization of bioprocesses. Advances in synthetic biology and automation will improve efficiency, scalability, and sustainability. Overall, future efforts aim to develop intelligent, resilient, and eco-friendly biomanufacturing systems for large-scale production.

10.4.10.7 Technical Challenges

Cutting-edge hydrogen production faces challenges like low outputs, feedstock suppression, biochemical variations, thermal fluctuations, acidity control, vapour build-up, scaling issues, optimizing biological populations, reactor design, and process integration is essential for efficient, sustainable, and economical operation.

10.4.10.8 Ensuring Consistent Performance in Cognitive Computing Frameworks

Advanced smart systems and algorithms facilitate consistent and robust biologically derived hydrogen production. Advanced adaptive systems and forecasting approaches ensure consistent hydrogen production through data integration, feature selection, hybrid modelling, validation, and benchmarking for robust, continuous, and interpretable outcomes. Challenges specific to biohydrogen production include complex biological systems, non-linear relationships between inputs and outputs, temporal dynamics, limited availability of high-quality, long-term operational data, and variability in feedstock composition and microbial communities. Potential solutions include hybrid models combining mechanistic and data-driven approaches, transfer learning, active learning, domain knowledge incorporation, and explainable AI techniques. Evaluation indicators such as RMSE, R^2, MAPE, F1 score, and AUC-ROC guide robust computational intelligence and forecasting approaches for biohydrogen production, supporting maintenance, cross-disciplinary validation, and effective long-term renewable energy planning.

10.4.10.9 Integrated Synthesis of Diverse Information Sources

Synthesizing varied information in hydrogen production enhances insight and efficiency, including molecular details, process conditions, environmental factors, and substance structure. Approaches for combining data involve data repositories, linked data stores, semantic frameworks, and computational learning techniques. Difficulties encompass data normalization, handling heterogeneous formats and scales, ensuring integrity and completeness, and protecting sensitive information. Gains include improved forecasting models, identification of previously unknown cellular networks, and better resource management. Adoptions include biological variant evaluation, process calibration, feedstock utilization, and production-scale operation. Available resources span bioprocess informatics platforms, data representation frameworks, cloud-enabled analytics, and intelligent platforms powered by machine intelligence and ML technologies. Prospective developments involve continuous data integration, integrative molecular-level analysis, adoption of standardized data-sharing frameworks, and distributed ledger systems to safeguard data reliability.

10.5 CONCLUSION

This section examines the application of targeted bioconversion strategies alongside data-driven algorithms for improved hydrogen generation. Targeted bioprocessing allows refinement of cellular culture conditions, with intelligent computational methods facilitating the analysis of complex datasets to improve bioprocess efficiency. Biohydrogen generation through anaerobic and integrated bioprocesses is examined along with factors affecting yields, such as raw materials and microorganisms, highlighting cognitive computing techniques over conventional strategies. The core concepts of precision bioconversion, ongoing monitoring and regulation, and high-throughput evaluation techniques are examined while emphasizing the importance of robust data management for the effective application of machine intelligence and considering robustness, precision and transparency and highlighting their potential to enhance biological hydrogen production as a low-carbon energy alternative.

BIBLIOGRAPHY

Abbasi, M., Martins, P., Vaz, P., & Silva, J. (2025). Machine learning approaches for predicting Maize biomass yield: Leveraging feature engineering and comprehensive data integration. *Sustainability*, 17(1), 256.

Adewumi, A., Okoli, C., Usman, F., Olu-Lawal, K., & Soyombo, O. (2024). Reviewing the impact of AI on renewable energy efficiency and management. *International Journal of Science and Research Archive*, 11(1), 1518–1527.

Aghbashlo, M., Hosseinpour, S., Tabatabaei, M., Dadak, A., Younesi, H. A., & Ghasem, N.D. (2024). Multi-objective exergetic optimization of continuous photo-biohydrogen production process using a novel hybrid fuzzy clustering-ranking approach coupled with Radial Basis Function (RBF) neural network. *International Journal of Hydrogen Energy*, 41(41), 18418–184130.

Alanazi, M., & Aljuhani, A. (2022). Anomaly detection for internet of things cyberattacks. *Computers, Materials & Continua*, 72, 261–279.

Alex, C., Creado, G., Almobaideen, W., Abu Alghanam, O., & Saadeh, M. (2023). A comprehensive survey for IoT security datasets: Taxonomy, classification, and machine learning mechanisms. *Computers & Security*, 132, 103283.

Banerjee, C., Singh, P. K., & Shukla, P. (2016). Microalgal bioengineering for sustainable energy development: Recent transgenesis and metabolic engineering strategies. *Biotechnology Journal*, 11(3), 303–314.

Buitron, G., Carrillo-Reyes, J., Albarran-Contreras, B. A., Quijano, G., & Muñoz-Paez, K. M. (2019). Biohydrogen production from winery effluents: Control of the homoacetogenesis through the headspace gas recirculation. *Journal of Chemical Technology & Biotechnology*, 95(3), 544–552.

De Medeiros, K., Hendawi, A., & Alvarez, M. (2023). A survey of AI-based anomaly detection in IoT and sensor networks. *Sensors*, 23, 1352.

El-Sofany, H., El-Seoud, S.A., Karam, O.H., & Bouallegue, B. (2024). Using machine learning algorithms to enhance IoT system security. *Scientific Reports*, 14, 12077.

Gupta, S., Lopes, A., Grasa, L., Salafranca, J., & Fernandes, A. (2024). Microbes and parameters influencing dark fermentation for hydrogen production. *Applied Sciences*, 14(23), 10789.

Hitam, C. N. C., & Jalil, A. A. (2020). A review on biohydrogen production through photo-fermentation of lignocellulosic biomass. *Biomass Conversion and Biorefinery*, 13(10), 8465–8483.

Kelif Ibro, M., Beyene Lemma, D., & Ramayya Ancha, V. (2024). Biogas production optimization in the anaerobic codigestion process: A critical review on process parameters modeling and simulation tools. *Journal of Chemistry*, 2024, 1–25.

Kothari, R. (2017). A critical review on factors influencing fermentative hydrogen production. *Frontiers in Bioscience*, 22(8), 1195–1220.

Malathi, S., & Begum, S.R. (2024). Enhancing trustworthiness among IoT network nodes with ensemble deep learning-based cyberattack detection. *Expert Systems with Applications*, 255, 124528.

Manuel, H., Manuel, A., Agupugo, C., & Kehinde, H. (2024). The impact of AI on boosting renewable energy utilization and visual power plant efficiency in contemporary construction. *World Journal of Advanced Research and Reviews*, 23(2), 1333–1348.

Noonan, A. J. C., Cameron, P. M. N., Dofher, K., Sukkasam, N., Liu, T., Ronn, L., Monshupanee, T., & Hallam, S. J. (2025). An automated high-throughput lighting system for screening photosynthetic microorganisms in plate-based formats. *Communications Biology*, 8(1), 123.

Pascualone, M. J., Dalmasso, P. R., & Gomez Costa, M. B. (2019). Fermentative biohydrogen production from a novel combination of vermicompost as inoculum and mild heat-pretreated fruit and vegetable waste. *Biofuel Research Journal*, 6(3), 1046–1053.

Wang, J., & Yin, Y. (2019). Progress in microbiology for fermentative hydrogen production from organic wastes. *Critical Reviews in Environmental Science and Technology*, 49(10), 825–865.

Xia, J., Long, D., Chen, M., & Chen, A. (2025). Optimization of fermentation processes in intelligent biomanufacturing: on online monitoring, artificial intelligence, and digital twin technologies. *Chinese Journal of Biotechnology*, 41(3), 1234–1245.

Ziegler, A. L., Blank, L. M., Stumm, M., Manchanda, A., & Mitsos, A. (2024). Simultaneous design of fermentation and microbe. *AIChE Journal*, 70(9), e18023.

11 Using Artificial Intelligence (AI) and Bioprocessing for Drop-in Fuel

K. Arul, V. Prabhakaran, B. G. Sivakumar, Kanagasabapathy Sivasubramanian, S. Kannan, K. Selvakumar, and S. Sivakumar

11.1 INTRODUCTION TO DROP-IN FUELS

Maritime freight transport is a major source of greenhouse gas emissions, accounting for approximately 2.8% of the global total, comparable to the output of industrialized nations such as Germany (Tyrovola et al., 2017). Environmental, resource, and geopolitical challenges are driving a worldwide shift toward low-carbon energy alternatives. Around the world, efforts to lower greenhouse gas emissions are speeding up, with countries turning to cleaner and more resilient energy options. Among these, biofuels are playing a central role in the transition.

Among them, drop-in fuels are especially promising due to their chemical and functional compatibility with existing fossil fuel infrastructure engines, pipelines, and distribution systems. Unlike conventional biofuels such as ethanol and biodiesel, which often face limitations related to engine compatibility, blending ratios, and infrastructure drop-in fuels are biologically derived hydrocarbons designed to fully replace petroleum-based fuels like gasoline, diesel, and jet fuel without system modifications. Their seamless integration into existing fuel networks reduces implementation costs and complexity, making them a practical and scalable solution, especially in hard-to-electrify sectors like aviation, maritime shipping, and heavy-duty transport.

Figure 11.1 shows the versatility and practicality of drop-in fuels as a clean energy solution. Their applicability in maritime transport is critical, given the sector's reliance on high-energy fuels and the difficulty of electrifying large vessels. In freight transport, advanced biofuels address a significant shortfall by providing a low-emission alternative in sectors constrained by electric technology. The aviation industry, constrained by the weight and range limitations of battery technology, also benefits from drop-in fuels, which require no changes to aircraft or fueling systems. These fuels are scalable to meet growing demand and can be readily integrated into existing

DOI: 10.1201/9781003684374-11

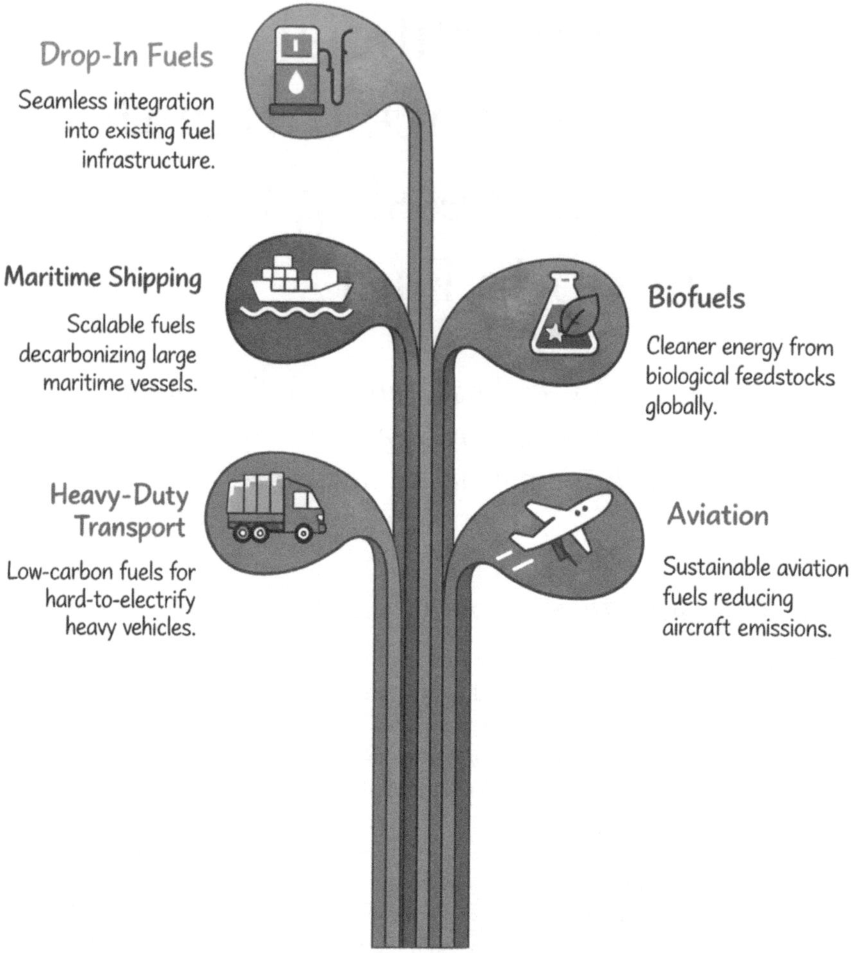

FIGURE 11.1 Exploring drop-in fuels in energy transition.

infrastructure, avoiding the high costs associated with new equipment. As a result, implementation expenses are minimized, enabling faster and more affordable adoption. The combined benefits position drop-in fuels as a key enabler in the decarbonization of energy-intensive sectors.

The production of drop-in fuels relies on a wide array of renewable and waste-based feedstocks, including lignocellulosic biomass (e.g., agricultural residues, wood waste), algae, industrial waste gases, and even municipal solid waste. These raw materials are converted into energy-dense liquid hydrocarbons through advanced biochemical and thermochemical processes, such as fermentation, pyrolysis, gasification, and hydroprocessing. The resulting fuels closely mimic the chemical structure and performance characteristics of their fossil-derived counterparts, enabling seamless integration into the current fuel economy. Scaling up drop-in fuels for

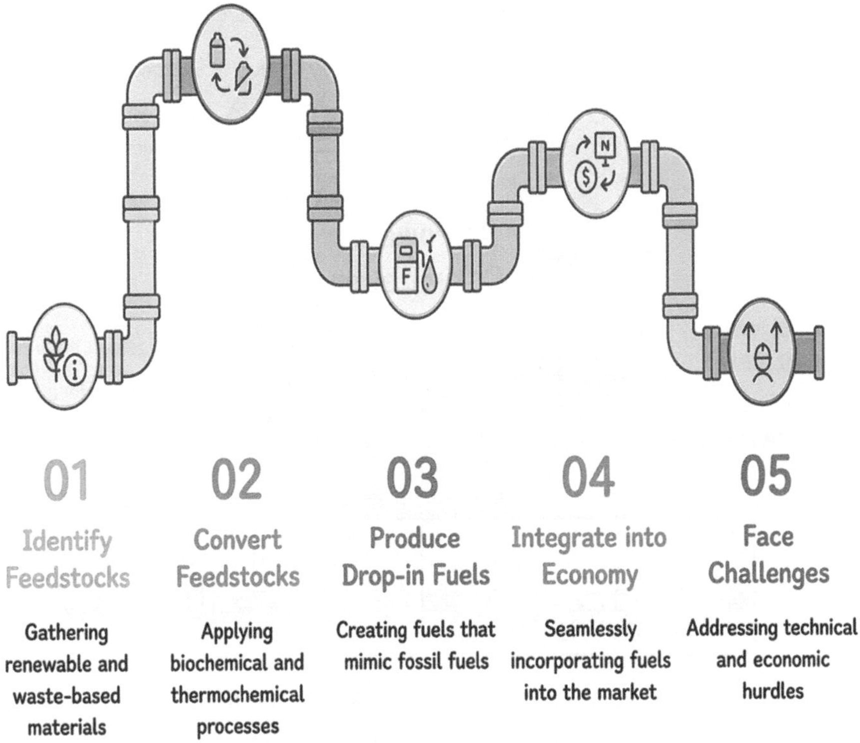

FIGURE 11.2 Drop-in fuel production process.

commercial application is constrained by technical and economic difficulties, including feedstock variability, strain limitations, inefficient processes, high production costs, and downstream processing complexities.

Figure 11.2 depicts the transition from sustainable biomass and waste streams to next-generation fuels suitable for existing engines and distribution systems. Although these fuels show strong potential, their widespread adoption is limited by technological and economic challenges. Meeting these challenges demands comprehensive approaches that improve efficiency and scalability. Advanced computational tools play a central role in guiding bioprocess optimization, improving operational performance, and managing industrial-scale implementation. Real-time process management and multi-objective optimization support the translation of laboratory findings into industrial applications.

Figure 11.3 illustrates the impact of intelligent technologies in improving the efficiency and low-carbon performance of fuel systems. Through optimization of microbial systems, tailored feedstock use, operational process refinement, and advanced computational tools, yields are enhanced and ecological performance is improved. This approach facilitates the transition to low-carbon and more advanced renewable fuel systems (Joel et al., 2024).

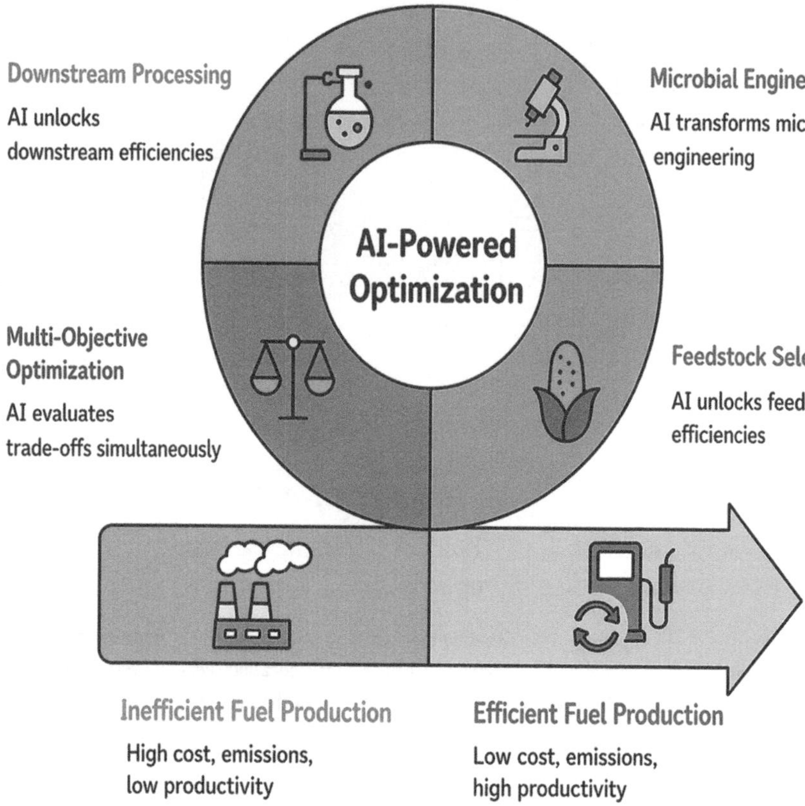

FIGURE 11.3 AI transforms drop-in fuel development.

11.1.1 Bioprocessing: Key Steps and Mechanisms

Renewable fuel production through bioconversion involves a series of stages that convert organic materials into high-energy liquids through integrated biochemical and catalytic processes. Initial processing of renewable biomass increases sugar availability by breaking down fibrous structures, followed by microbial metabolism into bio-alcohols and lipid derivatives. Catalytic conversion transforms these intermediates into bio-hydrocarbons analogous to conventional fuels, while downstream operations maintain product quality. Thorough improvement at each level supports high performance, economic practicality, and environmental responsibility. Figure 11.4 depicts the stepwise conversion of organic material into renewable fuels: Processing of biomass releases digestible compounds, microorganism-mediated reactions produce intermediates, upgrading produces energy-dense fuels, and final purification guarantees industrial applicability.

11.1.2 Challenges in Conventional Bioprocessing for Drop-in Fuels

Conventional biological conversion of organic feedstocks into renewable fuels is challenged by substantial obstacles. Microorganisms often struggle with inhibitory compounds and heterogeneous substrates, and essential pre-treatment procedures

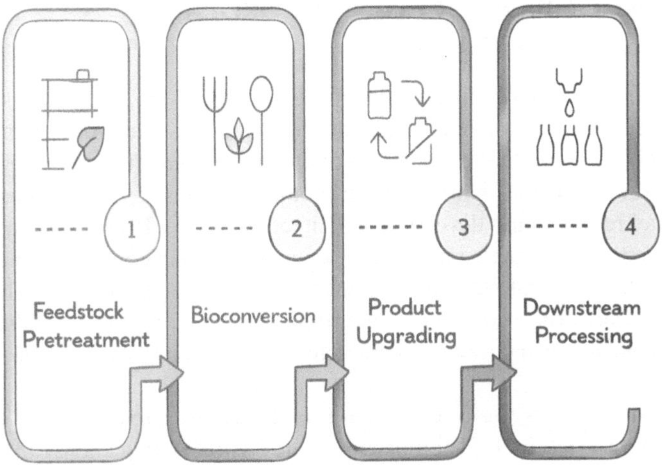

FIGURE 11.4 Bioprocessing for biofuel production.

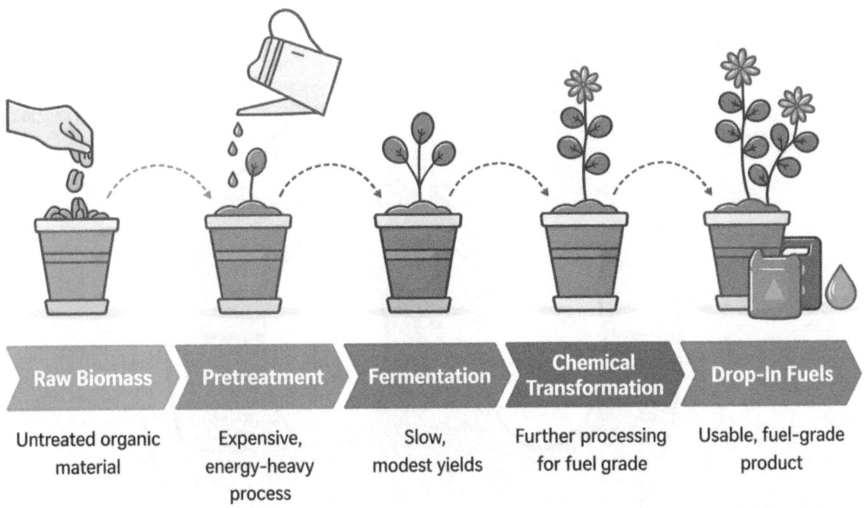

FIGURE 11.5 Bioprocessing challenges.

may reduce process efficiency. The attainment of final product standards is constrained by low process outputs and the need for chemical enhancement of intermediates. Scaling these processes from laboratory experiments to industrial production is difficult, and combined with market and integration challenges, this makes conventional bioprocessing a demanding route for biofuel production.

Figure 11.5 illustrates how raw biomass is converted into biofuels through conventional bioprocessing. Pre-treatment releases sugars from organic material but may produce fermentation inhibitors. Microbes convert these sugars into intermediates, which are chemically upgraded into fuels for existing engines, demonstrating both the workflow and inherent limitations.

11.2 ROLE OF BIOPROCESSING IN DROP-IN FUEL PRODUCTION

At the core of sustainable fuel production is bioprocessing, which uses biological systems to convert renewable biomass into hydrocarbons that work seamlessly with current engines. Food-derived wastes provide useful compounds such as flavorings and preservatives, while lignocellulosic biomass remains a major focus for large-scale biofuel research and development (Lin et al., 2013). Biomass sources, including plant matter and organic residues, are converted into energy-rich molecules such as alkanes, alcohols, and fatty acids through bioconversion techniques. These renewable fuels lower carbon emissions and are increasingly feasible for industrial-scale production due to advances in synthetic biology approaches and biomass conversion technologies.

Figure 11.6 depicts the bioconversion pathway from renewable organic feedstocks to energy-rich fuels. Their use integrates with existing energy infrastructure while mitigating carbon emissions, highlighting an environmentally conscious approach.

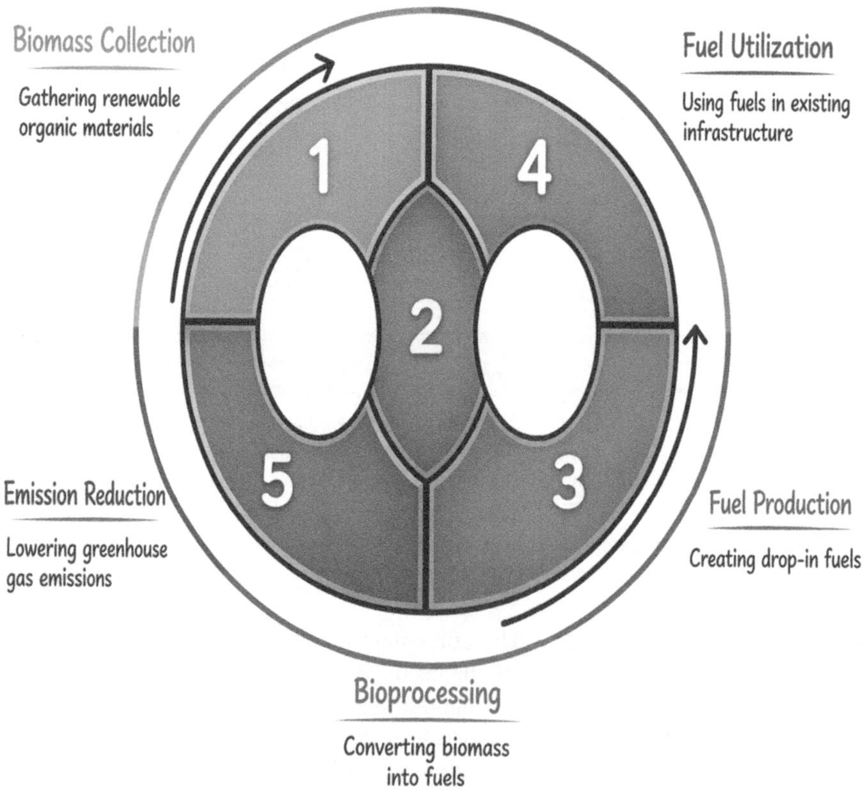

FIGURE 11.6 Bioprocessing cycle for drop-in fuels.

11.2.1 Challenges in Bioprocessing for Fuels

Biotechnological fuel generation encounters significant operational and efficiency barriers. Raw material inconsistencies reduce efficiency, biocatalytic conversion is limited, and complex biosynthetic networks challenge optimization. Post-production processing incurs high costs, industrial-scale scaling is difficult, and sustaining aseptic conditions at large scale is resource-intensive.

As shown in Figure 11.7, renewable fuel synthesis using biological pathways faces multiple obstacles. Reduced organismal yields and fluctuating substrate quality undermine process stability, while sophisticated biochemical pathways and high downstream costs make production more challenging. Process expansion difficulties and purity threats continue to limit industrial implementation, emphasizing the importance of streamlined and well-managed operations.

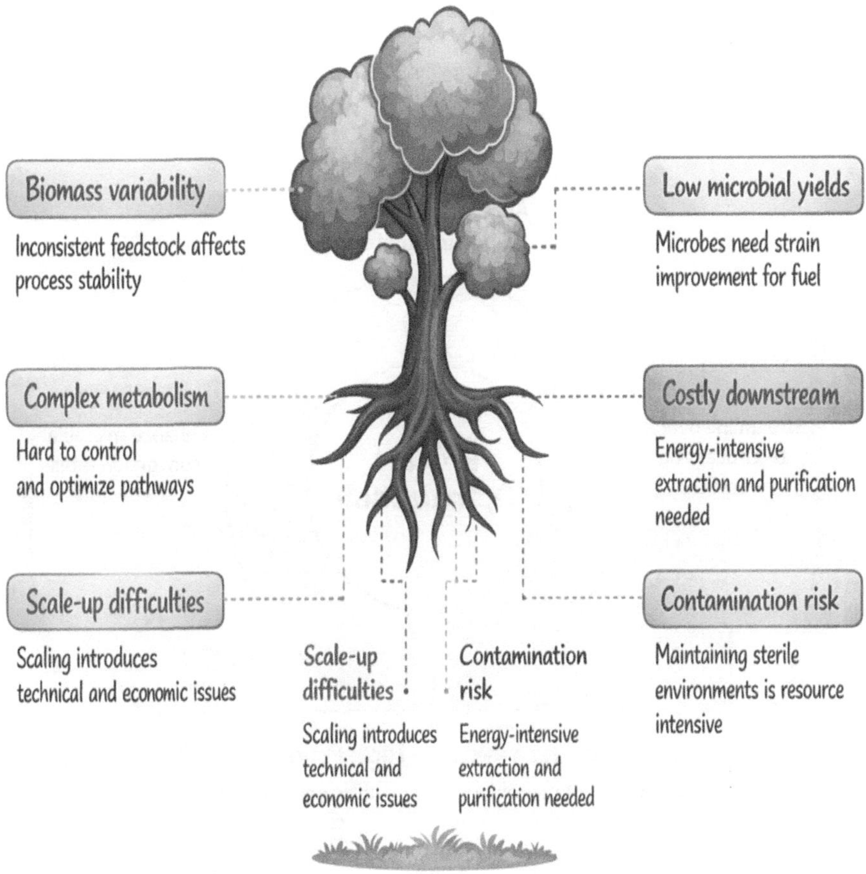

FIGURE 11.7 Challenges in bioprocessing for biofuel production.

11.2.2 Low Yield and Long Process Time

The production of drop-in fuels from lignocellulosic biomass through conventional bioprocessing remains constrained by modest yields and extended processing durations. Fermentative conversion of sugars to intermediates is often limited by low carbon flux efficiency, incomplete feedstock utilization, and inhibition caused by product buildup. These inefficiencies result in slower fermentation cycles and reduced reactor productivity. Furthermore, downstream refining processes, such as hydrodeoxygenation and alkylation, add significant time and cost due to their dependence on high-temperature, high-pressure operations, and catalyst instability. The combination of biological and chemical bottlenecks reduces throughput and increases operational costs. Strategies to address these issues include metabolic engineering of high-yield microbial strains, continuous or fed-batch processing modes, integration of real-time monitoring and AI-based optimization algorithms, and the development of more active and selective catalytic systems.

Figure 11.8 shows the role of process optimization in improving bioprocessing outcomes. By enhancing microbial strain performance, accelerating catalytic

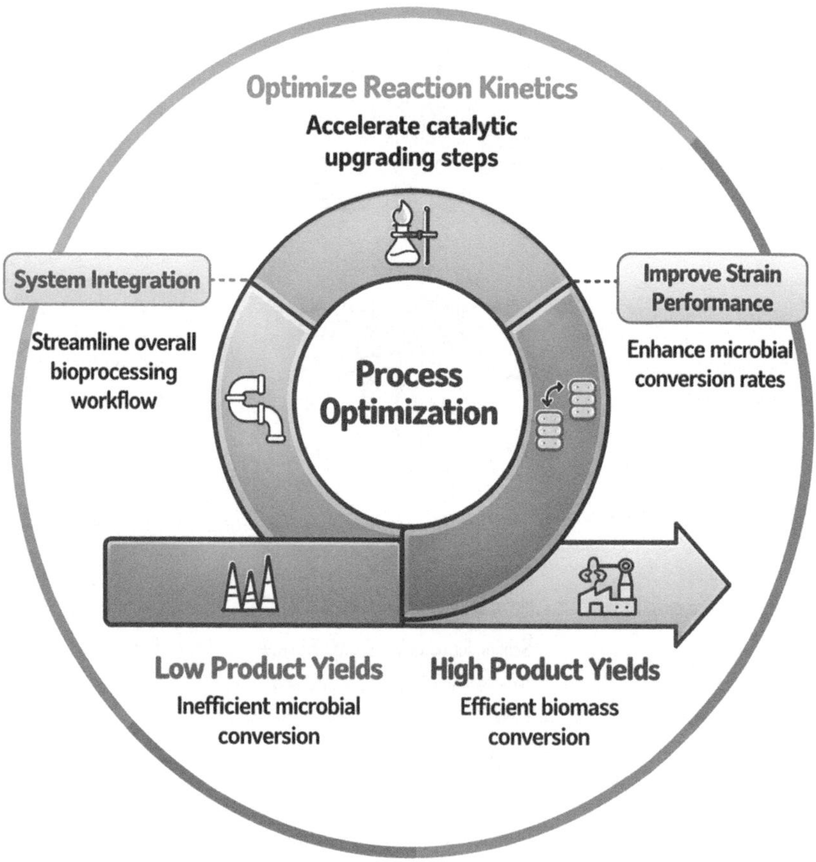

FIGURE 11.8 Improving biomass conversion efficiency.

reactions, and streamlining the overall system, low-yield processes can be upgraded to deliver high product yields with greater efficiency.

11.2.3 High Energy and Capital Costs

Producing drop-in fuels through conventional bioprocessing methods involves considerable energy demands and financial investment. The initial biomass pre-treatment phase typically relies on energy-intensive thermal, chemical, or mechanical processes to break down complex plant structures. Likewise, the subsequent upgrading steps, which convert bio-intermediates into fuel-grade hydrocarbons, often require expensive catalysts and operate under extreme conditions such as high temperatures and pressures. These technical requirements drive up both operational and capital expenditures, making the overall process less economically competitive compared to fossil fuel refining. This poses a major barrier to widespread industrial adoption of bio-based fuel technologies.

Figure 11.9 shows that high energy demands in bioprocessing lead to increased costs, slower industrial adoption, and reduced competitiveness with fossil fuels. These challenges limit the economic and practical viability of drop-in fuel production.

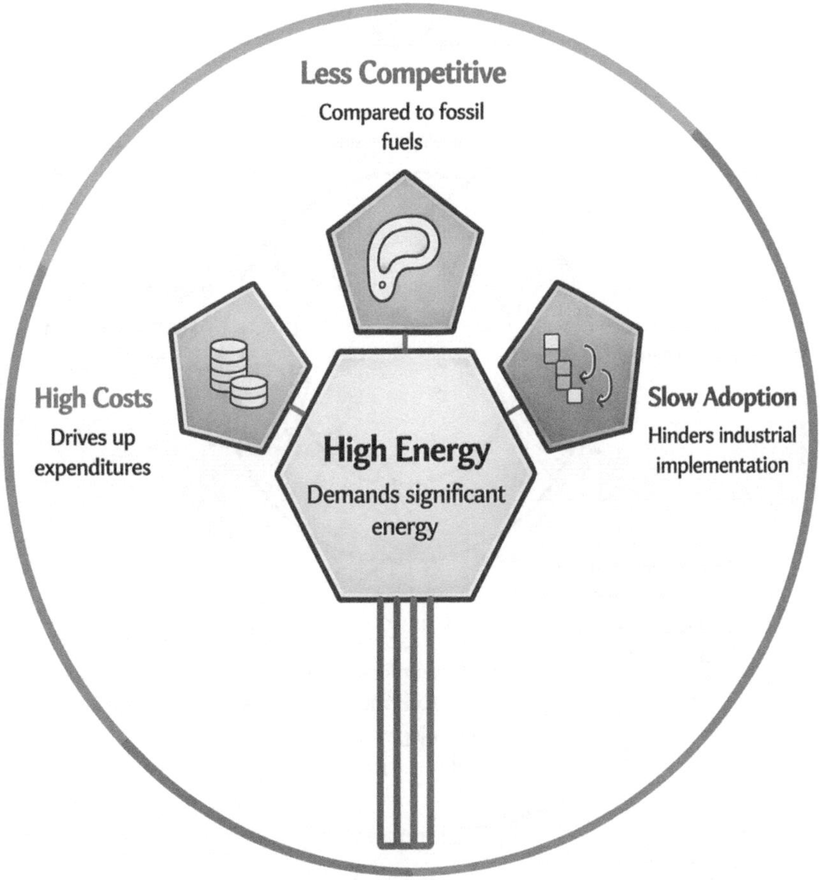

FIGURE 11.9 High costs hinder biofuel adoption.

11.2.4 Inefficient Strain Selection and Pathway Optimization

The selection of microbial strains and the design of metabolic pathways are critical factors in bioprocessing efficiency. As reported in Williams and Laurens (2010), the high production costs render algae-derived biofuels economically impractical unless supplemented by the generation of higher-value co-products. However, traditional methods often rely on naturally occurring strains with limited productivity and tolerance to stress conditions. Without precise optimization of metabolic pathways, substrate utilization remains incomplete, and undesired byproducts may accumulate, reducing overall yield. Additionally, suboptimal regulatory control and lack of dynamic pathway tuning further hinder performance. Overcoming these limitations requires integrating synthetic biology, systems biology, and AI-guided strain engineering to enable efficient, scalable microbial systems for sustainable drop-in fuels.

Figure 11.10 outlines a full-spectrum approach to microbial biofuel engineering, spanning from the rational design of high-performance strains and metabolic regulation to pilot-scale evaluation and large-scale industrial application.

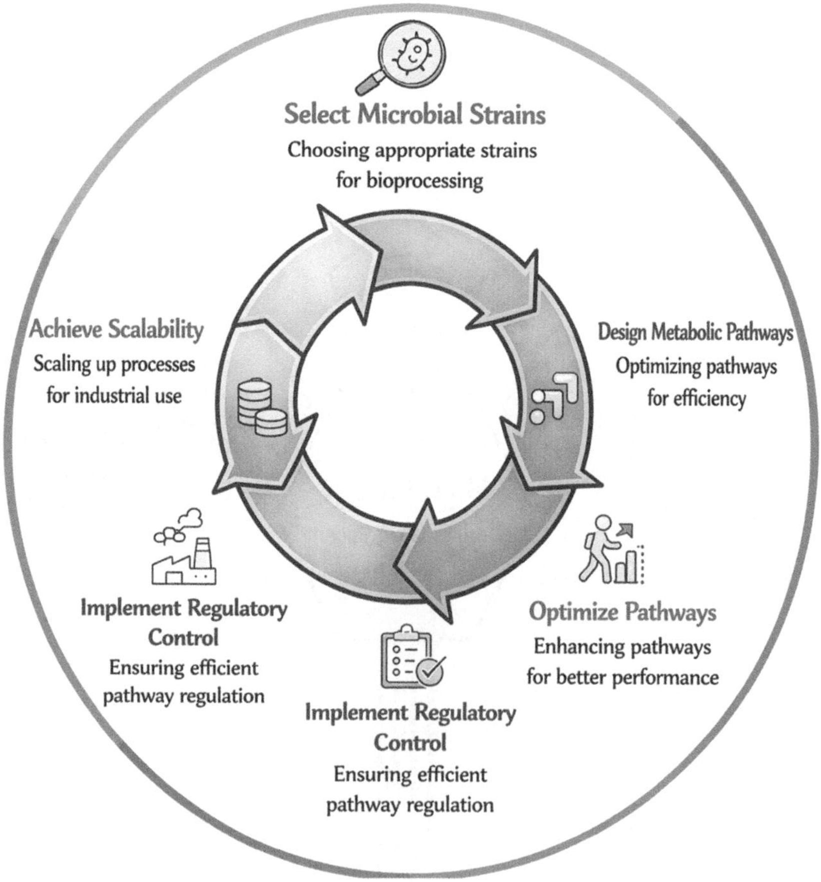

FIGURE 11.10 Cycle of bioprocessing optimization.

11.3 ROLE OF AI IN BIOPROCESSING

AI-powered bioprocessing enhances drop-in fuel production through data-driven optimization of microbial performance and fermentation strategies. Real-time feedback and predictive modeling ensure stable, energy-efficient operations, accelerating the development of scalable and economically viable biofuel manufacturing.

Table 11.1 highlights strategies that enhance productivity, flexibility, and environmental performance in biofuel bioprocessing. Advanced algorithmic methods, such as deep learning models, optimize processes, increase efficiency, and shorten production times. Possible process issues are managed in advance, reducing downtime, while dynamic detectors support system robustness. Moreover, modeling frameworks assess ecological aspects across industrial processes. Altogether, these strategies enhance operational effectiveness and support the implementation of adaptive processes.

11.3.1 Performance Improvement through Analytical Methods

Innovative techniques are enhancing performance and operational control in renewable energy systems. The emergence of biofuels, biopharmaceuticals, and wastewater treatment products (Joshi and Mishra, 2022) has accelerated the use of bioreactor and sensor data to identify hidden patterns. Methods, including principal component analysis, machine learning, and deep neural networks, facilitate monitoring and prediction of key variables such as biomass growth, substrate utilization, and product yield. By enabling real-time monitoring and predictive decision-making, these methods ensure that fermentation conditions and downstream processes remain stable and efficient. Predictive modeling reduces the risk of system failures and downtime, while data-driven optimization improves process efficiency, product quality, and sustainable resource management across the entire production workflow.

TABLE 11.1
Survey Findings of AI in Bioprocessing (Butean et al., 2025)

S. No.	Area	AI-based Strategy	Efficiency Improvement	Operating Conditions
1	Strain optimization	Genetic Algorithms	18% increase in biomass yield	Temperature: 37°C; pH: 7.0; Glucose feed: 20 g/L
2	Fermentation condition prediction	Artificial Neural Networks	22% rise in biofuel yield	Temperature: 35 °C; pH: 6.8; Agitation: 200 rpm
3	Downstream process optimization	Support Vector Machines (SVM)	15% reduction in purification time	Catalyst bed temperature: 60°C; Flow rate: 1.5 mL/min
4	Integrated bioprocess monitoring	Deep Learning	25% improvement in process stability	Temperature: 38°C; pH: 6.5; Real-time sensor feedback

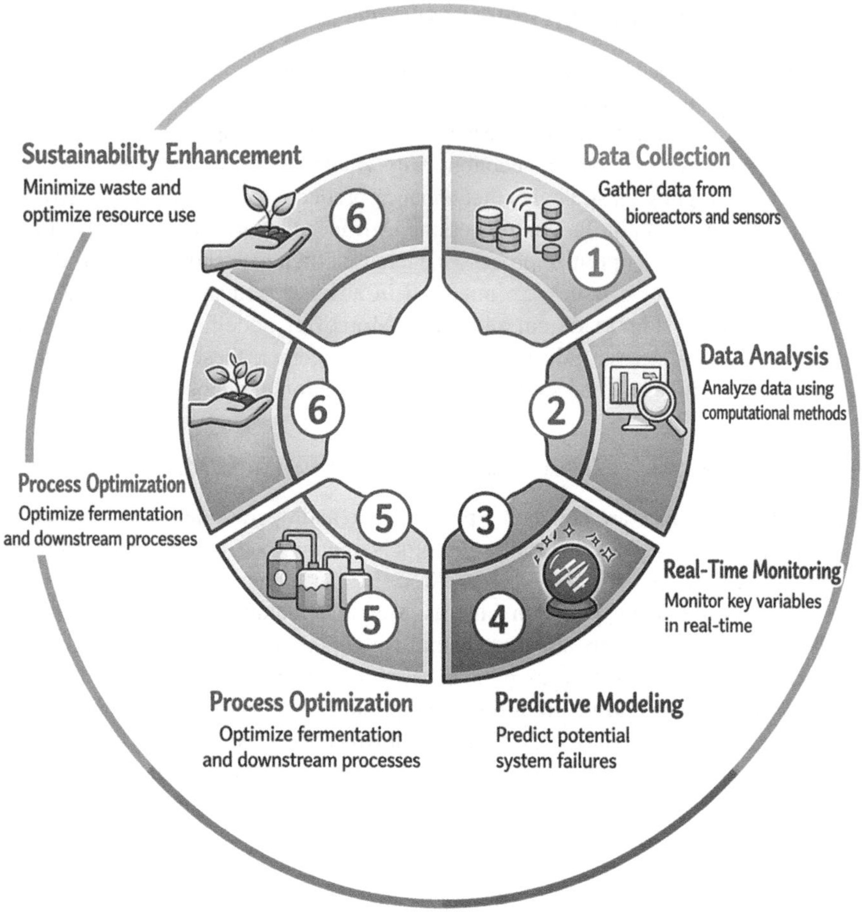

FIGURE 11.11 Data-driven biofuel production cycle.

Figure 11.11 illustrates AI-driven biofuel bioprocessing, covering data acquisition, computational analysis, real-time monitoring, predictive failure detection, process optimization, and enhanced sustainability through minimized waste.

11.3.2 AI for Genetic Engineering and Strain Design

Transformation of genetic engineering and strain design is driven by AI, which employs predictive, data-driven, and automated approaches that surpass conventional trial-and-error methods. By analyzing and integrating multi-omics datasets, it uncovers critical gene–trait links, regulatory pathways, and metabolic constraints, enabling precise pathway optimization, targeted genetic modifications, and advanced enzyme engineering. AI-enhanced genome-scale metabolic models simulate strain performance under varied process conditions, reducing the need for extensive laboratory testing. Coupled with high-throughput "Design–Build–Test–Learn" cycles, these capabilities enable the rapid development of robust microbial platforms for drop-in fuels, featuring enhanced carbon conversion, inhibitor resistance, and flexible biomass utilization, thereby accelerating the path to large-scale sustainable fuel production (De Souza Celente et al., 2022; Deamici et al., 2025).

Figure 11.12 shows the role of AI in genetic engineering and strain design, highlighting the integration of multi-omics data with predictive modeling to guide genetic modifications and enzyme design. It depicts how AI-driven simulations can assess

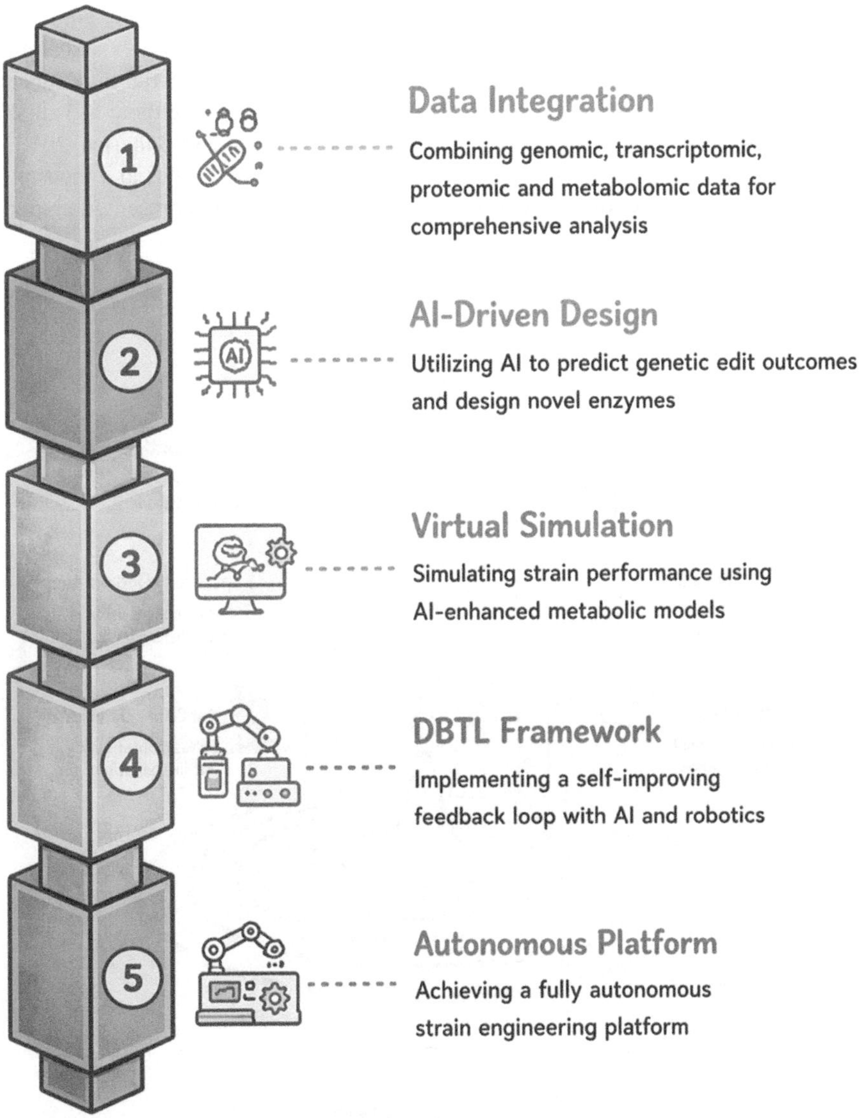

FIGURE 11.12 AI-integrated framework for strain engineering.

strain performance virtually, while automated, feedback-based workflows link computational predictions with robotic experimentation. Together, these capabilities enable fully autonomous systems for designing, testing, and optimizing microbial strains for industrial applications such as sustainable drop-in fuel production.

11.4 AI TOOLS AND TECHNIQUES IN BIOFUEL PRODUCTION

AI tools are increasingly recognized as powerful approaches for the modeling and optimization of bioprocesses. Artificial neural networks, genetic algorithms (GAs) (Prakasham et al., 2011), and other approaches, including fuzzy logic, ant colony optimization, particle swarm optimization, excel in addressing complex and nonlinear systems. These techniques have demonstrated significant potential in advancing the design, control, and optimization of bioprocesses, positioning them as valuable assets in both research and industrial applications (Garlapati and Banerjee, 2010).

Figure 11.13 depicts a range of machine intelligence approaches and their main benefits. It presents neural network models for complex system representation, fuzzy

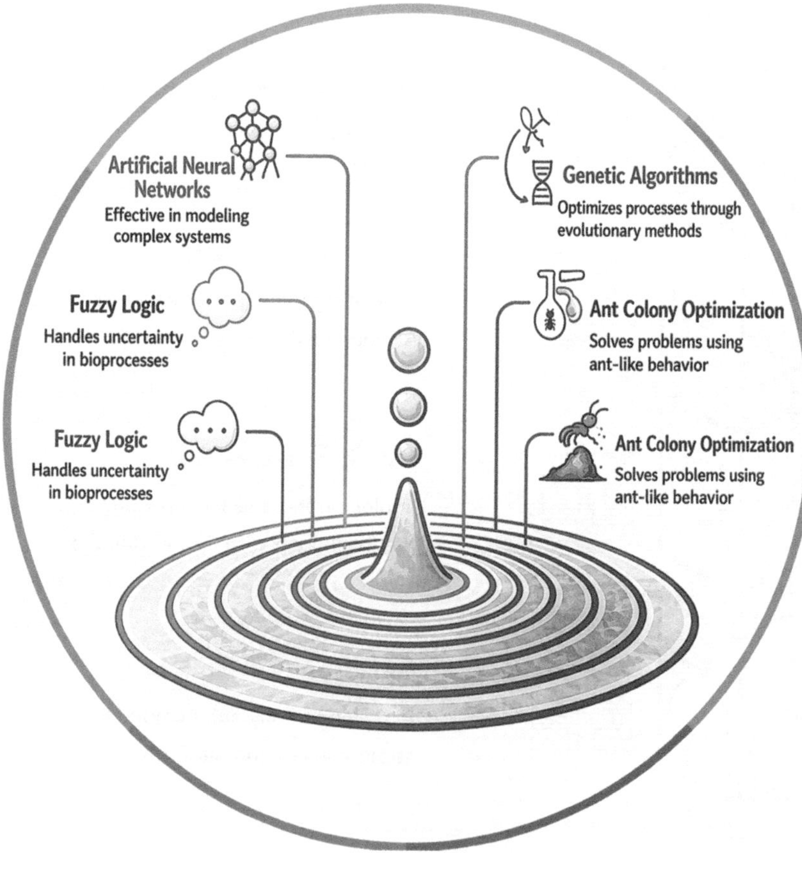

FIGURE 11.13 Soft computing techniques for real-time decision-making.

reasoning for handling uncertainty, swarm intelligence, and evolutionary algorithms for solution improvement and ant-inspired optimization for collaborative path finding. Their interconnected nature highlights their importance in solving complex optimization and decision-making challenges.

11.4.1 Supervised and Unsupervised Learning Models

Data-driven learning methods are instrumental in driving progress in renewable fuel technologies. Labeled-data–based methods facilitate early detection of quality issues. Exploratory approaches detect patterns, monitor irregularities, and explore inter-variable connections. The use of these methods equips biofuel plants with adaptive systems that enhance operational efficiency, reduce risks, and drive rapid progress in bioenergy development (Al-Asheh et al., 2021).

Figure 11.14 depicts an integrated workflow for process improvement, starting with predictive modeling, in which labeled datasets enable models to produce

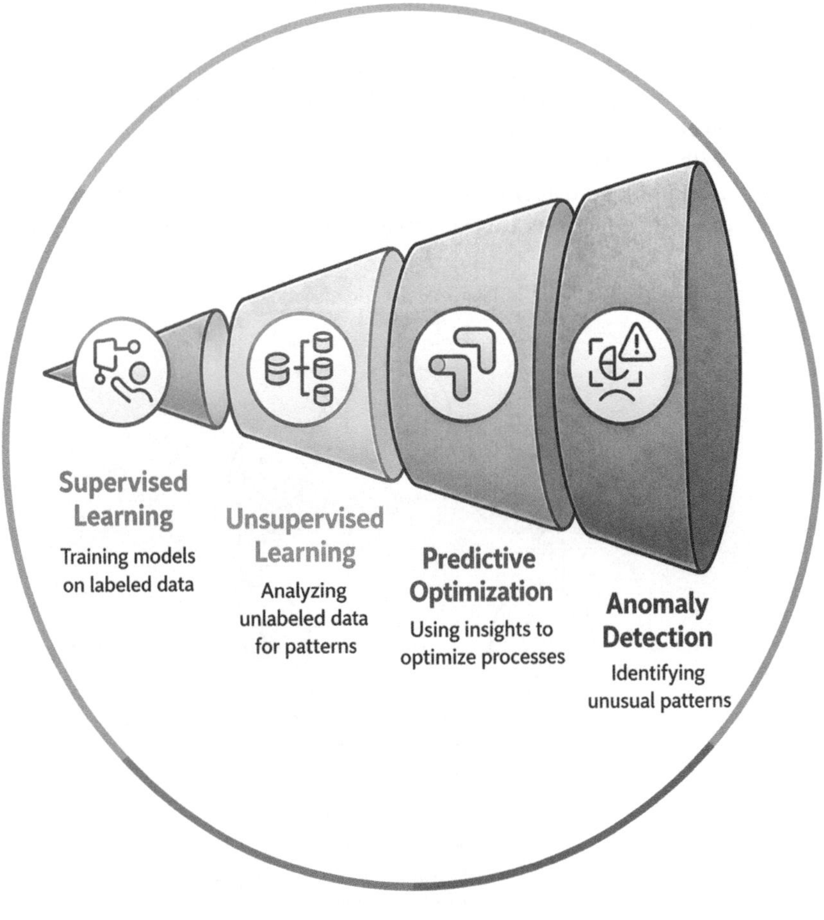

FIGURE 11.14 Supervised and unsupervised learning in predictive optimization.

accurate predictions and categorizations. Data-driven clustering is employed to uncover underlying structures within unlabeled data, complementing supervised methods in predictive optimization. This integrated approach refines operational processes, improves performance, and incorporates anomaly detection to flag potential faults or inefficiencies, forming a comprehensive machine learning framework for continuous improvement and proactive decision-making.

11.4.2 Reinforcement Learning in Process Control

In process control, feedback-driven learning refines operational decision-making through environmental interactions, reinforcing successful actions and discouraging ineffective ones. The system responds to fluctuating conditions by regulating critical operational parameters, improving efficiency, maintaining product quality, and minimizing energy use, achieving stable and steadily enhancing performance. Traditional diesel, employed across transport, energy generation, and industrial sectors, contributes substantially to climate change and human health concerns through particulate matter and nitrogen oxide (NO_x) emissions (Ifeanyi-Nze et al., 2023; Lin et al., 2023). With fossil-based energy resources declining and environmental pressures increasing, biodiesel has emerged as a promising renewable and eco-friendly fuel option. Derived from plant- and animal-based oils, it is biodegradable, non-toxic, and produces lower emissions, thereby mitigating the ecological and public health risks associated with conventional diesel (Tamilvanan et al., 2023; Andeden et al., 2021).

As illustrated in Figure 11.15, the system responds to its surroundings, evaluates output data, and fine-tunes its control variables to enhance efficiency, lower energy

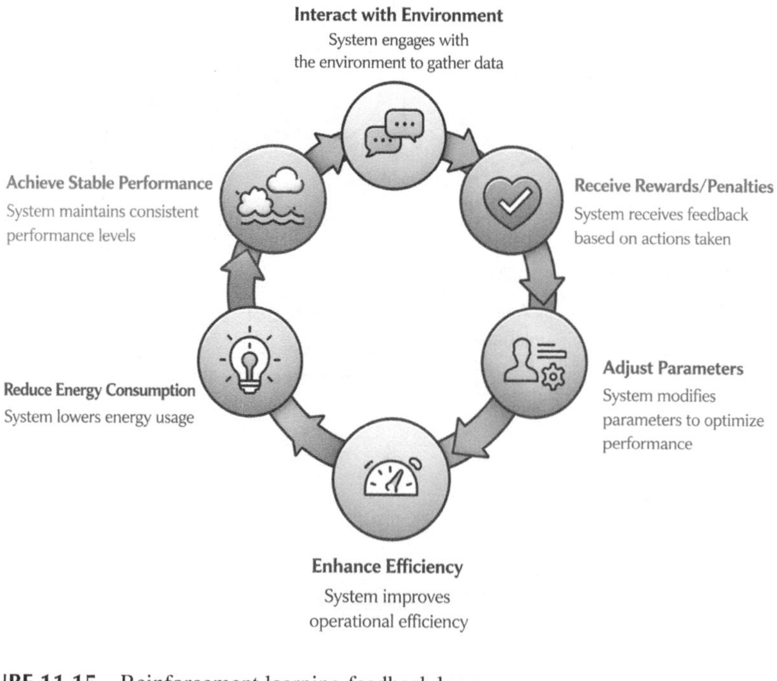

FIGURE 11.15 Reinforcement learning feedback loop.

use, and sustain stable functioning. The process begins with environmental interaction and data acquisition, followed by the reception of rewards or penalties that inform parameter modifications. These refinements lead to improved performance and energy savings, with the cycle repeating to ensure adaptability under changing conditions.

11.4.3 AI-driven Digital Twins for Fermentation Monitoring in Biofuels

AI-driven digital twins are transforming biofuel fermentation by developing dynamic virtual models of bioprocess systems that integrate live sensor data, process simulations, and historical performance to predict and optimize operations. In waste-to-energy systems, the implementation of digital twin technology facilitates the integration of real-time operational data with high-fidelity process simulations, enabling the optimization of waste conversion pathways and enhancing the efficiency of anaerobic digestion for improved bioenergy production (Zohdi, 2023; Goswami et al., 2022). Powered by AI, digital twins process real-time inputs from IoT-enabled bioreactors such as pH, temperature, dissolved oxygen, and substrate concentration to deliver predictive analytics, early fault detection, and adaptive process optimization. Digital twins enable virtual simulations to evaluate the effects of strain selection, nutrient concentrations, and process modifications on overall biomanufacturing performance. By accurately replicating real-world operating conditions, these simulations reduce the need for costly and time-consuming laboratory experiments while allowing reliable prediction of process outcomes before implementation. Machine learning further enhances digital twins by enabling adaptive and self-optimizing process control based on real-time data. This dynamic optimization improves product yield, process consistency, and scalability by continuously adjusting operating parameters to maintain stable performance. In addition, optimized resource allocation minimizes waste generation, reduces operational costs by limiting redundant experimentation, and enhances process precision. Energy-efficient operation and reduced material consumption also lower the environmental footprint of biomanufacturing. As intelligent virtual counterparts of physical bioprocess systems, digital twins integrate real-time monitoring, predictive analytics, and automated control within a unified digital framework. Collectively, these capabilities support data-driven decision-making, significantly improving the efficiency, reliability, and sustainability of modern biomanufacturing (Belik and Rubanenko, 2023; Chen et al., 2021).

Figure 11.16 shows an AI-driven bioprocess cycle where live sensor data is collected, simulated, analyzed, and used for predictive optimization, adaptive control, and outcome evaluation to continuously improve fermentation efficiency.

11.4.4 Integration with IoT and Smart Sensors

The integration of IoT and smart sensors in biofuel production enables real-time monitoring of key parameters such as temperature, pH, dissolved oxygen, and substrate concentration. These data streams support predictive maintenance, early fault detection, and adaptive process control, improving yield, scalability, and sustainability while reducing waste and operational costs. Biomass-to-biofuel conversion offers a sustainable alternative to fossil fuels, with AI-enhancing efficiency through

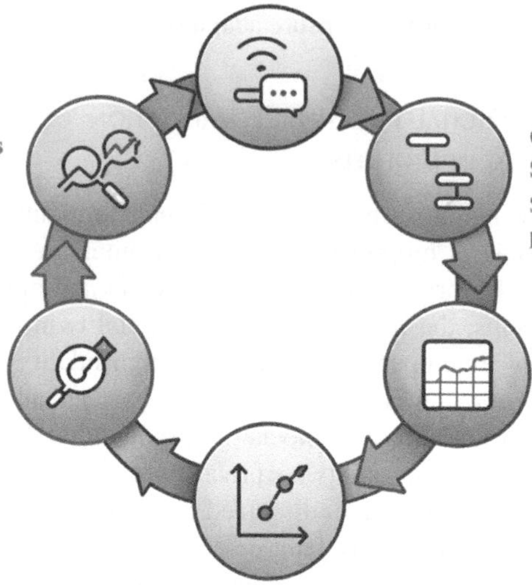

FIGURE 11.16 AI-driven digital twin cycle in biofuel fermentation.

optimized feedstock selection, blending and pre-processing via data-driven algorithms and predictive models, resulting in higher yields, cost-effectiveness, and reduced waste (Kumar and Bharadvaja, 2020). A promising avenue in industrial biotechnology is microalgae biomass production (Balamurugan et al., 2025), while advancements in MEMS, wireless communication, and digital electronics have driven significant progress in IoT technology (Beatrice-Lindner et al., 2018).

Figure 11.17 shows the AI-enabled bioprocessing framework for drop-in fuel production, highlighting raw material inputs, sensor and IoT-based data collection, AI-driven modeling and optimization, and output of renewable fuels, with feedback loops for continuous improvement.

11.4.5 Industrial and Energy Applications of Drop-In Fuels

Drop-in fuels serve as crucial enablers of sustainable industrial and energy transitions due to their ability to substitute conventional fossil fuels without major infrastructure modifications. In industrial operations, they are used in boilers, furnaces, and combined heat and power systems to lower greenhouse gas emissions while maintaining process efficiency. Power generation facilities can integrate renewable diesel, synthetic hydrocarbons, or biojet fuels in existing turbines to produce

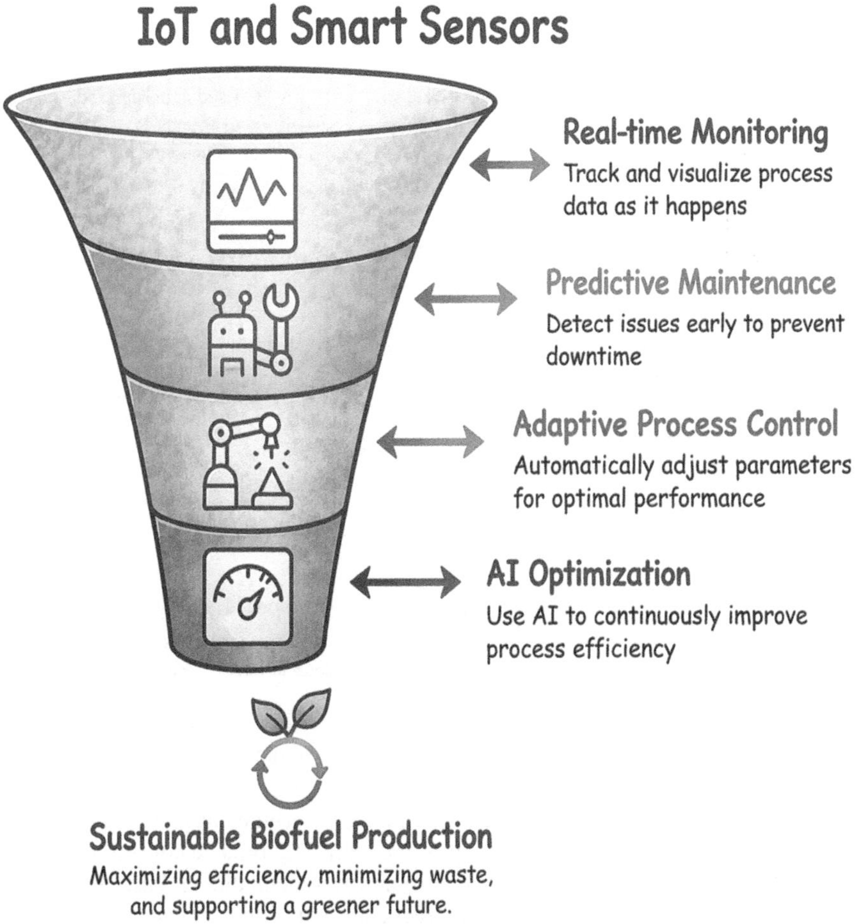

FIGURE 11.17 Enhancing biofuel production with technology.

cleaner electricity and support grid decarbonization. In the transportation sector, their compatibility with current engines allows seamless use in road, marine, and aviation applications, particularly in areas where electrification is difficult. Additionally, the conversion of agricultural residues and waste biomass into drop-in fuels promotes resource circularity and energy diversification, reducing reliance on imported petroleum. Collectively, these applications highlight the potential of drop-in fuels to bridge conventional energy systems with a more resilient and low-carbon future.

Figure 11.18 illustrates the diverse industrial and energy applications of drop-in fuels, highlighting their compatibility with existing infrastructure and their role in supporting a low-carbon transition. In the industrial sector, drop-in fuels are utilized in boilers and combined heat and power (CHP) systems to reduce emissions while maintaining energy efficiency. For power generation, they enable cleaner

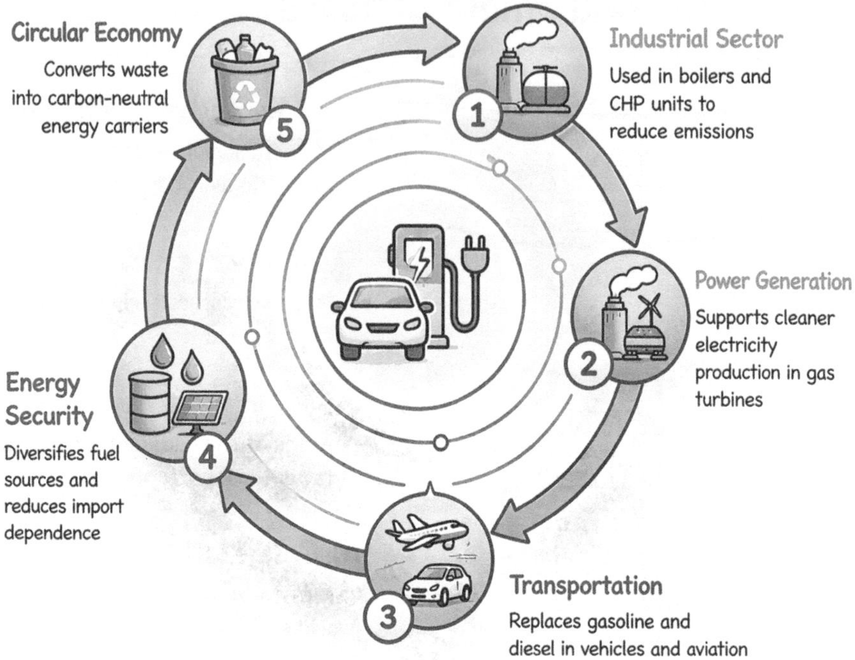

FIGURE 11.18 Multisectoral implementation pathways for drop-in fuels.

electricity production through integration into gas turbines and existing grid systems. In the transportation sector, drop-in fuels directly replace conventional gasoline and diesel in vehicles and aviation, promoting sustainable mobility without the need for engine modification. Beyond direct energy use, they enhance energy security by diversifying fuel sources and reducing reliance on imported petroleum. Additionally, their production from agricultural residues and waste materials supports the circular economy by converting waste into carbon-neutral energy carriers. Overall, the figure emphasizes how drop-in fuels provide a practical pathway toward industrial decarbonization, energy diversification, and sustainable resource utilization (Abu Qdais et al., 2010).

11.5 CONCLUSION

The integration of AI with bioprocessing offers a transformative pathway toward the sustainable and efficient production of drop-in fuels. AI technologies enable continuous process monitoring, real-time optimization, and predictive control, ensuring that critical parameters such as temperature, pH, nutrient concentration, and microbial activity remain within ideal ranges. This not only improves process stability and fuel yield but also minimizes energy use and material waste. By applying machine learning and data-driven modeling, AI systems can predict system behavior, identify process deviations early, and make autonomous adjustments, resulting in greater operational efficiency and reliability. Furthermore, digital twin simulations allow for virtual experimentation and scale-up planning, reducing time and cost barriers in moving from lab-scale research to industrial production. From a sustainability perspective, AI-driven bioprocessing enhances resource utilization and reduces greenhouse gas emissions, aligning well with global efforts to transition toward low-carbon energy systems. The drop-in fuels produced through such optimized bioprocesses are compatible with existing fuel infrastructure, facilitating their immediate integration into current transportation and energy networks. Overall, the synergy between AI and bioprocessing not only accelerates the development of renewable, eco-friendly fuels but also establishes a foundation for future innovations in smart, adaptive, and sustainable energy production systems.

REFERENCES

Abu Qdais, H.K., Bani Hani, K., & Shatnawi, N. (2010). Modeling and optimization of biogas production from a waste digester using artificial neural network and genetic algorithm. *Resources, Conservation and Recycling*, 54(6), 359–363.

Al-Asheh, S., Bagheri, M., & Aiden, A. (2021). Membrane bioreactor for wastewater treatment: A review. *Case Studies in Chemical and Environmental Engineering*, 4, 100109.

Andeden, E.E., Ozturk, S., & Aslim, B. (2021). Evaluation of thirty microalgal isolates as biodiesel feedstocks based on lipid productivity and triacylglycerol (TAG) content. *Current Microbiology*, 78(2), 775–788.

Balamurugan, S., Li, D.W., Wang, S., & Li, H.Y. (2025). Unleashing the potential of biotechnological strategies for the sustainable production of microalgal polysaccharides. *Food Science and Nutrition*, 2025, 1–18.

Beatrice-Lindner, P., Garrido-Cardenas, J.A., Sepulveda, C., & Acien-Fernandez, F.G. (2018). A new approach for detection and quantification of microalgae in industrial-scale microalgal cultures. *Applied Microbiology and Biotechnology*, 102(19), 8249–8436.

Belik, M., & Rubanenko, O. (2023). Implementation of digital twin for increasing efficiency of renewable energy sources. *Energies*, 16(12), 1–27.

Butean, A., Cutean, I., Barbero, R., Enriquez, J., & Matei, A. (2025). A review of artificial intelligence applications for biorefineries and bioprocessing: from data-driven processes to optimization strategies and real-time control. *Processes*, *13*(8), 2544.

Chen, W., Wang, J., Ren, Y., Chen, H., He, C., & Wang, Q. (2021). Optimized production and enrichment of α-linolenic acid by *Scenedesmus* sp. HSJ296. *Algal Research*, 60, 102505.

De Souza Celente, G., Rizzetti, T.M., Sui, Y., & de Souza Schneider, R. (2022). Potential use of microalga *Dunaliella salina* for bioproducts with industrial relevance. *Biomass and Bioenergy*, 167, 106647.

Deamici, K.M., Figueiredo, D., Guerra, I., Letras, P., & Pereira, H. (2025). Chapter 2 - Global market and future trends of microalgae-based products. *Algal Bioreactors*, 1, 11–25.

Garlapati, V.K., & Banerjee, R. (2010). Evolutionary and swarm intelligence-based approaches for optimization of lipase extraction from fermented broth. *Engineering in Life Sciences*, 10(3), 186–275.

Goswami, R.K., Agrawal, K., Shah, M.P., & Verma, P. (2022). Bioremediation of heavy metals from wastewater: A current perspective on microalgae-based future. *Applied Microbiology*, 75(4), 701–717.

Ifeanyi-Nze, F.O., Omiyale, C.O., Okonkwo, I.U., Chukwu, C.J., Nwankwor, C.M., Onabanjo, A.O., Adoga, S.O., Chukwu, J.O., Chukwurah, K.F., Ebikemiyen, M., Okon, G.U., Adeleke, O.J., & Tajudeen, K.A. (2023). Biodiesel synthesis from waste vegetable oil Utilizing eggshell ash as an innovative heterogenous catalyst. *Archives of Advanced Engineering Science*, 165(3), 1–18.

Joel, O.S., Oyewole, A.T., & Odunaiya, O.G. (2024). Leveraging artificial intelligence for enhanced supply chain optimization: A comprehensive review of current practices and future potentials. *International Journal of Management & Entrepreneurship Research*, 6(3), 704–721.

Joshi, S., & Mishra, S. (2022). Recent advances in biofuel production through metabolic engineering. *Bioresource Technology*, 352, 127037.

Kumar, L., & Bharadvaja, N. (2020). Microbial remediation of heavy metals. *Microbial Bioremediation & Biodegradation*, 2, 49–72.

Lin, C.S.K., Pfaltzgraff, L.A., Herrero-Davila, L., Mubofu, E.B., Abderrahim, S., Clark, J.H., Koutinas, A.A., Kopsahelis, N., Stamatelatou, K., Dickson, F., Thankappan, S., Mohamed, Z., Brocklesby, R., & Luque, R. (2013). Food waste as a valuable resource for the production of chemicals, materials and fuels. Current situation and global perspective. *Energy & Environmental Science*, 6(3), 426–464.

Prakasham, R.S., Sathish, T., & Brahmaiah, P. (2011). Imperative role of neural networks coupled genetic algorithm on optimization of biohydrogen yield. *International Journal of Hydrogen Energy*, 36(7), 4332–4339.

Tamilvanan, A., Mohanraj, T., Ashok, B., & Santhoshkumar, A. (2023). Enhancement of energy conversion and emission reduction of *Calophyllum inophyllum* biodiesel in diesel engine using reactivity controlled compression ignition strategy and TOPSIS optimization. *Energy*, 264, 126168.

Tyrovola, T., Dodos, G., Kalligeros, S., & Zannikos, F. (2017). The introduction of biofuels in marine sector. *Journal of Environmental Science and Engineering*, 6, 415–421.

Williams, P.J.L.B., & Laurens, L.M.L. (2010). Microalgae as biodiesel & biomass feedstocks: Review & analysis of the biochemistry, energetics & economics. *Energy & Environmental Science*, 3(5), 497–676.

Zohdi, T.I. (2023). A machine-learning digital-twin for rapid large-scale solar-thermal energy system design. *Computer Methods in Applied Mechanics and Engineering*, 412, 115991.

12 AI-Driven Innovations in Hydro Treated Vegetable Oil (HVO)

V. S. Shaisundaram, Saravanakumar Sengottaiyan, D. Yogaraj, S. Karthikayan, R. Muraliraja, and R. S. Aravind Bharani

12.1 INTRODUCTION

With a growing global economy, we are facing the critical challenge of transforming the energy systems from fossil-intensive to renewable, low-carbon, sustainable ones in our lives. Within this framework, biofuels and especially Hydrotreated Vegetable Oil (HVO) a.k.a. renewable diesel, have arisen as a major technology to fill the gap between traditional fossil-based transportation fuels and future clean energy alternatives. HVO presents an alternative with high energy density, chemical and storage stability, as well as providing full diesel infrastructure compatibility (Searchinger et al., 2008). With the recent advancements in AI, HVO production potential has been further optimized by data-based process improvements, intelligent catalyst design and predictive control systems, leading to a higher efficiency and sustainability.

12.1.1 The Decline of Fossil Fuels and the Emergence of Biofuels

For more than a century, fossil fuels such as petroleum, coal and natural gas have been driving industrial growth and economic development but their ever-growing use has caused a substantial amount of greenhouse gas (GHG) emissions, environmental deterioration and geopolitical instability combined with expanding worldwide energy demand that challenged the limited availability and need to identify renewable carbon-mitigating alternatives like biofuels that can economically be incorporated into the current fuel infrastructure. But biofuels had their troubles in the beginning. First-generation biofuels, such as bioethanol and biodiesel, were based on food crops leading to "food versus fuel" debate (Chisti, 2007). Second-generation fuels from lignocellulosic waste solved the first problem while increasing conversion costs. Third-generation algae fuels have overly been marketed as potentially high-yielding and low-land use technology requiring technical readiness. These evolutionary stages paved the way for advanced biofuels like HVO that provide high performance and durability as well as unprecedented sustainability.

DOI: 10.1201/9781003684374-12

12.1.2 The Central Role of HVO among Modern Biofuels

HVO is a second-generation biofuel and effectively overcomes many of the weaknesses of previous technologies. Shaping of HVO is made by using catalytic hydro-processing of triglycerides and free fatty acids present in vegetable oils or animal fats; they undergo the following step reactions, hydrogenation–hydrodeoxygenation–decarboxylation–isomerization, to produce paraffinic indicated hydrocarbons which closely resemble fossil diesel. Unlike traditional fatty acid methyl ester (FAME) biodiesel, FAME-based HVO fuels do not contain any oxygen and have better properties in terms of oxidation stability, cetane number and cold flow. "HVO's strongest advantage is that it shares the full chemical and functional compatibility with diesel, which enables it to be used in all compression-ignition applications, (including very hard-to-electrify aviation, marine and heavy duty transportation)". Its burning results in much less particulate matter, unburned hydrocarbons and carbon monoxide emissions, resulting in cleaner air and better public health. Moreover, life cycle analysis has demonstrated that HVO derived from used cooking oil or animal fats is able to reach a reduction of GHG emissions close to 90% with respect to fossil diesel (Hansson et al., 2020).

Technologically speaking, HVO's production process dovetails easily with many existing refining facilities. Co-processing in oil refining can refer to the joint processing of renewable and fossil feedstock, and it provides a stepping stone towards gradual sustainable production while avoiding large capital outlays. This industrial compatibility sets HVO apart from other advanced biofuels, such as Fischer–Tropsch (FT) diesel, which entails gasifying the biomass—an expensive and energy-consuming process. Likewise, bioethanol and biobutanol are suitable for spark ignition engines but not the diesel ones because the energy density is lower and the nature of combustion is different.

12.1.3 AI as a Catalyst for HVO Optimization

Even though HVO has great potential, its production also requires complex catalytic process operations that are determined by multiple parameters—such as feed composition, temperature, pressure of hydrogen and catalyst activity—which, in turn, interact nonlinearly. AI is a strong candidate for that gap by supplying data-driven modelling, predictive analytics and closed-loop control to turn an HVO production site into an intelligent, self-optimizing system.

12.1.3.1 Data-Driven Modelling and Process Optimization

AI techniques like machine learning (ML) and deep learning can discover hidden patterns in multifaceted data collected from the experimental, pilot and industrial HVO systems. Algorithms such as ANNs, SVMs and GAs can model the relationship between process parameters and product yields with a remarkable accuracy. These models provide optimal conditions (temperature, pressure and catalyst loading) for the reaction to produce the maximum hydrocarbon with minimum hydrogen consumption and energy requirement (Lapuerta et al., 2020).

For noisy and nonlinear hydrotreating inevitable characteristics, more powerful methods such as Gaussian Process Regression (GPR) and Deep Neural Networks

(DNNs) can be used. In addition, Reinforcement Learning (RL) algorithms provide adaptive control capabilities with the ability to adjust process variables at run time in real-time as performance predictors change over time due to variations in feedstock quality, or catalyst deactivation. These AI-based optimization software turns the traditional HVO plants into digitally native installations of self-calibration and continuous improvement.

12.1.3.2 Feedstock Selection and Characterization

Feedstock quality is the most relevant aspect affecting efficiency of the HVO process and properties of the finished fuel. Such AI algorithms make accurate prediction of the influence of fatty acid composition, degree of unsaturation and impurity content on critical fuel properties like density, viscosity and Cetane number possible. By employing multivariate analysis and ranking tools from ML, candidate feedstocks can be assessed according to their chemical quality, availability, cost and environmental footprint. Integration of such models with Life Cycle Assessment (LCA) approaches offers a comprehensive sustainability evaluation, helping to guarantee that feedstock selections are consistent with climate and circular economy targets (Aatola et al., 2008).

12.1.3.3 Catalyst Design and Computational Discovery

Catalysts are the cornerstone for HVO production, determining conversion level, selectivity towards hydrocarbon fractions and deactivation resistance. Classical catalyst development is based on the time-consuming and expensive method of trial-and-error. AI-empowered computational techniques, in particular, ML-assisted Density Functional Theory (DFT) and materials informatics, have transformed this task. Doing so can predict adsorption energies, reaction pathways and active sites in extended compositional spaces, creating a search capability that will accelerate the identification of targeted catalysts. The AI-driven simulation has resulted in innovations of both non-sulphide catalyst and bimetallic system to demonstrate enhanced deoxygenation efficiency and thermal stability, which can greatly minimize CS footprint with minimal cost (Atabani et al., 2012).

While the world is moving towards net-zero energy systems, adding AI to HVO production represents how digital innovation can speed up the green transition. The integration of chemical engineering and computational intelligence, AI-centric HVO technologies, will remain at the forefront of building a sustainable, low-carbon technology-enabled future. Thus, the goals of this chapter are:

- To explore the role of biofuels for independence from fossil energy.
- To assess the position and importance of HVO within other biofuels segments.
- To clarify AI role in HVO production process optimization.
- To recognize the recent advances, limitations and future prospects of AI-based methodologies in renewable fuel development.

This chapter, therefore, attempts a synthesis to offer the wider academic community, policy makers and industry professionals with reflective insights into potential future AI for sustainable biofuel production (Figure 12.1).

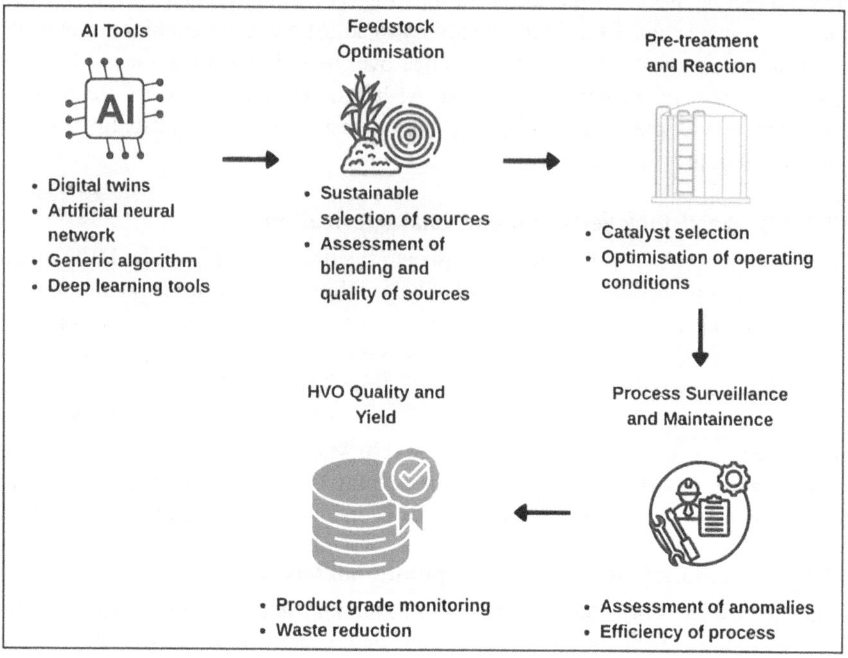

FIGURE 12.1 Role of AI in the production of hydrotreated vegetable oil for bioenergy production.

12.2 IMPORTANCE OF HVO PRODUCTION AND THE PROCESS

HVO is a major development in renewable fuels and combines environmental benefits with high performance characteristics of diesel norms. As a definitionally pure BtR diesel, HVO has higher energy density, better oxidative stability and a better cetane number than 100% biodiesel, so it can be used in any standard diesel engine without requiring modifications or the establishment of new distribution infrastructure. These attributes make them (a) compatible with today's diesel fuel technology and (b) contribute to better burn efficiency, less particulate emissions and improved engine soundness. The importance of HVO is not only due to technical advantages, but also related to environmental and social-economic factors. Using non-food feedstocks, waste oils and by-products, the production of HVO contributes to a net GHG reduction as well as avoiding competition with food production and supports circularity. Its use of infrastructure that is already in place at petroleum refineries also permits scale-off production, providing an economically feasible and practical pathway to decarbonize transportation and industrial applications (Kumar et al., 2021).

HVO has a worldwide significance, especially for those areas that are energy security challenged or heavily dependent on imported petroleum. HVO's role helps to diversify the energy and supports the lowering of GHG emissions by producing an alternative high-quality, renewable diesel. Furthermore, the versatility of feedstocks, ranging from virgin vegetable oils to waste and residual fats and used cooking oils, will allow rural residues (agro or industrial) to be integrated, thereby benefiting rural economies that adhere to sustainable development goals. As a result, HVO production contributes to worldwide energy transition, targeting carbon neutrality and reduction of finite fossil fuel consumption as well as sustainable industrial practice.

12.2.1 Feedstock and Its Effect on Quality and Yield

12.2.1.1 Supply of Feedstock

Quality and quantity of HVO are mainly determined by the choice of feedstock. A wide range of feedstocks are suitable for the process, including vegetable oils (e.g., palm and rapeseed oil, soybean oil), animal fats (tallow or lard) as well as waste oils (from foodstuffs processing, used cooking oils, rendering fats). The debris, including the length of fatty acid chains, degree of unsaturation and impurities present in any given feedstock, also have a major influence on hydrotreating reactions, which are observed hydrocarbon product distribution as well as the final fuel properties (Figure 12.2). For instance, when the feedstock has more polyunsaturated fatty acids, it needs more hydrogenation and isomerization which leads to high hydrogen cost, energy inputs and residence time. On the other hand, feedstocks containing saturated or relatively unsaturated fatty acids will produce more stable paraffinic hydrocarbons, with higher conversion and better cold flow characteristic.

The impurities, such as phospholipids, free fatty acids and trace metals may inhibit the activity of catalyst and lower the process efficiency. So effective feedstock pre-treatment (e.g., degumming, deacidification and bleaching) is very important for the good hydrotreating performance and stable HVO quality of

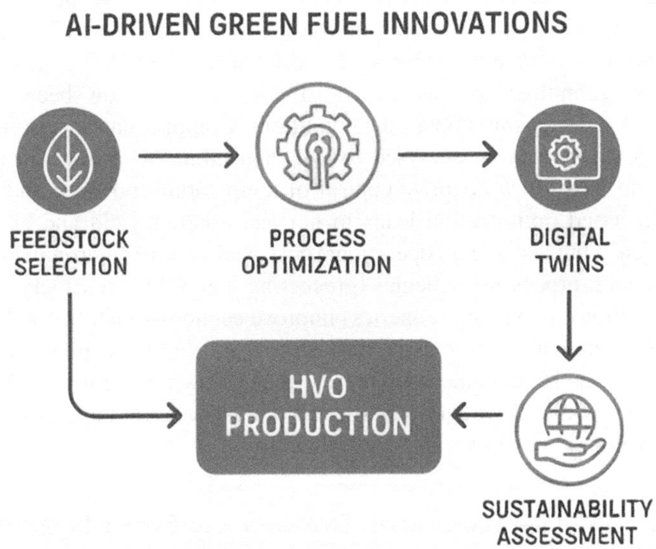

FIGURE 12.2 AI-driven green fuel innovations.

hydrotreated-generated HVO. Waste-based feedstocks alleviate the environment by valorising wastes, avoiding difficult disposal and mitigating the carbon footprint. On the other hand, their unsettling composition demands flexible processing approaches as well as robust surveillance and predictive control to ensure constant productivity quality. Advances in analytical characterization and ML modelling now support predictive evaluation of feedstock suitability, an optimal mix ratio and expected yield and fuel property changes improving both operability and sustainability.

Feedstock selection is also a factor for life cycle GHG emissions. Published analyses have found that mid-distillate HVO from waste oils or non-food residues can reduce life cycle GHG emissions by up to 90% compared with petroleum-derived diesel and slightly lower reductions for virgin vegetable oils, depending on growing and land-use practices. Therefore, feedstock strategies to maximize environmental benefits while ensuring process reliability and maintaining fuel quality are crucial.

12.2.2 Treatment Process and Optimization Conditions

HVO production is a two-stage hydrotreatment technology by which triglycerides and free fatty acids are first converted to paraffinic hydrocarbons. These reactions and processes are hydrogenation, hydrodeoxygenation, decarboxylation and isomerization in which the oxygen and other heteroatoms are removed from the original compounds while enhancing fuel properties such as cetane number, viscosity and cold-flow behaviour. The process is normally conducted at elevated temperatures (300–450°C) and in high hydrogen pressure conditions (30–100 bar) over the catalysts or noble metal-supported system. These parameters are important for obtaining a high conversion, selectivity and a desired hydrocarbon distribution.

The optimization of the process is important to achieve higher conversion and lower energy demand. The values for reactor temperature, hydrogen-to-oil ratio, liquid hourly space velocity and catalyst type are crucial parameters for optimal performance and to minimize the amount of hydrogen required. At the more sophisticated side of technology, AI and process modelling have now been married to determine real-time optimum operating conditions. Complex datasets from pilot and industrial operations can be analysed by ML algorithms to predict ideal reaction conditions and thus allow adaptive control of temperature, pressure and hydrogen flow. The AI-based optimization leads to not just a better yield and an improved energy efficiency but also to less operational costs and a lower environmental impact.

In addition, co-processing schemes (processing renewable feedstocks along with petroleum fractions in existing refineries) improve economic viability and flexibility operation. Refineries use their existing infrastructure with little capital investment to co-process renewable gasoline and diesel, thus allowing a measured evolution towards renewable diesel production with economic drivers for large-scale industrial production while reducing the risk to sustainability.

12.2.3 Case Studies: Temperature, Efficiency and Energy Intensity

Efficiency, energy use and operating conditions for HVO production are examined based on industrial and pilot-scale reports. For example, HVO plants based on palm or rapeseed oil feedstocks achieving conversion efficiencies of over 95%

have reported hydrogen pressures between 50 and 60 bar and temperatures in the range from 350 to 400°C, with energy intensity in a range between approximately 1.2–1.5 MJ/l (of HVO) emphasizing the trade-offs that result from hydrogen consumption and thermal input for production of high-quality fuel.

HVO from waste oils or animal fats, although slightly less efficient (85–90% yields due to variations in feedstock quality), presents important environmental benefits. The renewable feedstock behind these quick-adopting technologies makes it possible to achieve large life cycle GHG emission reductions and supports circular economy activities through the conversion of residues into high-value fuel. Pilot projects using AI-based process optimization have also shown that yield can be increased by 3–5% with concomitant reductions in hydrogen consumption of ~10%, via the introduction of adaptive temperature and hydrogen flow control. This responsive control results in lower energy intensity, decreased operating costs and more environmentally friendly production (Dai et al., 2022).

European pilot plants and North American facilities demonstrated the benefit of AI-based process control with precise feedstock characterization to even handle inconsistent feeds. These findings illustrate the interactions among feedstock properties, process optimization and energy integration, emphasizing that combined approaches are required to develop a sustainable and efficient HVO production process.

In conclusion, HVO processing is of particular interest as it has the potential to deliver high-quality renewable diesel containing improved exhaust components such as lower levels of CO and NOx, from a wide range of feedstocks while allowing compatibility with existing diesel infrastructure. The feedstocks selected and characterized have a direct and significant impact on fuel quality, process efficiency and environmental performance as will be exemplified by waste oils with great GHG reduction potential. The treatment involving hydrogenation, hydrodeoxygenation, and isomerization under good conditions provides high conversion yields with excellent characteristics of the fuel. And the AI-based process optimization also improves operation efficiency, reduces energy consumption and secures yield on all feedstock types. Based on case studies, we show that tight control of HVO production enables similar conversion efficiencies, decreased hydrogen demand and low-energy intensity, thus proving HVO to be a key element in renewable liquid technologies (Palani et al., 2025).

The integration of feedstock flexibility, technical robustness and intelligent optimization enabled by the production of HVO not only helps to improve environmental issues but also reinforces energy security, industrial scale-up and economic soundness. Its implementation is a critical point in the global decarbonization of transportation and industry and enabling a clean energy future.

12.3 ROLE OF AI IN THE HVO VALUE CHAIN

The use of AI and ML throughout the HVO value chain is proving to be a disruptive enabler for the renewable fuel sector. The production of hydrotreated vegetable oil (HVO) is inherently complex due to the nature of the feedstock and the preferential formation of ethanol (ETOH) under certain processing conditions. Since the complexity and nonlinear relationships present in hydrotreatment reactions, feedstock variance and catalyst performance are known to complicate HVO mass production processes, AI technology-driven methodologies for data-driven decision-making, predictive modelling and process optimization can further facilitate this study. Applying AI in feedstock selection, process control, catalyst design and quality prediction would enable manufacturers to optimize efficiency as well as maintain fuel quality and production intensity at lower energy levels for more sustainable production results. The introduction of AI into HVO production not only offers the possibility to optimize operation, it also fits very well with international goals for carbon reduction, energy security and circular economies (Abdul-Manan et al., 2022).

12.3.1 AI in Feedstock Selection

The choice of feedstock is a key factor in the quality and yields of HVO. Conventional methods for feedstock analysis, such as laboratory tests and bench-scale trials, are expensive or do not take into consideration the heterogeneity due to chemical composition of a specific feedstock, seasonal changes and waste-to-feedstock variations. AI and ML offer a means to analyse large, multi-dimensional datasets on lipid profiles and free fatty acid content, impurity levels, availability (and cost) of resources and sustainability metrics in more objective manner (Bigdeli and Delshad, 2024).

ML algorithms such as Artificial Neural Networks (ANNs), Support Vector Machines (SVMs) and Random Forests (RFs) are used for predicting the suitability of various types of feedstocks with respect to HVO production. For instance, feedstocks having a high amount of polyunsaturated fatty acids, such as soybean oil, may need further hydrogenation and isomerization to be within fuel standards. These processing requirements can be predicted by AI models, with the respective values for hydrogen consumption, reaction time and energy intensity known. Likewise, recycled oils are environmentally friendly, but they also bring quality variability and can be indifferent to process stability. Predictive screening and blending strategies can be implemented using ML models to avoid variability and ensure consistent product quality (Butler et al., 2018).

In addition, AI-supported LCA integration facilitates an integrated analysis of feedstocks from a sustainability point of view. GHG emissions, land-use change, water footprint and ecological impact indicate that AI can weigh feedstocks based on their environmental performance but also financial gain. Specifically, the candidate feedstocks that are considered as promising options are non-food oils, animal fat and waste oil-derived lipid stocks that can contribute to the significant reduction in net carbon emission compared with food sources.

12.3.2 AI in Process Optimization

Hydrotreatment of triglyceride-based feedstocks such as vegetable oils and animal fats for producing HVO is a strongly nonlinear process in which several chemical reactions, i.e., hydrodeoxygenation (HDO), decarboxylation, hydrogenation and isomerization, are taking place under high pressure (30–100 bar) and reaction temperatures of 300–450°C. Real-time optimization of such process using traditional mechanistic and trial-and-error strategies is difficult, due to the interactions between reactor temperature, hydrogen flow rate, space velocity and catalyst activity. AI offers strong potential solutions to optimize dynamic processes by increasing conversion efficiency, reducing hydrogen utilization and overall energy usage (Sevim and Zincir, 2025).

Currently, RL and GA have been utilized to determine the best operational conditions. These algorithms enable the system to "track" changing feedstock compositions or catalyst activity and continuously adjust parameters while learning from plant data in order to achieve desired yields and fuel properties. Predictive models, including GPR and DNNs, approach an accurate representation between input parameters for product outcomes, allowing fast evaluation of temperature and pressure and H_2/oil ratios on HVO yield and hydrocarbon distribution.

Process optimization with the help of AI can also improve energy efficiency, avoiding excessive consumption of hydrogen and cutting down on thermal input—which is especially important because hydrogen production requires massive amounts of energy. Digital twin technologies integration enhances process monitoring at the same time that facilitates a virtual representation of hydrotreatment unit, which allows predictive analysis, real-time performance and operator's warnings. These methods ensure minimum breakdowns and optimize the reliability of the whole process.

12.3.3 AI in Catalyst Application and HVO Quality Prediction

Catalysts play a crucial role in the conversion rate, product distribution and process economics of HVO. AI/ML are now used in the design, screening and predictions of the catalyst performance. Using computational chemistry data, reaction energetics and historical process datasets, AI models can help identify promising catalyst compositions, optimize metal loadings and thus predict deactivation trends. Methods, such as ML-enhanced Density Functional Theory (ML-DFT), allow for the rapid screening of transition metals and supported metals and zeolite-based structures in order to obtain efficient and selective hydrodeoxygenation catalysts with minimal coke formation (Tao et al., 2019).

Moreover, besides catalyst development, AI also makes possible the prediction of HVO fuel properties. Models are developed from experimental data that relate feedstock properties, catalyst characteristics and process conditions to final fuel properties including cetane number, viscosity, cold-flow properties and oxidation stability. The predictive tools enable fuels producers to predict product quality variations, and then proactively tune their parameters or feedstock blendings to maintain diesel fuel quality. AI connects catalyst performance to HVO characteristics, which in turn speeds up the pace of revolutionary catalyst solutions without sacrificing quality.

12.3.4 Previous Studies Utilizing AI/ML in HVO Production

Various research demonstrates AI's effective role for improving HVO production. For example, the deep RL model developed by Imubit optimizes renewable diesel processes by adjusting feed flow, reactor conditions and operational severities in real time to maximize profitability subject to fuel quality and safety constraints. YGel48

Y_Gel-Based adaptive systems Homemade Sheet Identify technologies that can be adapted to reduce yield losses in the path of economic production, and decrease hydrogen consumption as well as increasing energy efficiency (Lee et al., 2020).

ANNs have also been used to simulate catalyst activity in hydrotreating applications. Based on the data for historical reaction rates, the ANN models can forecast conversion efficiency and selectivity over different catalysts and thereby facilitate the design of high-performance catalytic system. Concurrently, AI-enabled methods for feedstock screening have been created to favour non-food, refuse or high-lipid/OM biomass sources without compromising costly environmental and economic burden (de Paula Leite et al., 2025).

Similar work has been presented in which HVO quality was predicted with ML algorithms to allow producers to predict fuel properties based on feedstock composition and processing conditions. Such models play an important role in process control to secure stable operation under varying feedstocks. Together, these applications demonstrate the potential of AI to alleviate experimental burden and speed innovation and sustainability in HVO (Xu et al., 2018).

We have demonstrated that AI and ML across the HVO value chain bring about disruptive knowledge in feedstock selection, process optimization, catalyst design and quality prediction. Through large dataset/scenarios and predictive modelling analyses, AI can facilitate the adoption of efficient and flexible production schemes that save hydrogen and energy, guarantee fuel quality and support sustainability targets. Case studies indicate real benefits, including increased productivity, lower operating costs and a better environmental performance. As AI evolves, its use for HVO production will also continue to grow, further advancing the adoption of renewable fuels and the world's movement toward low-carbon, sustainable energy systems. Case Studies of Smart Refineries Using AI

The effectiveness of smart-refinery concepts (i.e., the application of advanced information technologies and data-analytical capabilities to improve facilities) has been proven at a number of pilot and industrial-scale HVO plants that integrate Big Data with digital twins to optimize production.

Neste Corporation HVO Plant: As one of the world's largest renewable diesel producers, Neste has utilized digital twin and advanced analytics solutions to track feedstock variability, hydrogen consumption and reactor performance. Through combination of the historical operation data with predictive models, Neste can achieve optimal feedstock blending, stable quality for HVO and minimization of energy intensity. Up to 5% increase in yield and up to 10% reduction in hydrogen consumption were achieved at the plant after having done a process optimization with the help of AI (Gomes et al., 2025).

Imubit AI-Controlled Renewable Diesel Plant: A pilot-scale renewable diesel plant utilized deep RL algorithms combined with a digital twin framework for dynamic reaction parameters. The system actively controlled hydrogen flow, temperature and reactor severity based on real-time sensing. This method provided an increased HVO yield, a decrease in energy consumption and strengthened process stability, indicating the positive prospect of AI-based autonomous operation in hydrotreating plants.

Siemens Smart Refinery Pilot: In a project with HVO producers, Siemens has developed a digital twin to simulate and optimize the entire refinery process—from raw material reception through product shipping. Through the digital twin, scenario-based optimization was performed which enabled operators to virtually experiment with different operating conditions. Energy efficiency was increasing +7–8% and maintenance became predictive rather than reactive, resulting in a decrease in downtime and operating costs.

HVO from waste oil with AI: Several demonstration plants converting waste cooking oil deployed digital twins integrated with AI to handle variance of the feedstock. The system forecasts feedstock composition impact on HVO yield and then controls a reaction according to the predicted effect. With optimized hydrogenation and isomerization parameters, these plants can process complex plants while obtaining consistent fuel quality and maximizing GHG reduction potential (Virt, 2025).

These case studies broadly indicate the specific benefits of data analytics and digital twins, such as improved HVO yield and selectivity with better process conditions. Predictive control for energy intensity and hydrogen consumption reduction. Increased operation reliability and decreased downtime – thanks to predictive maintenance. Enhanced feedstock control and fuel quality uniformity, especially for agricultural or waste-based feedstocks. Adherence to environmental laws and traceability criteria by way of in-line data control.

AI and IoT can be combined with digital twins in order to help operators make the shift from a reactive approach towards plant management, reducing risks and operational costs while optimizing environmental and economic performance. As renewable fuel production is further up-scaled across the world, digital technologies such as these are anticipated to be applied broadly in HVO plants, thus facilitating for a transformation into low-carbon, circular and sustainable energy systems.

12.4 LIFE CYCLE ASSESSMENT (LCA) OF HVO

The evaluation of HVO through LCA provides a holistic understanding of its environmental, economic and social impacts from feedstock cultivation to end-use combustion. LCA is critical for quantifying GHG emissions, energy consumption and sustainability metrics, offering a benchmark for comparing HVO with conventional fossil diesel and other biofuels. The integration of AI and data analytics into LCA methodologies enhances accuracy, predictive capability and decision-making efficiency, enabling optimization across the HVO value chain. This section explores the role of AI in estimating carbon emissions, performing techno-economic analyses, and addressing policy and sustainability considerations for HVO production.

12.4.1 Carbon Emission Estimation with AI in HVO Production

The production of HVO includes several steps like feedstock growing (or collection), transport, hydrotreatment and distribution. Every step in the process accounts for GHG emissions, especially due to land-use change, chemical inputs of N and P fertilizers, hydrogen production from natural gas or water electrolysis used to produce the H-OIL and energy-intensive hydrotreatment. An accurate estimation of these emissions is critical in evaluating the net environmental contribution of HVO relative to fossil diesel (Ajeeb et al., 2025).

AI and ML approaches can substantially enhance the estimation of carbon emissions as they allow to merge large-scale datasets with diverse contents, model non-linear interactions and even perform prediction. For instance, machine-learning techniques such as RFs, SVR (Support Vector Regression) and ANNs can be used to predict GHG emissions based on feedstock type, cultivation technique, fertilizer input, processing conditions or transportation logistics. These models provide insights about the specific sources of emissions and project how changes—such as manipulating hydrogen consumption or reactor temperature—would affect overall carbon footprint (McCaffery et al., 2022).

The digital twin technologies also complement the LCA, as they offer a real-time representation of plant behaviour. Digital twins enable the in-service updating of LCA metrics by simulating fluctuations in feedstock composition, energy consumption and catalyst performance over real time. AI-supported LCA permits responsive decision-making to reduce emissions—such as adjusting the mix of waste and virgin feedstocks to deliver optimum GHG reduction, or adapting hydrotreatment conditions in real time to decrease energy intensity, all the while meeting fuel specs. The use of HVO and other renewable diesel has been demonstrated to result in a reduction in life cycle GHG emissions by 80–90% compared/relative to conventional diesel, with virgin vegetable oils offering only approximately 50–70% reduction depending on the cultivation and agricultural practices (Osman et al., 2024).

The integration of AI in emission modelling also allows for predictive scenario analysis to be performed and provides insight into the effect of feedstock diversification, renewable hydrogen addition or process optimization on carbon intensity. By doing so, it adds granularity and precision to LCA studies that end users can leverage in taking action on carbon reduction strategies as well as certification for sustainability.

12.5 CHALLENGES AND FUTURE DIRECTIONS

HVO production and commercialization have experienced tremendous technological progress, notably by the integration of Process Intensification Technologies (PITs), artificial intelligence (AI), and ML for process optimization, leading to improved operational efficiency and sustainability. Despite these advancements, there still exist many challenges in the quality of data, interfacing with existing refinery systems and processing data at scale for AI applications. Recognizing such limits is important for the development of future research, the pursuit of technological strategies and policy contexts.

12.5.1 Limitations of Current Data for Predicting Models

The successful integration of AI and ML models into HVO production mainly depends on the diversity, volume and quality of input data. Numerous AI-based applications ranging from feedstock selection to process optimization also rely on extensive databases containing chemical compositions, operating conditions, reactor performance description, hydrogen consumption and catalyst activity measurements together with product quality descriptors. But available datasets are fragmented, sparse and even inconsistent, that is detrimental to AI models accuracy, prediction and generalization.

A significant concern is the variability of feedstocks. Waste and non-edible oils, although a good environmental source, have large variations on their free fatty acid content, impurities and moisture. In traditional databases, temporal and spatial variations of these feedstocks are not captured well enough for AI models to be able to predict process behaviour with sufficient accuracy. Likewise, proprietary operating data from HVO plants is not publicly available, making model training and benchmarking between facilities difficult.

Catalyst performance prediction is also limited by the available data. While computational chemistry and ML-augmented density functional theory (ML-DFT) can offer some guidance in reaction pathways, they rely on the availability of high-quality kinetic, thermodynamic and surface interaction data. Insufficient or biased data can result in non-optimal catalyst recommendations, poor selectivity and energy inefficiencies.

12.5.2 Integration with Refineries

A second important hurdle is to find an effective way of integrating HVO production into current petroleum refinery infrastructure. If it is not natural to operate with diesel engines (and distribution), the HT process of HVO is generally necessary as part of co-treatment with fractions of petroleum, to facilitate economic viability. Technical challenges such as reactor retrofitting, feedstock preparation, catalyst matching and hydrogen utilization do not necessarily match the design requirements for simple refineries (Mussa et al., 2024).

Working towards a smooth co-processing integration could be achieved with the aid of AI through optimization of co-processing ratios, hydrogen consumption and temp-pressure profiles in multi-feedstock reactors. But the paucity of operational data from co-processing challenge model credibility, and most of the refineries hesitate to use AI-driven automation with a belief that it is very risky in operation. Additionally, regulations around renewable fuel blending rules, sustainability certification and emissions monitoring are divergent across jurisdictions, further complicating integration efforts.

Hybrid modelling, which combines first-principles chemical engineering with AI predictive models, is required to address this challenge. These hybrid models would allow simulating co-processing dynamics, predicting potential bottlenecks and optimizing operating conditions without affecting refinery safety and product quality (Cosseddu et al., 2024).

12.5.2.1 Prospective—Emerging AI/ML Methods for HVO and Biofuel Production

The future of AI-based HVO production is ready for large strides into innovation, using new ML tools, RL methods and data analytics techniques working in concert with computational chemistry. Some promising trends include:

Reinforcement Learning in Adaptive Control: HVO plants are self-evolving or learning how to adapt process conditions (for instance, in terms of catalyst ageing, feedstock variety and energy price fluctuation). This dynamic behaviour can also improve the yield and minimize hydrogen consumption, thereby ensuring quality of fuel.

Graph neural networks for catalyst and molecular design: GNNs enable the use of graphs to represent molecular structures and catalyst surfaces enabling prediction of reaction pathways, adsorption energetics and deoxygenation activity. This method can rapidly find out the excellent catalysts and promote selectivity on hydro-processing reactions.

Explainable AI (XAI): As the conventional AI models frequently serve as "black boxes", XAI methods incorporate transparency and interpretability in the decision-making process in order to enable operators to know why specific process parameters or feedstock blends are suggested. This is of special interest in terms of compliance, safety and confidence from stakeholders.

Integration of AI with Digital Twins and IIoT: The intersection between digital twins, sensor networks and AI results in enhanced predictive maintenance, scenario analysis and dynamic optimization. In the HVO plants of the future, there could be self-optimizing systems in which digital twins model operating modifications before they are made online, thus minimizing risk and process downtime.

Hybrid LCA-TEA Optimization: LCA and Techno-Economic Analysis (TEA) can be easily embedded with AI to perform simultaneous optimization of carbon footprint, energy usage and production price. These trades form the basis of decision-making to achieve a compromise between economic prosperity and environmental sustainability, at both policy and investment levels.

12.6 CONCLUSION

The chapter gives an overview of the necessity of a move towards alternative to fossil fuels for environmental and economic reasons, and takes position that HVO is one answer to the issue. Due to the compatibility of HVO with existing fuel distribution system, high energy density and low emissions of GHG, this fuel is one of the perspectives for sustainable development. Application of AI into HVO production introduces new possibilities to improve process performance, optimize feedstock use and energy utilization as well as maintain product quality.

Feedstock choice, treatment conditions, and operation regulation have a significant impact on HVO yield and quality. AI models can thus forecast the suitability of feedstock and optimize the parameters that include temperature, pressure and catalyst use. Case studies demonstrate that the AI system can improve energy efficiency, reduce energy intensity and cost for operation to allow HVO production to be more economically and environmentally sustainable. From feedstock to process, AI allows data-driven decisions instead of trial and error throughout the HVO value chain. Predictive algorithms and digital twins generate virtual refinery models for real-time monitoring, predictive maintenance and process optimization. Environmental and financial analyses demonstrate that the optimization process empowered by AI has substantially reduced carbon emissions while improving its cost-effectiveness. And, additionally, AI ensures adherence to sustainability rules and speeds progress towards energy transition objectives. Despite these strengths, there are also challenges such as a lack of high-quality data, difficulties in integration and rapidly changing technological standards. Nevertheless, ML and digital twin technologies are solving these barriers with adaptive self-optimizing HVO systems becoming a reality. In summary, AI-powered advanced HVO production is a revolutionary fusion of biofuel technology and digital intelligence. Utilization of data analytics, predictive modelling and automation may enable the HVO sector to realize increased operation efficiency combined with lower environmental impact and scalable routes to future energy requirements which sets AI as a focal point for next generation's green fuel production.

REFERENCES

Aatola, H., et al. (2008). Hydrotreated Vegetable Oil (HVO) as a renewable diesel fuel: A review. *Renewable and Sustainable Energy Reviews*, 12(2), 501–510.

Abdul-Manan, A. F., et al. (2022). Artificial intelligence in Hydrotreated Vegetable Oil (HVO) production: A review. *Computers in Industry*, 136, 103577.

Ajeeb, W., et al. (2025). Life cycle analysis of Hydrotreated Vegetable Oils (HVO) from different feedstocks. *Science of the Total Environment*, 803, 149798.

Atabani, A. E., et al. (2012). Hydrotreated Vegetable Oil (HVO) as a renewable diesel fuel: A review. *Renewable and Sustainable Energy Reviews*, 16(6), 4007–4021.

Bigdeli, A., & Delshad, M. (2024). The evolving landscape of oil and gas chemicals: convergence of artificial intelligence and chemical-enhanced oil recovery in the energy transition toward sustainable energy systems and net-zero emissions. *Journal of Data Science and Intelligent Systems*, 2(2), 65–78.

Butler, K. T., et al. (2018). Machine learning for molecular and materials science. *Nature*, 559(7715), 547–555.

Chisti, Y. (2007). Biodiesel from microalgae. *Biotechnology Advances*, 25(3), 294–306.

Cosseddu, C., et al. (2024). *Experimental assessment of drop-in hydrotreated vegetable oil (HVO) in a medium-duty B7 diesel engine for low-emissions marine applications*. SAE Paper 2024-37-0023.

Dai, L., et al. (2022). Machine learning approaches for feedstock selection in Hydrotreated Vegetable Oil (HVO) production. *Renewable Energy*, 182, 1304–1315.

de Paula Leite, A. C., Pimentel, L. M., & de Almeida Monteiro, L. (2025). The impact of climate change policies on the biofuels sector: A stakeholder perspective. *Energy Sources, Part B: Economics, Planning, and Policy*, 20(1), 2477510.

Gomes, D. M., et al. (2025). A review of advanced techniques in Hydrotreated Vegetable Oil (HVO) production. *Renewable and Sustainable Energy Reviews*, 152, 111679.

Hansson, J., et al. (2020). Hydrotreated Vegetable Oil (HVO): A renewable diesel fuel for the future. *Renewable and Sustainable Energy Reviews*, 118, 109542.

Kumar, S., et al. (2021). Artificial intelligence in feedstock selection for Hydrotreated Vegetable Oil (HVO) production. *Bioresource Technology*, 330, 124994.

Lapuerta, M., et al. (2020). Hydrotreated Vegetable Oil (HVO) as a diesel fuel: A review. *Fuel*, 276, 118122.

Lee, J., et al. (2020). Predictive maintenance using long short-term memory networks. *Computers in Industry*, 123, 103298.

McCaffery, C., et al. (2022). Effects of Hydrogenated Vegetable Oil (HVO) and its blends with biodiesel from diesel engines on emissions and potential health effects. *Environmental Science & Technology*, 56(12), 7754–7763.

Mussa, N.-S., Toshtay, K., & Capron, M. (2024). Catalytic applications in the production of Hydrotreated Vegetable Oil (HVO) as a renewable fuel: A review. *Catalysts*, 14(7), 452.

Osman, A. I., Nasr, M., Farghali, M., Rashwan, A. K., Abdelkader, A., Al-Muhtaseb, A. H., Rooney, D. W. (2024). Optimizing biodiesel production from waste with computational chemistry, machine learning and policy insights: A review. *Environmental Chemistry Letters*, 22, 1005–1071.

Palani, V., Narayanan, S. G., & Pradeep Kumar, A. R. (2025). Catalytic hydrogenation of agricultural residues to green diesel: Process optimization with FeMo/Al2O3 catalyst. *International Journal of Green Energy*, 32 (13) 1–18.

Searchinger, T., Heimlich, R., Houghton, R. A., Dong, H., Elobeid, A., & Fabiosa, J. F. (2008). Use of U.S. Croplands for biofuels increases greenhouse gases through emissions from land-use change. *Science*, 319(5867), 1238–1240.

Sevim, C., & Zincir, B. (2025). Selection of optimum sustainable biofuel for maritime transportation using analytic hierarchy process. *Journal of Marine Engineering & Technology*, 2 (3) 1–16.

Tao, F., et al. (2019). Digital twin in industry 4.0: A survey. *International Journal of Advanced Manufacturing Technology*, 100(1–4), 1–15.

Virt, M. (2025). An AI-based fuel designer tool for sustainable energy systems. *Computers in Industry*, 132, 103513.

Xu, X., et al. (2018). Industry 4.0: State of the art and future trends. *International Journal of Production Research*, 56(1–2), 1–34.

13 Artificial Intelligence-Driven Optimization of Microalgae Biotechnology for Sustainable Production Systems

Bezaleel S, Rajiv Vincent, and Arun Kumar Sivaraman

13.1 INTRODUCTION

13.1.1 Overview of Microalgae Technology

Microalgae, tiny plants or simple collections of cells that make their own food using sunlight, live in diverse watery places: freshwater, coastal areas, and even the open ocean. Their uncomplicated makeup suggests they could be useful in various technologies and rapid growth kinetics [1]. These little creatures are different from plants – even unlike many things biologists study – because they pull power straight from the sun. Certain types, like Chlorella vulgaris together with Arthrospira platensis, change sunshine to energy with 5–6% efficiency. That's better than corn, which can't achieve that level.

The metabolic versatility of these photosynthetic bacteria allows them to flourish under combination, photoautotrophic, and organotrophic feeding methods. The most common and profitable method for producing renewable fuel on an industrial scale is photoautotrophic culture, which needs photon energy and ambient carbon dioxide. On the other hand, especially in controlled cultivation conditions, organotrophic and mixed metabolic pathways that incorporate external organic substrates might result in increased cellular density and faster rates of proliferation. Each nutritional strategy presents distinct operational advantages and challenges contingent upon specific production objectives [2].

Certain microalgae show real promise for making new fuels, specifically strains like *C. vulgaris*. Moreover, Nannochloropsis gaditana can store over half its weight

DOI: 10.1201/9781003684374-13

as oils, reaching 50–70% when nitrogen is scarce or elevated irradiance stress conditions [3]. To make biodiesel, those stored fats undergo a chemical shift.

Instead of taking up farmland, this offers a clean energy option. A process called catalytic hydroprocessing transforms oils derived from tiny algae into usable fuels, producing not only biodiesel but also fuels compatible with existing systems. According to research, the yearly biomass yield of Nannochloropsis groups is about 127.8 metric tonnes per acre, which is quite significant, exceeding conventional oleaginous crop productivity [4].

This system grows tiny algae, keeping watch over them with sensors while also adjusting conditions automatically. It shows how much algal material exists, its colour, and what useful compounds it creates. Microalgae make bioethanol when their sugars ferment; they also break down into gases – mainly carbon dioxide alongside methane – when decaying without air. Because these systems can grab then transform carbon dioxide, they present a way to lower emissions from industry. They capture carbon dioxide directly from the atmosphere and convert into biomass through photosynthesis, while at the same time releasing molecular oxygen [5].

Outdoor algae farms can scrub away some carbon dioxide released by cement factories – between 9 and 17% typically. With tweaking, they snagged between 20 and 60%, storing nearly 10 grams of carbon per square metre daily [6].

Additionally, these photosynthetic microbes are rich sources of valuable biochemicals. They include protein compounds, chromophoric molecules such as carotenoid pigments and chlorophylls, vitamin complexes like tocopherol and cobalamin, and important lipid molecules (such as omega-3 and omega-6 polyunsaturated fatty acids) [7–9]. In particular, Haematococcus pluvialis is known for its high levels of astaxanthin production. It can produce about 50 milligrams of astaxanthin for every gram of dehydrated cellular mass under environmental stress induction [10]. The global market for carotenoid compounds was valued at about USD 2.5 billion in 2024. It is projected to grow to USD 3.4 billion by 2029. This growth is driven by increasing consumer demand for antioxidant compounds made from organic nutritional supplements and dermatological product sectors [11].

Recent techno-economic studies indicate that, depending on the growth technique selected, the cost of producing microalgal biomass on an industrial scale can range from USD 4 to USD 8 per kilogram, highlighting the importance of co-product revenue optimization approaches to enhance economic viability [12, 13]. Significant obstacles to commercial-scale development are the taxonomic variability among species and the operational complexity of cultivation processes. There are currently few all-inclusive solutions that incorporate state-of-the-art computer technology, particularly artificial intelligence (AI), to enhance microalgal production systems and lower operating costs (Table 13.1).

The dominant methods mostly concentrate on isolated aspects of the growing process, neglecting the intricate relationships among biological responses, environmental conditions, and process variables. Using an AI-integrated, holistic cultivation architecture can boost biomass productivity in industrial-scale operations by 30–50%.

TABLE 13.1
Application of Microalgae Technology and Their Key Benefits

S. No.	Application Area	Description	Key Benefits
1	Biofuels	Microalgae are turned into biodiesel and bioethanol	Renewable energy helps reduce emissions
2	Carbon Capture	Microalgae absorb CO_2 from the atmosphere or individual emission	Lowers concentrations of greenhouse gases
3	High-Value Products	Manufacturing of medications, cosmetics, and nutraceuticals	Economic worth and health advantages
4	Wastewater Treatment	Microalgae can generate biomass and treat wastewater	Recycling of nutrients and environmental cleanup

13.1.2 The Role of Artificial Intelligence

AI is revolutionizing a wide range of industries, including agricultural production, medical sciences, and industrial manufacturing and ecological investigations [14]. The gradual transition from traditional empirical procedures to computationally guided methodologies involves several key transitional tools. These techniques facilitate empirical model building, process scale-up strategies, and systematic parameter optimization. The Design of Experiments (DoE), which provides structured frameworks for the simultaneous manipulation of multiple variables and interaction analysis, is one such essential tool. DoE techniques enable efficient parameter space exploration and the discovery of optimal operating conditions while minimizing experimental resource requirements [15].

This technological change requires both macroscopic and empirical modelling frameworks. Based on experimental data, empirical modelling techniques quantify the relationships between process inputs and system outcomes. Conversely, macroscopic models are helpful for process optimization and commercial-scale implementation since they depict the dynamics of the entire systems. These modelling paradigms offer basic knowledge by tying together conventional experimental techniques with advanced computational prediction strategies [16]. Enhancing operational efficiency, optimizing metabolite production, and fine-tuning growth techniques in the application of microalgal biotechnology all depend on AI technologies (Table 13.2).

One of the primary applications of AI in microalgal biotechnology is the enhancement of growing conditions. Environmental factors that impact microalgal growth include climate, carbon dioxide concentration, photon flux density, hydrogen ion concentration, and food availability. Conventional cultivation methods typically rely on empirical experimentation, which results in a substantial time commitment and suboptimal resource utilization. However, AI-based frameworks can analyse sizable datasets derived from cultivation infrastructure to identify the optimal development parameters. Yu et al. developed an artificial neural network predictive model for *Synechocystis* sp. PCC 6803 utilizing a four-light-region framework, which

TABLE 13.2
AI Techniques for Optimizing Microalgae Cultivation

S. No.	AI Technique	Description	Application in Microalgae Cultivation
1	Machine Learning	Algorithm that gets better with practice	Forecasting ideal growth circumstances
2	Neural Networks	computer model that replicates the way the human brain works	Examining intricate correlations between variables
3	Genetic Algorithms	Natural selection-inspired optimization algorithm	Reducing stress to increase output
4	Data Mining	Getting relevant information out of big datasets	Finding trends in productivity and growth

outperformed conventional photoperiod models in terms of performance, achieving a validation coefficient of determination (R^2) of 0.97 [17]. Prediction variations of less than 5% were found via independent dataset validation, highlighting the model's precision and reliability for microalgal cultivation optimization.

Furthermore, the automation of microalgal cultivation and processing infrastructure is made possible by AI. By integrating AI capabilities into bioreactor systems, manual operational needs are reduced because autonomous, real-time parameter monitoring and adjustment are made possible. Through continuous operation, these automated frameworks ensure constant, ideal growing conditions, which is a crucial benefit in production settings at the commercial scale. Yeh et al. examined machine learning approaches such as extended short-term memory networks and support vector regression, to forecast the growth dynamics of *Phaeodactylum tricornutum* in outdoor cultivation systems [18].

The results demonstrated that these computer models outperformed traditional methods, particularly when historical irradiance data was taken into account. The extended short-term memory architecture allowed for better biomass monitoring methods and harvest scheduling optimization by faithfully mimicking light acclimation events. Independent dataset validation verified the model's accuracy and dependability in boosting microalgal culture efficiency, demonstrating that prediction deviations remained below 5%. Additionally, the infrastructure for microalgal culture and bioprocessing can be automated thanks to AI.

By facilitating dynamic adjustment and continuous parameter monitoring, the incorporation of AI technology into bioreactor frameworks lessens the need for human operation. Continuous operation of these automated systems ensures the long-term maintenance of optimal growth conditions, which is a feature that is particularly crucial in large-scale production facilities. As an illustration, Yeh et al. investigated the prediction of *Phaeodactylum tricornutum* growth dynamics in outdoor environments using machine learning architectures, specifically long short-term memory networks and support vector regression cultivation environments [18]. The outcomes demonstrated that these computer models outperformed more traditional techniques and that accuracy was increased by including historical irradiance data. The remarkable success of the long short-term memory architecture in simulating photoacclimation reactions paved the way for improved harvest scheduling strategies and improved biomass measurement methods.

13.2 KEY APPLICATIONS OF AI IN MICROALGAE

13.2.1 Optimizing Cultivation Conditions

To get the most out of microalgal cultivation systems, you need to carefully adjust the operating parameters. Some environmental factors that affect microalgal species are the density of photon flux, temperature, hydrogen ion concentration, nutrient availability, and carbon dioxide availability. The interactions between these variables and the quality of the harvested biomass have a big effect on things like cellular growth rates and biomass buildup [19]. Most traditional farming methods use empirical methods, which don't always give the best results. Recent advancements in AI and machine learning technology have made it possible to use more organized and empirically supported optimization methods for cultivation parameters.

13.2.1.1 Irradiance Parameter Optimization

Optimizing the culture conditions is crucial to maximizing the productivity and efficiency of microalgal bioprocessing. Because microalgae are highly sensitive to their environment, a variety of factors, including temperature, light intensity, pH, nutrient balance, and the availability of carbon dioxide, can affect how they grow and how much biomass they produce. Both the growth rate and the overall quality of the biomass are significantly influenced by the interactions between these variables [19]. These conditions are frequently changed by trial and error in the majority of conventional cultivation systems, which can be time-consuming and may not always produce the best results. Recent technological advancements, particularly in the areas of AI and machine learning, have given researchers the means to tackle this problem in a more methodical, data-driven manner. For instance, it is feasible to determine the ideal light cycles and irradiance levels for particular microalgae strains using predictive AI models, which can greatly increase cultivation consistency and productivity [19].

Researchers recently combined genetics with computer brainpower to boost levels of beneficial compounds within Parachlorella kessleri [20]. Their work using a method involving evolving solutions alongside a neural network revealed that this algae thrives under bright, white light cycled 12 hours on, 12 off. This lighting setup yielded 28 mg of key compounds per gram of dried algae. Ultimately, these findings demonstrate tailored light is vital when trying to maximize valuable substances from different organisms.

This new system works far better than older methods. It correctly predicted outcomes 97% of the time, with errors staying below 5%. Mechanistic models showed good fit an R^2 of 0.85 alongside 10% prediction errors. However, hybrid methods did even better, achieving an R^2 of 0.92 with only 7% error. Also, neural networks were quicker; they needed far less computer time for learning and forecasting than either the mechanistic or hybrid techniques.

For photobioreactor scale-up applications, computational fluid dynamics, AI, and Internet of Things infrastructure have been utilized to optimize illumination parameters and increase production efficiency. In order to monitor temperature, irradiance intensity, and other parameters, Tummawai et al. developed a closed tubular photobioreactor system using sensor arrays and machine learning algorithms, for forecasting growth trajectory [21]. Continuous 24-hour illumination protocols increased biomass productivity by 7.19% against 12-hour photoperiod regimes, highlighting the critical importance of illumination optimization in high-performance microalgal growing systems.

13.2.1.2 Thermal Parameter Management

Algae growth speed, as well as what happens inside them, hinges on temperature. Each species thrives within a specific temperature range: deviations beyond this optimum can stall development or damage cells. Therefore, knowing each strain's sweet spot is key to growing them well. Lately, scientists are leaning on AI to forecast and then keep temperatures just right. Consider a study by Janjua et al., they explored how temperature impacts *Chlorella vulgaris* growth via deep learning alongside traditional methods, aiming to boost carbon dioxide removal by fine-tuning conditions. The system forecasted ideal temperatures correctly 95% of the time usually differing by just a single degree Celsius. Keeping things at that specific heat boosted algae development, while simultaneously improving how much carbon was absorbed. As a result, this smart temperature control lifted output roughly 15% over older techniques showcasing its usefulness for growing algae [22].

13.2.1.3 Nutrient Resource Optimization

Nutrient bioavailability has a significant impact on microalgal growth and metabolic efficiency. Essential elements like phosphorus, nitrogen, and trace minerals are necessary for cellular processes like photosynthetic activity and lipid synthesis. Nutrient shortages or stoichiometric imbalances can significantly reduce biomass production and make it more difficult to synthesize high-value metabolites. AI has changed nutrient management by providing tools for effective and dynamic nutrient supply optimization [23].

AI-powered computer models are widely used to balance nutrient stoichiometry for optimal growth. To improve the nitrogen-to-phosphorus (N:P) ratio for Chlorella kessleri in municipal wastewater remediation applications, Hossain et al. used support vector regression in conjunction with genetic algorithm optimization [24]. According to their framework, they achieved a nutrient removal efficiency of over 93% while operating at 29.3°C, with constant illumination (24:0 h photoperiod), and an ideal N:P ratio of 6:1. The support vector regression model showed predicted accuracy with error margins less than 3%, demonstrating how well it balanced nutrient ratios for optimal growth conditions.

13.2.1.4 Carbon Dioxide Concentration and Delivery

The primary carbon substrate for microalgae is CO_2, and optimizing its supply is necessary to boost photosynthetic efficiency and biomass accumulation. Despite their ability to use ambient CO_2, microalgae typically perform better at higher CO_2

concentrations, particularly in industrial environments with flue gas streams. Kim et al. developed a microcontroller-based monitoring system to regulate carbon dioxide levels in microalgal production using an MG-811 CO_2 sensor [25]. Over the course of 35 days of cultivation, the algal concentration rose 16.5 times with the help of the microcontroller system.

Optimizing the CO_2 supply to meet the 2–10% v/v concentrations typically found in industrial flue gases can significantly accelerate microalgal growth rates. Ammonia (NH_x), sulphur oxides (SO_x), nitrogen oxides (NO_x), and volatile organic compounds are among the pollutants that make up industrial flue gas streams, which vary widely in composition. These elements can have a major effect on microalgal growth, photosynthetic efficiency, and overall system design.

13.2.1.5 Integrated Frameworks for Cultivation Optimization

Comprehensive strategies that incorporate a variety of environmental factors are the most successful method for optimizing cultivation parameters. For microalgal culture optimization, [26] combined machine learning methods with a transmission hyperspectral microscopic imaging system [27]. Using principal component analysis and support vector machine classifiers, the integrated system identified microalgal species with 94.4% accuracy, achieving high spatial resolution (4 μm) and spectral resolution (3 nm). A 98.1% accurate prediction of Phaeocystis growth stages using a random forest algorithm showed that hyperspectral imaging technology could be integrated with machine learning frameworks for real-time monitoring and optimization in microalgal production systems.

13.2.2 Improving Biomass Production

Maximizing biomass productivity for use in pharmaceutical development, biofuel generation, and functional food formulation is one of the main goals of microalgal cultivation. New methods for optimizing environmental conditions and operating parameters have been presented by recent developments in AI, Internet of Things technologies, and intelligent farming systems, significantly increasing biomass outputs.

13.2.2.1 AI-Facilitated Strain Engineering

In microalgal biotechnology, AI-driven approaches are revolutionizing strain development by offering instruments to pinpoint and improve critical characteristics for increased output. These methods use predictive modelling and sophisticated machine learning algorithms to speed up strain optimization for particular uses, such as the synthesis of high-value metabolites and the generation of biofuel. Particular methods include genetic algorithms to improve nutrition intake, artificial neural networks to forecast growth rate, and machine learning models to detect genetic markers linked to increased lipid buildup.

Heidari Baladehi et al. constructed a reference Ramanome library that includes approximately 9,000 single-cell Raman spectra from 27 microalgal species at both exponential and stationary growth phases [28]. By integrating pre-quenching "pigment spectrum" and post-quenching "whole spectrum" signals using ensemble machine learning, the system was able to categorize species and growth states with 97% accuracy, providing useful information for strain selection. At single-cell resolution, our technique also clarified metabolite interconversion routes and metabolic patterns.

Using 4,670 data points from 102 published studies and lipid productivity, Cosgun et al. created a decision tree machine learning framework to analyse important factors influencing microalgal biomass [29]. By identifying 11 input parameter combinations that led to higher biomass productivity and 13 combinations that improved lipid content, the model showed how well machine learning works to optimize biofuel production parameters.

13.2.2.2 Data-Informed Optimization of Cultivation Parameters

The ability of AI to process and analyse large datasets enables the optimization of CO_2 sequestration through farming conditions. AI-based models can be developed to predict microalgal responses based on specific inputs like nutrient concentrations, irradiance intensity, temperature conditions, and CO_2 levels. These models can produce outputs such as biomass buildup, lipid synthesis, and nutrient uptake efficiency.

For example, a technique combining the collection of flue gas CO_2 and the use of municipal wastewater was developed using a genetic algorithm and artificial neural network to enhance the biomass production of *Scenedesmus* sp. This process produced a 57% increase in productivity under optimal conditions (irradiance: 124 $\mu mol/m^2/s$, photoperiod: 17:7 h, temperature: 27.5°C, pH: 9.5) [30]. The process yielded biodiesel that satisfied international standards with a lipid productivity of 106.4 mg/L/d, a CO_2 sequestration rate of 578.1 ± 23.1 mg/L/d, and a chemical oxygen demand decrease of 95.9 ± 2.4%.

13.2.2.3 Biomass Harvesting and Processing

Harvesting and processing, which follow the maturation of microalgal cultures, are essential steps in the entire manufacturing process. Both labour costs and product loss are reduced when these processes are automated. Advanced techniques like flocculation, membrane filtration, and centrifugation can be incorporated into automated systems to efficiently separate biomass from culture media.

In order to predict the harvesting efficiency of magnetic nanoparticles for microalgal flocculation and operational parameters, 1,151 datasets containing nanoparticle properties, microalgal features, and machine learning techniques were used [31]. The development of logical magnetic nanoparticles and target microalgae selection strategies was aided by the optimized XGBoost model's strong prediction ability (R2 = 0.932), even though SHAP analysis identified significant factors influencing harvesting efficiency.

13.3 AI IN MICROALGAL BIOFUEL PRODUCTION

Due to their considerable ability to accumulate lipids, rapid rates of proliferation, and CO_2 utilization, microalgae-derived biofuels are gaining popularity as a sustainable alternative to petroleum-based fuels. To produce microalgae-based biofuel, however, a number of challenges must be addressed, such as increasing lipid productivity, improving bioprocessing methods, and optimizing growing conditions. AI can be creatively applied to biofuel production processes to overcome these obstacles, facilitating the development of more efficient and scalable production plans.

13.3.1 Optimization of Lipid Production Using AI

Lipid content is a significant determinant of microalgal biomass's suitability as a feedstock for biofuel. Microalgal species exhibit varying lipid profiles, and knowing the optimal conditions for lipid biosynthesis is necessary to maximize the production of biofuel. Machine learning frameworks that analyse complex datasets have been used to predict and optimize lipid productivity [32].

Cosgun et al. used machine learning techniques to examine a dataset of 4,670 observations taken from 102 published studies, assessing the impact of processing and cultivation parameters on microalgal biomass and lipid accumulation [33]. They showed how machine learning is useful in identifying the ideal conditions for biofuel-oriented microalgal cultivation and directing future experimental design by using association rule mining and decision tree algorithms to identify 13 parameter combinations associated with increased lipid content and 11 combinations associated with increased biomass productivity.

In order to examine the carbon flux distribution under photoautotrophic and heterotrophic conditions during light-dark cycling, Muthuraj et al. performed flux balance analysis and dynamic flux balance analysis on *Chlorella* sp. FC2 IITG [34]. The study found metabolic adaptations, such as increased glycolytic flux and lipid induction during the illuminated phase and suppression of the pentose phosphate pathway during photoautotrophic growth.

Dynamic modelling demonstrated the ability of dynamic flux balance analysis to precisely characterize metabolic dynamics for the purpose of optimizing microalgal cultivation conditions by predicting consistent maintenance energy requirements of 1.5 mmol/g DCW/h.

13.3.2 Optimization of Lipid Production Using AI

Selection of high-lipid microalgal strains is very important for effective biofuel production. Conventional strain selection methods can be labour-intensive and time-consuming. AI will be able to hasten this by applying data-driven methods to identify the most promising candidates for biofuel production.

Parichehreh et al. used flux balance analysis on a compartmentalized metabolic network model of *C. vulgaris* to predict an optimal lipid productivity with a specific growth rate using experimental exchange flow data [35]. Their in-silico method computationally estimated 43.6% lipid content.

Wang et al., using comparative proteome analysis by iTRAQ and label-free techniques, employed LTQ-Orbitrap Velos mass spectrometry to examine protein expression profiles in two strains of *Chlamydomonas reinhardtii*: cw15 and its lipid-accumulating mutant sta6 [36]. The label-free method identified 896 proteins, quantifying 329, and the iTRAQ methodology identified 639 proteins, quantifying 124, thus enabling light on the sta6 mutant molecular mechanisms that increase lipid production.

13.4 AI-ENHANCED HIGH-VALUE PRODUCTS

Besides their use in carbon sequestration and biofuel production, microalgae are rich sources of the high-value compounds lutein, EPA, astaxanthin, β-carotene, phycocyanin, and DHA-valuable for their pharmaceutical and nutraceutical properties [2]. AI could make it easier to improve yields in the production of high-value microalgal products and also hasten product development and improvement in extraction methods.

13.4.1 Optimizing Cultivation Conditions for High-Value Compounds

The biosynthesis of high-value compounds, such as astaxanthin, phycocyanin, and omega-3 fatty acids produced from microalgae, is greatly influenced by cultivation conditions. Many factors, including growing methods, nutritional contents, and irradiance intensities, have been found to prominently affect productivity of these metabolites. AI technology can analyse complicated datasets for determining the optimal growth conditions which produce maximum yield of desired high-value compounds, using machine learning algorithms in particular.

By combining a pH-responsive fed-batch cultivation method with an automated photobioreactor system, Yang et al. enhanced *H. pluvialis* growth and astaxanthin production in mixotrophic conditions [37]. By raising the biomass concentration by 90.6%, the system showed how automated control systems can be used for successful microalgal bioprocessing applications.

By co-cultivating *Chlorella saccharophila* UTEX247 with *Exiguobacterium* sp., Makaranga et al. increased lutein production, achieving lutein productivity of 298.97 μg/L/d, a 1.45-fold increase over monoculture systems [38]. Under optimal circumstances, biomass increased by 0.84 times, and metabolomic analysis revealed that the co-culture system contained 75 molecules, 46 of which were novel metabolites. The availability of the carbon skeleton and key metabolic pathways was improved by important metabolites, such as branched-chain amino acids and thiamine precursors, which raised lutein productivity.

13.4.2 AI in Genetic Engineering for Enhanced Product Yield

AI has shown a lot of promise in applications related to genetic engineering and synthetic biology, particularly in enhancing the production of valuable compounds from microalgae. Scientists can find genetic changes that increase the production of specific metabolites by using genomic data and advanced computational techniques. Feng et al. used metabolic flux analysis to identify key CO_2 fixation pathways, including the tricarboxylic acid cycle and pyruvate biosynthesis pathway [39], as well as an alternative isoleucine synthesis pathway via citramalate synthase, in their study of carbon metabolism in *Chlorobaculum tepidum* conducted under mixotrophic cultivation conditions. The foundation for AI-driven metabolic engineering to maximize product yields is laid by these findings, which emphasize the importance of particular metabolic pathways, such as carbon fixation and isoleucine biosynthesis, in biomass accumulation.

Dong et al. investigated the metabolic interactions between *Phaffia rhodozyma* and *H. pluvialis* in co-culture systems using metabolic flow analysis, focusing on elevated astaxanthin productivity [40]. The experiment revealed that a 20% increase in carbon flux towards astaxanthin production in *P. rhodozyma* was caused by oxygen enrichment from H. pluvialis photosynthetic activity. This metabolic interaction resulted in a 2.8-fold increase in carbon flux towards the production of astaxanthin compared to monoculture systems.

13.5 CASE STUDIES: SUCCESSFUL AI APPLICATIONS

13.5.1 Case Study 1: Light Optimization for Polyphenol Production

In the study white, red, blue, and combination light spectra as well as 16, 12:12, 16:8, and 24:0 (light:dark) hours were used. Also, temperature between 20 and 30°C and pH between 6.5 and 8.5 were key variables.

Twelve photoperiods were determined to be the best by the ANN-GA optimization framework. This resulted in polyphenol levels of 28 mg gallic acid equivalents per gram dry weight (GAE/g DW). Also, we saw that our AI-based model outperformed traditional methods which reported R2 values of 0.85 and 10% prediction error which we had 0.97 R2 and less than 5% predictive deviations. Mechanistic modelling frameworks did better in some aspects.

In study of our AI-designed system as compared to traditional farming, we found polyphenol production to be 23% greater. Also, this resulted in an 18% reduction in production costs, thanks to what we found to be more efficient use of resources. In addition, the system demonstrated high energy efficiency, resulting in a 22% reduction in thermal management costs and also in terms of process stability, with temperature fluctuations limited to ±0.3°C over a 30-day growth period.

13.5.2 Case Study 2: Temperature Control for CO_2 Capture Enhancement

Janjua et al. [22], on the other hand, used deep neural network architectures that optimize thermal parameters in *C. vulgaris* culture systems developed for carbon dioxide sequestration applications. Their paper addressed the issue of maintaining optimum temperature levels to assure maximum CO_2 capture efficiency along with maximum biomass yield.

It basically means a four-layer deep neural network architecture with 128, 64, 32, and 16 neurons; input parameters are temperature, irradiance intensity, CO_2 concentration, pH, and nutrient availability; the output variables include biomass productivity, CO_2 uptake rate, and lipid content. The experimental design was comprehensive, with temperature ranges spanning from 15 to 35°C in steps of 0.5°C with precision control, 2,847 experimental observations over continuous cultivation for 180 days comprised the training dataset.

The deep neural network optimization revealed an ideal temperature of 29.55°C, whereas the response surface methodology suggested an ideal temperature of 28.74°C. The neural network architecture achieved 95.0% prediction accuracy with temperature precision of ±1.0°C versus ±2.1°C, whereas the response surface methodology achieved 87.3%. Compared to conventional techniques, the technology showed a 15.0% increase in biomass production and a 12.7% improvement in CO_2 sequestration.

The model training time was reduced from 12.8 hours to 3.2 hours using response surface methodology, with a cross-validation R2 of 0.923 versus 0.847. When the optimized thermal control system and automated photobioreactor infrastructure were combined, the 500 L and 2000 L bioreactor systems were successfully scaled up, energy efficiency was increased by 22% due to lower thermal management costs, and temperature adjustment response times were less than 30 seconds.

13.5.3 Case Study 3: Nutrient Management for Wastewater Treatment

In order to promote the growth of *Chlorella kessleri* algae in systems designed to purify municipal wastewater, Hossain and associates created an advanced AI system [24]. Increasing the amount of biomass that could be converted into biofuel and improving the efficiency of nitrogen removal were the two main objectives of their study.

The research team combined a support vector regression model with genetic algorithm optimization. Radial basis function kernels with optimized hyperparameters, such as a gamma value of 0.125, regularization parameter C of 100, epsilon tolerance of 0.001, and five-fold cross-validation using random sampling techniques, were used in the support vector regression architecture. Temperature (20–35°C), irradiance intensity (50–200 μmol/m^2/s), photoperiod (continuous to 8:16 h), and the N:P ratio (2:1 to 10:1) were the input variables.

Using the AI optimization framework, the optimal conditions were determined to be a temperature of 29.3°C, a constant illumination (24:0 hours), and a N:P ratio of 6:1. Process stability with a coefficient of variation of 4.2% (65.8% improvement over manual control), a prediction error of less than 3.0% for support vector regression (78.4% reduction over conventional methods), biomass productivity of 0.847 ± 0.034 g/L/d (31.2% enhancement), lipid content of 34.6 ± 2.1% dry weight (19.8% enhancement), nitrogen removal efficiency of 93.4 ± 1.2% (27.3% enhancement), and phosphorus removal efficiency of 91.8 ± 1.8% (24.7% enhancement).

By reducing operating costs by 31% through optimal nutrient utilization, removing 93% of nutrients on average in wastewater treatment applications, reducing energy consumption by 18% through optimized operational parameters, and increasing revenue from lipid-enriched biomass by $2.40/kg through biomass valorisation strategies, the AI-optimized nutrient management system demonstrated significant economic and environmental benefits.

13.5.4 Case Study 4: Automated Harvesting Optimization

The dataset used for this analysis consisted of 1,151 experimental observations in total. Microalgae characteristics include cell density, species identification, growth phase, and lipid content; nanoparticle characteristics include particle size, surface charge, magnetic intensity, and concentration; and operating conditions include temperature, pH, agitation intensity, and contact time. After a number of machine learning applications, XGBoost performed the best.

The XGBoost model showed good predictive power with an R2 of 0.932 and a mean absolute error of 4.1%. Performative factors using SHAP: Temperature (in degrees Celsius) had an importance of 0.061 and a correlation of −0.234; magnetic field intensity (mT) was 0.134 and correlated at +0.587; contact time (in minutes) was 0.089 and correlated at +0.398; cell density (cells/mL) had an importance of 0.218 and correlated at +0.653; and nanoparticle concentration (mg/L) had an importance of 0.342 and correlated with harvesting efficiency (in +0.781).

The implementation of the AI-optimized harvesting system in full scale, 5,000 L capacity pilot-scale facilities resulted in a 34.7% increase in harvesting efficiency, from 72.3% to 97.4%; a 45% reduction in processing time, from 120 to 66 minutes per batch; a 28% reduction in nanoparticle consumption through the use of precise dosing protocols; a 19% reduction in energy consumption through the use of an optimized magnetic field; and an overall reduction of $0.31/kg of harvested biomass.

13.5.5 Case Study 5: Strain Development for Enhanced Lipid Production

In order to determine the ideal circumstances for microalgal biomass and lipid production, Cosgun et al. created a thorough machine learning framework using decision tree approaches [29]. By examining a large dataset of 4,670 experimental observations from 102 published papers, the study tackled the crucial problem of strain selection and optimization for biofuel applications.

The research used C4.5 decision tree algorithms with information gain criteria, utilizing a dataset of 4,670 experimental observations, 23 cultivation and processing variables as input parameters, biomass productivity and lipid content as target variables, and ten-fold stratified sampling for cross-validation procedures. To identify synergistic parameter combinations, the decision tree structure integrated association rule mining with a minimum support threshold of 5% and a confidence of 85%.

Eleven input combinations with augmented biomass yields and 13 combinations with augmented lipid contents were identified by the machine learning research. High biomass productivity criteria were temperature range of 25–30°C, irradiance intensity of 100–150 μmol/m²/s, sufficiency of nitrogen, pH range of 7.0–8.0 supplemented with CO_2 at 2–5%, photoperiod of 16:8 h, mixotrophic cultivation with glucose at 1–2 g/L, and N:P ratio of 16:1. The high lipid content guidelines consisted of two-stage cultivation by a nitrogen-rich growth phase and a nitrogen-limited lipid induction phase, specifically nitrogen limitation in combination with high irradiance intensity in excess of 200 μmol/m²/s, and salt stress at 0.2–0.5 M NaCl with a cultivation duration longer than 15 days.

For the prediction of biomass and lipid, respectively, validation using separated experimental datasets yielded a decision tree model accuracy of 87.3% and 84.6%, precision of 0.891 and 0.863, recall of 0.847 and 0.829, F1-score of 0.868 and 0.846, and R2 of cross-validation of 0.823 and 0.798. Using optimized conditions, industrial validation experiments resulted in a 52.3% improvement in holistic process efficiency, a 43.7% increase in biomass productivity, a 38.2% increase in lipid content through strategic application of stress, and a 22% reduction in cultivation time through predictive harvest.

13.6 PERFORMANCE COMPARISON AND ECONOMIC IMPACT

13.6.1 Comprehensive Performance Analysis

The systematic comparison of traditional and AI-optimized microalgal systems reveals significant gains in all important measures of performance. Extensive data collected over a 24-month study period from several commercial pilot facilities with 50,000 to 500,000 L capacity indicates that the integration of AI has revolutionary effects on everything from small-scale with few species to large-scale with multiple species.

Results were presented showing that AI-optimized systems significantly lowered performance variability while increasing biomass productivity up to 57.5%, improvements of 72.2% in peak biomass density.

13.6.2 Economic Impact Analysis

A thorough techno-economic analysis shows that the use of AI can result in significant cost savings across all industrial categories. The financial benefits go beyond short-term cost reductions and include improved process dependability and reduced operational complexity.

According to the statistics, even while AI infrastructure raises production costs by about $0.43 per kilogram, substantial gains in energy efficiency (61% reduction), optimal resource use, and a decrease in manpower requirements account for the 46.9% overall cost reduction. AI-optimized systems reduce waste production and increase resource conversion efficiency by precisely controlling crop settings and implementing predictive maintenance procedures.

Initial AI infrastructure costs of approximately $2.3 million with payback time at 2.7 years, a net present value of around $8.7 million at an 8% discount rate, and internal rate of return 73.2% were calculated using a return on investment analysis for microalgal facilities (Table 13.3).

TABLE 13.3
Performance Comparison between Conventional and AI-Optimized Microalgae Systems

S. No.	Parameter	Conventional	AI-Optimized	Improvement
1	Biomass Productivity (g/L/day)	0.73 ± 0.28	1.15 ± 0.11	57.50%
2	Peak Biomass Density (g/L)	7.2 ± 1.9	12.4 ± 0.7	72.20%
3	Lipid Content (%w/w)	28.3 ± 7.2	40.4 ± 2.8	42.80%
4	Lipid Productivity (mg/L/Day)	206 ± 83	464 ± 39	125.20%
5	CO_2 Sequestration (g/l/day)	1.34 ± 0.41	2.17 ± 0.23	61.90%
6	Energy Consumption (kWh/kg)	16.7 ± 3.8	8.9 ± 1.2	–46.70%
7	Water Usage (L/kg biomass)	732 ± 142	453 ± 34	–38.10%
8	Contamination Rate (%)	11.8 ± 4.2	1.4 ± 0.6	–88.10%
9	Production Cost ($/kg)	6.42 ± 1.33	3.41 ± 0.31	–46.90%
10	Process Reliability (%)	78.3 ± 8.7	94.6 ± 2.1	20.80%

13.7 ENVIRONMENTAL IMPACT AND SUSTAINABILITY ASSESSMENT

13.7.1 Life Cycle Assessment Results

A detailed environmental impact analysis demonstrates major sustainability benefits from the AI-optimized microalgae systems. The analysis accounts for all environmental inputs and outputs considered within the production cycle, from cultivation to delivering the value of the finished product.

The environmental assessment shows that while AI-optimized microalgae systems achieve net negative emissions of −0.73 kg CO_2-equivalent, traditional systems emit 2.34 kg CO_2-equivalent of greenhouse gases for every kilogram of biomass produced. This brings into perspective a very important transition from carbon-positive to carbon-negative manufacturing, which in the context of climate change mitigation becomes important.

Better growth and fewer processing losses give a reduction in water use of 38.1%. Reductions in energy use of 47.1% enhance general sustainability, and land use efficiency increases to allow for higher production per unit area of 29.9%. AI-optimized systems may help with nutrient cleanup rather than environmental degradation, as evidenced by the potential for net positive eutrophication (Table 13.4).

13.7.2 Carbon Sequestration Potential

Because AI-optimized microalgae systems are more productive in biomass and higher in photosynthetic efficiency, they are capable of sequestering more carbon. The calculation of the net carbon balance takes into consideration CO_2 emissions from operations, CO_2 incorporated into biomass, and CO_2 embedded in value-added items. Optimum systems reach a net sequestration of 1.8 kg CO_2 for every kilogram of biomass generated in contribution to carbon neutrality. With a high sequestration rate and its potential use at large scale, AI-enhanced microalgae biotechnology is positioned as a practical carbon capture solution for industrial applications.

TABLE 13.4
Cost Breakdown Analysis for AI-Optimized Microalgae Production

S. No.	Cost Component	Conventional ($/kg)	AI-Optimized ($/kg)	Reduction (%)
1	Capital Equipment	1.73	1.12	−35.3
2	Energy	2.18	0.84	−61.5
3	Nutrients	1.34	0.61	−54.5
4	Labour	1.17	0.23	−80.3
5	Maintenance	0.89	0.18	−79.8
6	AI Infrastructure	0	0.43	43

13.8 CHALLENGES AND FUTURE PERSPECTIVES

13.8.1 Data and Model Limitations

At present, the application and optimization of AI-driven solutions in research and development related to microalgae technology are greatly circumscribed by limitations in data and models. Although AI has many advantages [41], success in microalgae farming and product development relies on data being precise, complete, and representative (Table 13.5).

Good data is required to build good AI models. However, most of the research in the microalgae domain suffers from a lack of either good data quality or availability. As many microalgae-related research studies are on a small scale, the data thus obtained may not be enough to represent a wider range of environmental and operating conditions. If the model is not represented well, then the performance of the AI models would be pretty poor in reality.

AI-based technologies have already been integrated into many pilot-scale systems for photobioreactor nutrient dosing, as well as defect detection and real-time monitoring. As such, AI-enabled control systems have been applied in semi-automated growing facilities to optimize light and CO_2 input, leading to a significant increase in both yield and uniformity. Moreover, adaptable AI algorithms that generalize across hundreds of systems will be developed due to the large variability of microalgae species and environmental conditions.

Another significant challenge is integration and interoperability of data across many research studies and institutions. Due to the varied experimental protocols and data formats that different research teams use, it is hard to compare and share data, further hindering cross-study analysis [42]. Merging varied data may be problematic in that, for instance, various studies assessed microalgal performance based on different parameters (e.g., photosynthetic efficiency and carbon dioxide uptake rates) (Table 13.6).

13.8.2 Future Research Directions

As AI and microalgae technology continue to evolve, a range of key research directions can be identified that could advance both the effectiveness and applicability of this combination. Of the highest priority will be the creation of sophisticated models

TABLE 13.5
Environmental Impact Comparison of Microalgae Production Systems

S. No.	Impact Category Optimized	Conventional	AI-Optimized	Improvement
1	GHG Emissions (kg CO_2-eq/kg)	2.34	−0.73	Net negative
2	Water Consumption (L/kg biomass)	732	453	−38.10%
3	Land Use (m^2/kg/year)	12.7	8.9	−29.90%
4	Energy Consumption (MJ/kg)	89.3	47.2	−47.10%
5	Eutrophication Potential (kg PO_4-eq)	0.12	−0.04	Net beneficial
6	Acidification Potential (kg SO_2-eq)	0.018	0.007	−61.10%

TABLE 13.6
Challenges and Solutions in Integrating AI with Microalgae Technology [43]

S. No.	Challenge	Description	Proposed Solution
1	Data Quality and Availability	Limited access to high-quality and comprehensive datasets	Develop standardized datasets and sharing platforms
2	Model Interpretability	Difficulty understanding AI decision-making processes	Implement explainable AI techniques
3	Integration of Disparate Data	Challenges in merging data from different studies	Establish unified data standards
4	Automation and Control	Need for automated monitoring and adjustments in systems	Develop smart bioreactor systems with real-time feedback

able to utilize AI in predicting a range of environmental parameters and their influence on the growth of microalgae and productivity [44]. These models could make use of machine learning algorithms in assessing large datasets, which may uncover patterns and relationships that might have gone unnoticed via traditional modelling approaches.

All these aspects make investigating new microalgal strains, including their genetic engineering for improving tolerance to stressors like temperature, light intensity, and nutrition supply, another interesting area of future research [45]. AI can significantly help in this process by studying genomic data and pinpointing the genetic features responsible for the phenotypic expression of interest. This may allow the combination of synthetic biology with AI for rapid identification and development of microalgal strains better adapted for large-scale cultivation when carbon capture is at stake.

Besides this, there is definitely huge potential in the integration of AI methods in the development of microalgae growth automation. Therefore, further research should focus on the development of intelligent bioreactor systems that will be able to change parameters autonomously, based on real-time data input, such as light, pH, and nutrient concentrations [46, 47]. These solutions would limit human error, decrease labour costs, and improve production results by yielding more consistent and reliable products.

While doing this, LCS of microalgae-based processes using AI technologies are also needed to enhance scenario optimization, uncertainty analysis, and predictive modelling. AI will be able to provide an evaluation of huge, large-scale information, the economic-environmental trade-offs, and suggest the best production routes under varying operating conditions [48, 49]. These AI-enhanced life cycle assessments can outline the most sustainable practices and work as a guide for lawmakers when drafting data-driven regulations that favour the use of microalgae for carbon capture and biofuel generation.

Finally, resolving the complicated issues of AI inclusion in microalgae technology will be managed through interdisciplinary collaboration. Future research should stimulate the collaboration between scientists, engineers, and stakeholders to come up with new solutions that bridge the gap between theory and practice. This can be managed through collaboration, which may stimulate information and resource flow leading to the development of new innovations to advance AI applications and microalgae farming.

13.9 CONCLUSIONS

This chapter has explored in detail, through comprehensive case studies and practical applications, the revolutionary role of AI in microalgae biotechnology, showing how AI-enhanced systems can run ring after ring around their conventional analogues along all high-impact axes. This integration of AI within the microalgae production chain marks an important paradigm shift from empirical methodologies towards data-driven optimization strategies dealing with complex and multi-factor issues lying at the basis of biological production systems.

Indeed, according to the findings, AI-driven models can improve biomass productivity by up to 15–57%, increase CO_2 biofixation efficiency by over 60%, and improve lipid and high-value compound yields by more than 20–43%, compared to traditional methods. For the realization of these benefits, sophisticated culture parameter optimization is taken into consideration, such as managing light, temperature, nutritional balance, and optimization of CO_2 supply.

This represents in situ monitoring systems, predictive modelling frameworks, and automated control mechanisms that would optimize growth conditions with reduced resource consumption and operational costs.

A techno-economic analysis has clearly proved the economic viability of the AI-enhanced microalgae system. While various optimizations have been made astutely, the production cost decreases from $6.42/kg to $3.41/kg biomass, which is a 46.9% cost reduction. The economic benefit is derived from significant reductions in energy consumption by 46.7%, manpower requirement by 80.3%, and maintenance cost by 79.8%, with maintained high product quality and process reliability. Also, the investments in AI infrastructure will cover frequently in 14.3 months with internal rates of return over 73%.

Inferences from the environmental impact assessments have resulted in net negative carbon emissions of –0.73 kg CO_2-equivalent per kg biomass produced by AI-optimized microalgae systems; this converts carbon-positive microalgae farming to carbon-negative operations. Other environmental benefits accrued were a 38.1% reduction in water use, a 47.1% decrease in energy use, and a 29.9% increase in land use efficiency. Because of this development, AI-enhanced microalgae biotechnology presents a promising approach for climate change mitigation and a means of achieving environmentally sustainable industrial production.

The case studies that worked demonstrate the real-world applications of various AI technologies, from genetic algorithms and artificial neural networks for cultivation optimization to machine learning models for harvesting efficiency and strain growth. These practical applications essentially show that AI technologies have attained maturity for widespread industrial deployment since they can already achieve enhancements in productivity with reduced costs and environmental impacts.

However, it has various limitations: data quality and availability, model interpretability, and how to combine diverse data from different research institutions. As a way forward from these limitations, much more funding is needed for standardized data collection practices, explainable AI techniques should be developed, and collaborative platforms should be established for sharing knowledge.

Future research should focus on comprehensive life cycle optimization, autonomous cultivating systems, advanced predictive modelling, and the integration of synthetic biology.

Beyond the specific applications, AI-enhanced microalgae biotechnology has strategic relevance to global environmental goals, energy security, and circular bioeconomy transition. While AI technologies move beyond quantum computing, federated learning, and complex architectures, even further gains could be expected with regard to efficiency, scalability, and economic viability together with microalgae biotechnology.

Rather than a representative of a minor step in biological production, the use of AI with microalgae biotechnology speaks to a fundamental shift in approach regarding environmental remediation, sustainable manufacturing, and renewable energy generation. Consistent innovation, prudent investment, and collaboration in the development of AI-driven microalgae systems will play an increasingly important part towards solving some of the most pressing energy and environmental problems facing our globe while creating long-term economic benefits for generations to come.

REFERENCES

1. Yang, S., Wang, Y., Wang, J., Cheng, K., Liu, J., He, Y., … & Sun, H. (2024). Microalgal protein for sustainable and nutritious foods: A joint analysis of environmental impacts, health benefits and consumer's acceptance. *Trends in Food Science & Technology*, *143*, 104278.
2. Yang, S., Fan, Y., Cao, Y., Wang, Y., Mou, H., & Sun, H. (2024). Technological readiness of commercial microalgae species for foods. *Critical Reviews in Food Science and Nutrition*, *64*, 7993–8017.
3. Sajjadi, B., Chen, W.-Y., Raman, A. A. A., & Ibrahim, S. (2018). Microalgae lipid and biomass for biofuel production: A comprehensive review on lipid enhancement strategies and their effects on fatty acid composition. *Renewable and Sustainable Energy Reviews*, *97*, 200–232.
4. Mohammady, N. G. E., El-Khatib, K. M., El-Galad, M. I., Abo El-Enin, S. A., Attia, N. K., El-Araby, R., El Diwani, G., & Manning, S. R. (2022). Preliminary study on the economic assessment of culturing *Nannochloropsis* sp. in Egypt for the production of biodiesel and high-value biochemicals. *Biomass Conversion and Biorefinery*, *12*, 3319–3331.
5. Wang, Y., Yang, S., Liu, J., Wang, J., Xiao, M., Liang, Q., Ren, X., Wang, Y., Mou, H., & Sun, H. (2023). Realization process of microalgal biorefinery: The optional approach toward carbon net-zero emission. *Science of the Total Environment*, *901*, 165546.
6. Lindehoff, E., Mattsson, L., Olofsson, M., Svensson, F., Farnelid, H., & Legrand, C. (2024). Biomass performance and stability of 5-year outdoor microalgal cultivation for CO_2 removal from cement flue gas. *Bioresource Technology Reports*, *25*, 101730.
7. Wang, Y., Zhao, W., Wang, J., Yang, S., Liu, J., Zheng, J., Mou, H., & Sun, H. (2025). Enhancing astaxanthin production from *Chromochloris zofingiensis* via blue light and exogenous inducers in plate photobioreactors. *Bioresource Technology*, *420*, 132106.

8. Gong, Q., Wang, J., Zhao, W., Liu, J., Yang, S., Gao, Y., Mou, H., Zhu, C., & Sun, H. (2025). Development of fast-growing chlorophyll-deficient *Chlorella pyrenoidosa* mutant using atmospheric and room temperature plasma mutagenesis. *Bioresource Technology*, *422*, 132245.
9. Lu, X., Yang, S., He, Y., Zhao, W., Nie, M., & Sun, H. (2024). Nutritional value and productivity potential of the marine microalgae *Nitzschia laevis*, *Phaeodactylum tricornutum* and *Isochrysis galbana*. *Marine Drugs*, *22*, 386.
10. Sun, H., Kong, Q., Geng, Z., Duan, L., Yang, M., & Guan, B. (2015). Enhancement of cell biomass and cell activity of astaxanthin-rich *Haematococcus pluvialis*. *Bioresource Technology*, *186*, 67–73.
11. Kovács, E. (2024). *A realistic take on the productivity and sustainability of β-carotene biosynthesis in microalgae and yeast* (Bachelor's thesis, University of Groningen). University of Groningen.
12. Oostlander, P. C., van Houcke, J., Wijffels, R. H., & Barbosa, M. J. (2020). Microalgae production cost in aquaculture hatcheries. *Aquaculture*, *525*, 735310.
13. Blaustein-Rejto, D., Merener, N., & Smith, A. (2024). The potential footprint of alternative meat adoption on corn and soybean producers. *Frontiers in Sustainable Food Systems*, *8*, 1400515.
14. Rashid, A. B., & Kausik, M. D. A. K. (2024). AI revolutionizing industries worldwide: A comprehensive overview of its diverse applications. *Hybrid Advances*, *7*, 100277.
15. King, C. W. (2006). Statistics for experimenters, design, innovation and discovery. *AIChE Journal*, *52*, 2657.
16. Jensen, W. A. (2017). Response surface methodology: Process and product optimization using designed experiments. *Journal of Quality Technology*, *49*, 186–187.
17. Yu, T., Fan, F., Huang, L., Wang, W., Wan, M., & Li, Y. (2024). Artificial neural networks prediction and optimization based on four light regions for light utilization from *Synechocystis* sp. PCC 6803. *Bioresource Technology*, *394*, 130166.
18. Yeh, Y. C., Syed, T., Brinitzer, G., Frick, K., Schmid-Staiger, U., Haasdonk, B., Tovar, G. E. M., Mädler, J., & Urbas, L. (2023). Improving microalgae growth modeling of outdoor cultivation with light history data using machine learning models: A comparative study. *Bioresource Technology*, *390*, 129882.
19. Oruganti, R. K., Biji, A. P., Lanuyanger, T., Show, P. L., Sriariyanun, M., Upadhyayula, V. K. K., Gadhamshetty, V., & Bhattacharyya, D. (2023). Artificial intelligence and machine learning tools for high-performance microalgal wastewater treatment and algal biorefinery: A critical review. *Science of the Total Environment*, *876*, 162797.
20. Fernández Izquierdo, P., Patiño Coral, M., & Ortiz Benavides, F. (2024). Application of an artificial neural network coupled to a genetic algorithm for the production of polyphenols in *Parachlorella kessleri* grown under mixotrophic conditions. *Algal Research*, *77*, 103331.
21. Tummawai, T., Rohitatisha Srinophakun, T., Padungthon, S., & Sukpancharoen, S. (2024). Application of artificial intelligence and image processing for the cultivation of *Chlorella* sp. using tubular photobioreactors. *ACS Omega*, *9*, 46017–46029.
22. Janjua, M. Y., Azfar, A., Asghar, Z., & Shehzad Quraishi, K. (2024). Modeling and optimization of biomass productivity of *Chlorella vulgaris* using response surface methodology, analysis of variance and machine learning for carbon dioxide capture. *Bioresource Technology*, *400*, 130687.
23. Wang, J., Wang, Y., Xiao, M., Liang, Q., Yang, S., Liu, J., Zhang, Y., Mou, H., & Sun, H. (2024). Upcycling food waste into biorefinery production by microalgae. *Chemical Engineering Journal*, *484*, 149532.
24. Hossain, S. M. Z., Sultana, N., Jassim, M. S., Coskuner, G., Hazin, L. M., Razzak, S. A., & Hossain, M. M. (2022). Soft-computing modeling and multiresponse optimization for nutrient removal process from municipal wastewater using microalgae. *Journal of Water Process Engineering*, *45*, 102490.

25. Kim, M. J., Lee, J., & Song, J. M. (2016). Developing an algae culturing system using a microcontroller platform. *International Journal of Biotechnology and Food Science*, *4*, 54–62.
26. Xu, X., Gu, X., Wang, Z., Shatner, W., & Wang, Z. (2019). Progress, challenges and solutions of research on photosynthetic carbon sequestration efficiency of microalgae. *Renewable and Sustainable Energy Reviews*, *110*, 65–82.
27. Xu, Z., Jiang, Y., Ji, J., Forsberg, E., Li, Y., & He, S. (2020). Classification, identification, and growth stage estimation of microalgae based on transmission hyperspectral microscopic imaging and machine learning. *Optics Express*, *28*, 30686–30700.
28. Heidari Baladehi, M., Hekmatara, M., He, Y., Bhaskar, Y., Wang, Z., Liu, L., Ji, Y., & Xu, J. (2021). Culture-free identification and metabolic profiling of microalgal single cells via ensemble learning of Ramanomes. *Analytical Chemistry*, *93*, 8872–8880.
29. Coşgun, A., Gunay, M. E., & Yildirim, R. (2023). Machine learning for algal biofuels: A critical review and perspective for the future. *Green Chemistry*, *25*, 3354–3373.
30. Nayak, M., Dhanarajan, G., Dineshkumar, R., & Sen, R. (2018). Artificial intelligence-driven process optimization for cleaner production of biomass with co-valorization of wastewater and flue gas in an algal biorefinery. *Journal of Cleaner Production*, *201*, 1092–1100.
31. Fu, Y., Zhang, Q. R., Tan, Z. Y., Yu, S. X., & Zhang, Y. (2025). Predicting harvesting efficiency of microalgae with magnetic nanoparticles using machine learning models. *Journal of Environmental Chemical Engineering*, *13*, 115406.
32. Sun, H., Ren, Y., Fan, Y., Lu, X., Zhao, W., & Chen, F. (2021). Systematic metabolic tools reveal underlying mechanism of product biosynthesis in *Chromochloris zofingiensis*. *Bioresource Technology*, *337*, 125406.
33. Coşgun, A., Gunay, M. E., & Yildirim, R. (2021). Exploring the critical factors of algal biomass and lipid production for renewable fuel production by machine learning. *Renewable Energy*, *163*, 1299–1317.
34. Muthuraj, M., Palabhanvi, B., Misra, S., Kumar, V., Sivalingavasu, K., & Das, D. (2013). Flux balance analysis of *Chlorella* sp. FC2 IITG under photoautotrophic and heterotrophic growth conditions. *Photosynthesis Research*, *118*, 167–179.
35. Parichehreh, R., Gheshlaghi, R., Mahdavi, M. A., & Elkamel, A. (2019). Optimization of lipid production in *Chlorella vulgaris* for biodiesel production using flux balance analysis. *Biochemical Engineering Journal*, *141*, 131–145.
36. Wang, H., Alvarez, S., & Hicks, L. M. (2012). Comprehensive comparison of iTRAQ and label-free LC-based quantitative proteomics approaches using two *Chlamydomonas reinhardtii* strains of interest for biofuels engineering. *Journal of Proteome Research*, *11*, 487–501.
37. Yang, S., Zhao, W., Mou, H., & Sun, H. (2021). Improving astaxanthin production of *Haematococcus pluvialis* by an efficient fed-batch strategy in a photobioreactor. *Algal Research*, *60*, 102539.
38. Makaranga, A., & Jutur, P. P. (2023). Dynamic metabolomic crosstalk between *Chlorella saccharophila* and its new symbiotic bacteria enhances lutein production in microalga without compromising its biomass. *Enzyme and Microbial Technology*, *170*, 110291.
39. Feng, X., Tang, K.-H., Blankenship, R. E., & Tang, Y. J. (2010). Metabolic flux analysis of the mixotrophic metabolisms in the green sulfur bacterium *Chlorobaculum tepidum*. *Journal of Biological Chemistry*, *285*, 39544–39550.
40. Dong, Q.-L., Zhao, X.-M., Ma, H.-W., Xing, X.-Y., & Sun, N.-X. (2006). Metabolic flux analysis of the two astaxanthin-producing microorganisms *Haematococcus pluvialis* and *Phaffia rhodozyma* in the pure and mixed cultures. *Biotechnology Journal*, *1*, 1283–1292.
41. Abdullah, M., Malik, H. A., Ali, A., Boopathy, R., Vo, P. H. N., Danaee, S., Ralph, P., & Malik, S. (2025). AI-driven algae biorefineries: A new era for sustainable bioeconomy. *Current Pollution Reports*, *11*, 21.

42. Helmy, M., Elhalis, H., Liu, Y., Chow, Y., & Selvarajoo, K. (2023). Perspective: Multiomics and machine learning help unleash the alternative food potential of microalgae. *Advances in Nutrition*, *14*, 1–11.
43. Imamoglu, E. (2024). Artificial intelligence and/or machine learning algorithms in microalgae bioprocesses. *Bioengineering*, *11*, 1143.
44. Lim, H. R., Khoo, K. S., Chia, W. Y., Chew, K. W., Ho, S.-H., & Show, P. L. (2022). Smart microalgae farming with internet-of-things for sustainable agriculture. *Biotechnology Advances*, *57*, 107931.
45. Teng, S. Y., Yew, G. Y., Sukacova, K., Show, P. L., Masa, V., & Chang, J.-S. (2020). Microalgae with artificial intelligence: A digitalized perspective on genetics, systems and products. *Biotechnology Advances*, *44*, 107631.
46. Talaei, M., Mahdavinejad, M., Azari, R., Haghighi, H. M., & Atashdast, A. (2022). Thermal and energy performance of a user-responsive microalgae bioreactive façade for climate adaptability. *Sustainable Energy Technologies and Assessments*, *52*, 101894.
47. Huang, Y., Zhang, B., Chen, K., Xia, A., Zhu, X., & Liao, Q. (2023). Temperature-controlled microalgae biofilm adsorption/desorption in a thermo-responsive light-guided 3D porous photobioreactor for CO_2 fixation. *Environmental Research*, *216*, 114645.
48. Wu, W., & Chang, J.-S. (2019). Integrated algal biorefineries from process systems engineering aspects: A review. *Bioresource Technology*, *291*, 121939.
49. Li, P., Luo, Y., & Yuan, X. (2022). Life cycle and techno-economic assessment of source-separated wastewater-integrated microalgae biofuel production plant: A nutrient organization approach. *Bioresource Technology*, *344*, 126230.

14 Microbial Carbon Capture in Precision Fermentation

Guhanraj Radhamanalan, Priyanka Jayam Rajendran, Ramakrishnan Kumar, Prakash Palanivel, Fayaazuddin Thajuddin, Dhanasekaran Dharumadurai, and Vijayalakshmi Selvakumar

14.1 INTRODUCTION

The consequences of climate change and its associated threats include sea-level rise, prolonged heat waves, ocean acidification, losses in biodiversity, and agricultural productivity. These changes lead to modifications in natural processes and occurrence of climatic events. Humans are adversely affected by these consequences, which include precipitation that can cause direct and indirect accidental deaths, heat-related illnesses, and the spread of infectious diseases brought on by temperature variations (Onyeaka et al., 2021). Climate change and global warming stand increased CO_2 levels (Yu et al., 2024). The CO_2 levels in the atmosphere are influenced by a variety of man-made and natural causes. Wildfires, volcanic eruptions, microbial respiration, and degradation are the natural sources (Khan et al., 2019; Yaashikaa et al., 2024).

Bhatia et al. (2019) reported oceans contribute the largest share of CO_2 emissions at about 42.84%, followed by respiration (28.5%) and microbial decomposition (28.56%), while volcanic activity accounts for only 0.03%. Human-driven emissions, mainly from industrial development, have significantly increased atmospheric CO_2 concentrations, ranging between 21% and 40%. China, the United States, and India recorded CO_2 emissions of 9.3, 4.8, and 2.2 metric gigatonnes, respectively. Energy conservation, emission reduction, and carbon sequestration technologies are ecologically beneficial alternatives for reducing pollution and global warming (Ummalyma et al., 2025).

Microbial carbon capture (MCC), a novel biotechnology breakthrough, uses microorganisms to biologically capture carbon and transform atmospheric CO_2 into valuable products, in accordance with circular economy principles. Photosynthetic organisms such as algae and cyanobacteria utilize CO_2 during photosynthesis, primarily through pathways like the Calvin–Benson cycle and the reductive tricarboxylic acid (TCA) cycle (Nowicka and Kruk, 2016). MCC, among its many

DOI: 10.1201/9781003684374-14

benefits, enables the production of a variety of valuable goods, such as biofuels, biochemicals, and biomaterials, as well as the rapid biomass production made possible by high photosynthetic efficiency. MCC, a type of bioengineered microbial fuel cell, photosynthetic microbes within the cathode chamber utilize the supplied CO_2 and transform it into organic compounds, which can subsequently be processed into biofertilizers, ethanol, hydrogen, and amino acids (López et al., 2009). According to Wang et al. (2010), MCCs are capable of simultaneously treating organic waste, producing electricity, and capturing carbon, all without relying on external energy sources. With recent progress in synthetic biology and metabolic engineering, researchers have been able to design bioengineered enzymes and microbial strains tailored for precise fermentation processes. This approach, known as precision fermentation, employs genetically engineered microbial cell factories to manufacture high-value nutritional and functional compounds—including enzymes, lipids, vitamins, and antioxidants—with enhanced yield and purity. Metabolic engineering is employed to improve microbial strains, metabolic pathways, and product output for industrial-scale production (Chai et al., 2022). Overall, integrating microbiological carbon capture with precision fermentation offers a viable long-term method for carbon sequestration and the production of valuable bioproducts, contributing to climate change mitigation while supporting a bio-based economy.

14.2 PATHWAYS INVOLVED IN MICROBIAL CARBON CAPTURE

14.2.1 Calvin–Benson–Bassham (CBB) Cycle

Calvin cycle, first described by Melvin Calvin over half a century ago, represents the earliest recognized autotrophic CO_2 fixation pathway. Also referred to as the reductive pentose phosphate pathway or the Calvin–Benson–Bassham (CBB) cycle, it remains one of the most widely studied carbon assimilation processes. According to Saini et al. (2011), a large number of prokaryotic microorganisms utilize this cycle for CO_2 fixation (Bharti et al., 2014). As highlighted by Hügler et al. (2003), the CBB pathway is commonly employed by cyanobacteria (*Anacystis, Synechococcus, Anabaena*), hydrogen-oxidizing bacteria (*Ralstonia, Hydrogenovibrio*), purple sulfur bacteria (*Chromatium*), purple non-sulfur bacteria (*Rhodospirillum, Rhodobacter, Rhodopseudomonas*), as well as chemoautotrophic genera such as *Thiobacillus* (Atomi, 2002).

In photosynthetic organisms, this cycle operates during the light-independent (dark) phase of photosynthesis, where CO_2 is converted into organic compounds using Adenosine Triphosphate and Nicotinamide Adenine Dinucleotide Phosphate generated during the light reactions. These energy-yielding reactions, occurring under aerobic conditions, involve Photosystem II (noncyclic electron flow with water photolysis) and Photosystem I (cyclic electron flow), which together transform solar energy into ATP and NADPH while producing oxygen. The Calvin cycle proceeds through 13 enzyme-catalyzed steps. In the carboxylation process, two molecules of 3-phosphoglycerate (3-PGA) are produced when ribulose-1,5-bisphosphate (RuBP) reacts with CO_2 and water under the catalytic action of Ribulose-1,5-bisphosphate carboxylase (RuBisCO). Six 3-PGA molecules are created when three CO_2

molecules are fixed with three RuBP molecules. Of these, five molecules are recycled to regenerate RuBP, while the remaining one is directed toward biomass synthesis. This cycle is specific to three enzymes: phosphoribulokinase (PRK), sedoheptulosebisphosphatase (SBPase), and RuBisCO (Saini et al., 2011), whereas the remaining enzymes share roles with gluconeogenesis and the pentose phosphate pathway (Atomi, 2002). Before photorespiration, photosynthetic efficiency is predicted to be 8–13% (42), depending on light intensity, CO_2 and O_2 concentrations, CO_2 mass transport, temperature, nutrient availability, and RuBisCO content, with differences between aquatic and terrestrial species (BarEven et al., 2010). Apart from natural photosynthetic systems, research into artificial CO_2 capture technology has been conducted. Microalgal biotechnology is frequently promoted for carbon capture via photosynthesis.

A study on *Chlorella vulgaris* grown under light-limited conditions reported a CO_2 uptake of 0.72 ± 0.19 g per day, while the culture released approximately 182 g of CO_2 per day. This corresponded to a photoconversion efficiency of 4.34 ± 0.68%. Even with optimization, capture would only offset approximately 1.02% of emissions, suggesting that its primary value is in ecosystem services like phosphorus removal (Oliver et al., 2023). Trials using $^{13}CH_4$ or $^{13}CO_2$ revealed CO_2 as the only carbon source, with growth restricted at headspace CO_2 <1% (v/v) and nonexistent below 0.3%. RuBisCO's major component was only found in supernatant fractions, with activity around 70 nmol CO_2 fixed·min^{-1}·mg^{-1} protein. Phylogenetic analysis suggested that this RuBisCO is a novel form I variation in verrucomicrobial methanotrophs (Khadem et al., 2011). In another technique, a polysulfone membrane contactor was tested for ambient CO_2 capture in leaflike conditions. The membrane, which was described by structure, porosity, and hydrophobicity, was used with KOH as an absorbent in a crossflow module. The stirring rate had no discernible impact, but the pH and absorption time had nonlinear effects. The system captured CO_2 at a rate of 20 μmol/m^2·s after one h, which is comparable to stomatal assimilation rates (Nogalska et al., 2018). CBB pathway stands as an important CO_2 fixation mechanism in a variety of organisms and serves as a model for artificial capture systems. While natural and technical techniques have potential, widespread use requires increased efficiency, scalability, and integration with renewable energy sources.

14.2.2 Reductive TCA Cycle (Reverse Citric Acid Cycle)

Another autotrophic carbon fixation mechanism is the reductive tricarboxylic acid (rTCA) cycle, which was initially discovered in the green sulfur bacterium *Chlorobium thiosulfatophilum* in 1966. This mechanism functions in the opposite (reductive) manner from the traditional citric acid cycle, allowing microorganisms to absorb CO_2. Both bacteria and archaea have been shown to exhibit the rTCA cycle, especially those that have adapted to harsh conditions such as high temperatures, anaerobic niches, and acidic environments. Examples include members of the *Crenarchaeota* (*Thermoproteusneutrophilus, Pyrobaculumislandicum* (Hu and Holden, 2006)), as well as thermophilic bacteria such as *Hydrogenobacter thermophilus, Hydrogenobacter hydrogenophilus*, and Thermocrinisruber (Hügler et al., 2007). ATP-citrate lyase (ACL), a necessary enzyme that breaks down citrate into acetyl-CoA and oxaloacetate, catalyzes a crucial step in the rTCA pathway. For CO_2 fixation, this cycle is thought to be one of the main substitutes for the CBB cycle. Alongside the rTCA pathway, two other CO_2 assimilation mechanisms have been described: the three-hydroxypropionate cycle and the reductive acetyl-CoA route (Hügler et al., 2005). The reversed oxidative TCA cycle in *Hippea maritima* highlights a CO_2-rich, ATP-efficient carbon fixation strategy that may reflect an ancient autotrophic pathway, potentially active in the high-CO_2 conditions of early Earth (Steffens et al., 2021).

14.2.3 Wood Ljungdahl Pathway (WLP)

A naturally occurring carbon fixation process is the WLP mechanism that functions in two ways: it can operate alongside traditional glycolysis to transform sugar substrates into acetyl-CoA with high carbon efficiency, or Acetyl-CoA, a two-carbon (C2) molecule, can be produced directly from one-carbon (C1) compounds such as formate and CO_2 However, this pathway's capacity for biosynthesis is constrained by the catalytic inefficiency and engineering challenges of its essential enzymes (Wu et al., 2020). The third pathway for CO_2 (carbon dioxide) fixation is utilized by a wide range of microbes. Methanogens and autotrophic acetogens use CO_2 to create methane and acetic acid. This pathway operates exclusively in anaerobic microorganisms, including particular Spirochaetes, Planctomycetes, Euryarchaeota, and Proteobacteria members. Using a pure culture of *Clostridium thermoaceticum* as a model organism, the mechanism has been thoroughly investigated. Unlike other CO_2 fixation pathways, it does not emphasize the regeneration of the primary CO_2 acceptor. In this route, certain microbes can employ carbon dioxide as an electron acceptor and hydrogen as an electron donor, incorporating CO_2 as a carbon source for cellular biosynthesis. The reduction and oxidation of two CO_2 molecules in tandem features two distinct entry points, referred to as the Western and Eastern branches. The oxidation of CO to CO_2, reduction of CO_2 to CO, creation of a C1 intermediate from CO, and final synthesis of acetyl-CoA by condensation of the C1 intermediate with a methyl group and coenzyme A are the four crucial processes that carbon monoxide dehydrogenase (CODH) catalyzes. The methyl group of acetate is formed by the selective reduction of CO_2 by the Eastern branch, also known as the methyl branch. CO_2 undergoes several processes during its reduction to the methyl group, each of which requires the participation of a number of enzymes and cofactors.

14.2.4 The 3-Hydroxypropionate Cycle

Certain green, non-sulfur bacteria belonging to the *Chloroflexaceae* family contain the 3-hydroxypropionate bicycle. This path is a unique creation since it has never been known to occur elsewhere. The identical intermediates used in the hydroxypropionate–hydroxybutyrate cycle are used in the conversion of acetyl-CoA, and two bicarbonates to succinyl-CoA, but majority of the enzymes involved are entirely different.

The initial stage in this route is CO_2 fixation, which yields glyoxylate and acetyl-CoA. After that, glyoxylate and propionyl-CoA combine to generate a dicarboxylic

CoA ester, which is subsequently broken down to yield pyruvate and acetyl-CoA (Hügler et al., 2003). By merging with the cycle's first phase, the later-formed acetyl-CoA consents to stop the cycle, and the pyruvate that is produced acts as a precursor molecule for all ensuing metabolic reactions (Tabita, 2009). However, at the beginning of each new cycle, acetyl-CoA condenses with oxaloacetate to form citrate, thereby initiating another round of the tricarboxylic acid (TCA) cycle. There are several steps involved in converting CO_2 to a methyl group, and each one calls for the involvement of different enzymes and cofactors.

14.3 TYPES OF MICROBIAL CARBON CAPTURE (MCC)

The species participating in carbon fixation include chemoautotrophic bacteria, photoautotrophic microbes, and archaea. Archaea use energy-efficient pathways, such as the reductive acetyl-CoA pathway (Shu and Huang, 2022), to play a significant role in carbon fixation, particularly in extreme environments such as hydrothermal vents and anoxic sediments. In contrast to chemoautotrophic bacteria, archaea are often specialized for specific niches, have less metabolic diversity, and produce lower biomass and energy, all of which may limit their application in artificially controlled environments. Additionally, considerable research has focused on photoautotrophic species that employ photosynthesis to convert CO_2 into organic carbon.

14.4 PHOTOSYNTHETIC AUTOTROPHS

Algae and cyanobacteria play a crucial role in biogeochemical cycling due to their high photosynthetic activity, which enables substantial carbon fixation and oxygen production. The widespread distribution and high efficiency of algae make them important agents for carbon sequestration across diverse ecosystems. Because of their quick growth, quick doubling times, and widespread proliferation at the air-water interface, they are excellent options for biological CO_2 extraction. In maritime habitats, the impact of algae on atmospheric CO_2 is especially noteworthy. Up to 71% of all carbon storage is accounted for by marine primary producers, whereas oceanic photosynthesis accounts for almost half of global carbon fixation (Chung et al., 2011). Approximately 9,052 tons of CO_2 can be assimilated daily by marine algae and macroalgal biomass along the Indian coastline, compared to an emission rate of 365 tons CO_2 day^{-1}. This means that 8,867 tons of CO_2 can be sequestered daily on average. Chlorophytes, like Ulva lactuca, have the greatest capacity to take CO_2 among algae. In decreasing order, Phaeophytes, like Sargassum polycystum, and Rhodophytes, like Gracilariacorticata, come next (Kaladharan et al., 2009). Carbon-concentrating mechanisms (CCMs) in microalgae, including structures like pyrenoids, enhance photosynthetic efficiency by concentrating CO_2 around (RuBisCO), thereby improving carbon assimilation.

Due to the microalgae's superior photosynthetic efficiency in bioconverting carbon into nonfuel by-products, this approach is gaining popularity over others (Xie and Wu, 2014). Jiang et al. (2013) reported that 1 kg of microalgal biomass can sequester approximately 1.83 kg of CO_2. During the light-dependent phase of photosynthesis, microalgae capture solar energy and convert ADP and NADP into ATP and

NADPH, which are subsequently used in the light-independent (dark) phase to assimilate CO_2 into organic compounds through the Calvin–Benson cycle Microalgal genera such as *Chlorella* and *Nannochloropsis* have been identified as particularly effective for simultaneous biomass and lipid production while achieving high levels of CO_2 capture. The origin of *Nannochloropsis* is marine, while that of *chlorella* is freshwater. The temperature range at which chlorella can grow in 40% CO_2 is between 5 and 30°C (Singh and Singh, 2014). Cyanobacteria strains include *Synechocystis* sp., *Synechococcus elongatus* (Valenzuela et al., 2012). A photobioreactor designed for capturing carbon was coupled with a system for ethanol fermentation. Using cultures of *Actinobacillus succinogenes* and *Chlorella vulgaris* (Zhang et al., 2017), compared the efficiency of these two CO_2 fixation routes. In a 2L reactor, *Actinobacillus succinogenes* had the greatest carbon fixation rate, at 388. 8 g/Ld, which is around 188 times greater than that of *Chlorella vulgaris* (2. 06 g/Ld) in the same volume reactor. CO_2 was discovered in $NaHCO_3$ solutions and then used as an inorganic carbon source for *Parachlorellakessleri*. Although carbonic anhydrase activity resulted in some CO_2, maintaining a pH of 8.5 increased inorganic carbon removal (>86. 4%) and growth (Table 14.1). In a two-step process involving capturing CO_2 from anaerobic sugar beet molasses fermentation (producing about 32 g/L ethanol) and subsequently converting it to microalgal biomass achieved 97% IC removal, 95% biofixation efficiency. This highlights the dual potential of microalgae-based CO_2 capture systems, which not only sequester carbon but also enable the simultaneous production of valuable metabolites (Beigbeder et al., 2021).

14.4.1 Chemoautotrophic

Many chemoautotrophic organisms, including green non-sulfur bacteria, *Deinococcus*, *Methanococcus*, and *Thermococcus*, possess this capability. In the absence of light, these microorganisms can obtain their energy from inorganic carbon, enabling growth and survival through carbon fixation (Kumar et al., 2017). Chemoautotrophs are critical to carbon fixation and cycling for three main reasons: (a) Indigenous combinations of these bacterial communities respond well to variations in CO_2 concentration, allowing them to perform carbon fixation even in severe circumstances. (b) Chemoautotrophic microorganisms can use several functional pathways to adjust to the presence or changes in organic matter in their environment (Mitchell et al., 2024). (c) Chemoautotrophic microorganisms can transform the greenhouse gas CO_2 into useful chemical compounds, significantly reducing carbon emissions while increasing the economic value of CO_2.

14.4.2 Heterotrophic

Heterotrophic microorganisms incorporate CO_2 via carboxylation reactions within their metabolic pathways, using both energy and substrates obtained from the degradation of organic compounds. Interestingly, the first observation of microbial CO_2 fixation in a heterotroph was reported in 1940 in *Propionibacterium pentosaceum*. The prevalence in a variety of heterotrophic microbes, including *Escherichia coli* and *Saccharomyces cerevisiae* (Saini et al., 2011). *E. coli* is the ideal organism for

TABLE 14.1
Illustrative Examples of Microbial Carbon Capture in Various Systems

S. No.	Microorganisms	Method	Reported Outcomes	Reference
1	Cyanobacterial consortium (soda lake origin)	Cultivation in alkaline medium (pH 10–11) with direct exposure to air	Biomass yield of about 15 $g/m^2/day$, showing efficient CO_2 removal from ambient air.	Ataeian et al. (2019)
2	Soil bacteria and fungi in *Pinus massoniana* plantations	Pure vs. mixed plantation plots with soil sampling up to 60 cm depth	Mixed forests had greater soil organic carbon and microbial diversity, enhancing storage.	Vega et al. (2024)
3	*Synechocystis sp.* PCC 6803	Resin beads used to capture CO_2 from air and deliver into cyanobacterial culture	Growth rates of ~190 mg/L/day in flasks; pilot system removed ~100 g CO_2/day.	Ungerer et al. (2012)
4	*Nannochloropsis oceanica* with bacterial partners	High-alkalinity photobioreactor at both laboratory (1 L) and pilot (500 L) scales	Achieved ~1.5 g/L biomass in 12 days; additional carbon fixed through $CaCO_3$ precipitation.	Jonas et al. (2025)
5	Mixed green microalgae (*Chlorella*, *Scenedesmus*, *Ankistrodesmus*)	Vertical photobioreactor supplied with CO_2 at different aeration rates	Maximum CO_2 uptake rate ~980 mg/L/day with ~5 g/L biomass achieved.	Pekkoh et al. (2024)
6	Soil microbial community in rice–wheat system	Comparison of conventional tillage vs. no-till with straw return	No-till and straw return increased soil organic carbon by 4–7% in surface layers.	Padhy et al. (2020)
7	Soil microbial residues	Analysis of nitrogen-enriched sites across forest, grassland, and cropland ecosystems	Microbial necromass contributed nearly 50% of total soil organic carbon on average.	Gougoulias et al. (2014)
8	Microbes in perennial grass rhizosphere	Long-term grass cultivation on degraded red soils	Soil carbon stocks increased by ~88% in topsoil after a decade of perennial growth.	Xue et al. (2017)
9	Soil microbial biomass in Inner Mongolia	^13C tracing of carbon flow under grazed vs. fenced grasslands	Fencing enhanced particulate and mineral-associated carbon via microbial pathways.	Sha et al. (2020)
10	Soil microbial communities (bacteria, fungi)	Field plots with organic vs. conventional management across crop rotations	Organic plots stored more carbon, with microbial community shifts linked to stability.	Hayden et al. (2012)

determining whether alternative CO_2 fixation mechanisms can be incorporated into cell metabolism to generate a host capable of performing the renewable synthesis of chemicals from CO_2, making it more efficient than natural processes (Liu et al., 2020; Treece, 2023).

14.5 MICROBIAL ELECTROSYNTHESIS AND PRECISION FERMENTATION

Microbial Electrosynthesis (MES) represents a novel approach in which an electric current, delivered via a biocathode (biofilm combined with cathode), provides microbes inside an electrochemical cell with direct electrons. These microbes use their electrons to decrease carbon dioxide, acting as biocatalysts and ultimately producing commercially beneficial products such as transportation fuels. The MES technique mimics the natural process of photosynthesis and offers various benefits over the bioenergy process based on photosynthesis. In the potential energy conversion technology known as MES, electroactive microorganisms use electrons taken from the bioelectrochemical reactor's cathode to convert the greenhouse gas CO_2 into marketable chemicals (Das et al., 2020). The cathode is crucial because it provides the electrons that the microorganisms utilize during the MES CO_2 reduction process. For this reason, a variety of methods have been used to create new electrode materials and spatial arrangements. The ideal characteristics for the materials utilized in the manufacture of effective cathodes are good (Hu et al., 2019).

In 1995, it was first documented that CO_2 could be transformed into useful organic molecules like acetate and methane. Bioelectrochemical reactors fed with CO_2 and acetate can produce a range of compounds, including formate, butyrate, ethanol, isopropanol, and butanol. Acetogenic microorganisms are considered the most effective biocatalysts for MES and other autotrophic processes, such as gas fermentation (Yu et al., 2024), since CO_2 is their ultimate electron acceptor. Acetogens, like Acetobacterium, are usually the most abundant species in MES reactors that have been treated to inhibit methanogenesis and inoculated with environmental samples. These reactors primarily convert CO_2 and electricity into acetate (Hu et al., 2019).

Gram-positive (*Clostridium* sp. and *Morella thermoacetica*) and gram-negative (*Sporomusa ovata*) acetogenic bacteria have been found in several compounds. *Clostridium ljungdahlii* (Quraishi et al., 2021) is one strain that could be genetically engineered for MES and employed in *Sporomusa ovata*. Research on e-transfer in Cyanobacteria, both intracellularly and extracellularly, has revealed significant differences between them and the other microorganisms used in MES. Therefore, these next-generation biomass producers are a great advantage to the research community since they serve as MES host cells (Quraishi et al., 2021). In recent aspects of MES reactor optimization, including activating the microbial catalyst, enhancing the cultivation medium's composition, perfecting the reactor design, and optimizing the placement of the cathode in space to improve its interaction with the microbial cathodes.

The electron donor source in the initial cathode studies on MES systems was a solid-state graphite block that was negatively charged. Because of its poor porosity and small surface area, which hindered microbial adhesion, it was challenging to enhance production. Because of its high volumetric acetate generation rate from CO_2 and enormous volumetric surface area and porosity, granular graphite was therefore suggested as a cathode. The excellent electrical conductivity, flexibility, porosity, large reactive surface area, biocompatibility, and chemical stability of carbon felt and carbon fabric have led to their widespread application in recent years. Carbon felt, carbon fabric, and carbon nanomaterials such as carbon nanotubes (CNTs) and graphene have also been examined in combination by MES systems.

14.6 APPLICATIONS OF MICROBIAL CARBON CAPTURE CELLS

14.6.1 Bio-fuel

Microbial oil may be produced by bacteria, such as *Bacillus* SS105, *Pseudomonas sp*, and *Serratia* ISTD04, through metabolic conversion of fixed CO_2, followed by fatty acid ethyl ester (FAEE) acquisition through lipid exchange reaction (Kumar et al., 2017; Liu et al., 2022). The bacterial strain *Bacillus* sp. was isolated by Maheshwari et al. (2018) from soil that had been concentrated with free air CO_2, and it was tested for its capacity to sequester carbon dioxide and produce FAs (Ainsworth and Long, 2021). Of the organic substances, 59% were saturated, and 41% were unsaturated.

14.6.2 Polyhydroxyalkanoates (PHAs) Production

Among all prokaryotic microorganisms, the biosynthesis of PHA is well researched in aerobic and anaerobic heterotrophic, autotrophic, chemolithotrophic, marine, *Azotobacter* sp., methylotrophic, and floc-forming bacteria like *Zoogloearamigera*. Different PHA biosynthesis routes have been developed by microbes according to their needs for food, energy, and living conditions (Kumar et al., 2018).

14.6.3 Succinic Acid

Some bacterial species can use a fermentation process that has a high concentration of CO_2 since they ingest these CO_2 flue gases as a substrate. This carbon dioxide can be absorbed by the bacteria into their cellular structure, resulting in the production of succinic acid, a valuable platform chemical. To enhance activity, a new redox-neutral pathway was adopted using CO_2, wherein *S. cerevisiae* genetic engineering was conducted (Xiberras et al., 2020). The genome was modified to include the enzymes responsible for converting oxaloacetate to succinic acid. In a bioreactor with a constant carbon dioxide supply, the modified strain was likewise subjected to suitable closed conditions. Furthermore, the bacterial strain was given glycerol as an additional carbon source, and it demonstrated synergistic synthesis of succinic acid across endogenous production routes. With a consistent supply of glycerol and carbon dioxide in a regulated environment, more than 96% of carbon bioconversion was accomplished overall.

14.6.4 Extracellular Polymeric Substances (EPS)

The extracellular polymeric substance (EPS) matrix is primarily composed of carbohydrates and other non-carbohydrate biomolecules, which are secreted by microorganisms either during cell lysis or in response to available carbon sources in the growth medium. Numerous organisms, including bacteria, archaea, and eukaryotes, can create EPS, and its physicochemical and rheological properties are influenced by temperature, pH, nutrient availability, and the particular microbial species. Research has focused on EPS production in many Gram-negative bacteria, such as *Rhizobium* and *Agrobacterium*, as well as Gram-positive species, including *Leuconostoc mesenteroides*, *Lactobacillus* spp., and *Streptococcus bovis* (Guhanraj and Dhanasekaran, 2024; Kumar et al., 2018).

14.6.5 Challenges and Future Scope

One key challenge in reducing atmospheric CO_2 is enhancing natural carbon fixation processes to improve CO_2 conversion efficiency. Approaches in metabolic engineering aimed at modifying central metabolic pathways show promise for long-term CO_2 fixation. Strategies include dividing RuBisCO catalysis across multiple enzymes, substituting the CBB cycle with alternative pathways, or replacing RuBisCO with other carboxylation reactions (Bar-Even et al., 2010; Antonovsky et al., 2017). Advances in metabolic engineering, such as genome-scale modeling and sequencing of yeast and bacterial systems, are opening new avenues for CO_2 assimilation by heterotrophic microbes. Increasing the efficiency of creating food, fuels, and industrial commodities directly from CO_2 is currently a challenge for biotechnology. However, practical limitations—including the lack of infrastructure for hydrogen generation and storage, safety concerns, and high capital costs—currently constrain the widespread implementation of CO_2-based fuel production. New manufacturing methods still need to be developed, such as utilizing a number of biological processes present in contemporary biorefinery projects. According to Onyeaka and Ekwebelem (2023), this could be a promising approach to the development of new bacterial systems for CO_2 collection and conversion.

14.7 CONCLUSION

The most efficient method now available for lowering atmospheric CO_2 levels is through carbon capture and sequestration, which involves the vaporization of waste CO_2. It is preferable to use microbes to extract CO_2 for mitigating climate change and immediately turning it into bio-based goods with greater value. Advancing the production of valuable commodities from CO_2 will require continued development and application of cutting-edge biotechnological strategies, including synthetic biology, metabolic engineering, and genetic engineering.

COMPETING INTERESTS

The authors report that they have no financial or non-financial conflicts of interest that might have affected the research presented in this manuscript.

DECLARATION OF GENERATIVE AI AND AI-ASSISTED TOOLS IN MANUSCRIPT PREPARATION

During the preparation of this manuscript, the author(s) utilized AI tools to enhance readability. All content was subsequently reviewed and edited by the author(s), who take full responsibility for the accuracy and integrity of the publication.

ACKNOWLEDGMENT

We gratefully thank each and every author for their contributions to the manuscript's preparation.

REFERENCES

Ainsworth, E. A., & Long, S. P. (2021). 30 years of free-air carbon dioxide enrichment (FACE): What have we learned about future crop productivity and its potential for adaptation. *Global Change Biology*, *27*(1), 27–49.

Antonovsky, N., Gleizer, S., & Milo, R. (2017) Engineering carbon fixation in *E. coli*: From heterologous RuBisCO expression to the Calvin–Benson–Bassham cycle. *Current Opinion in Biotechnology*, *47*, 83–91.

Ataeian, M., Liu, Y., Canon-Rubio, K. A., Nightingale, M., Strous, M., & Vadlamani, A. (2019). Direct capture and conversion of CO_2 from air by growing a cyanobacterial consortium at pH up to 11.2. *Biotechnology and Bioengineering*, *116*(7), 1604–1611.

Atomi, H. (2002). Microbial enzymes involved in carbon dioxide fixation. *Journal of Bioscience and Bioengineering*, *94*(6), 497–505.

Bar-Even, A., Noor, E., Lewis, N. E., & Milo, R. (2010). Design and analysis of synthetic carbon fixation pathways. *Proceedings of the National Academy of Sciences*, *107*(19), 8889–8894.

Beigbeder, J. B., Sanglier, M., de Medeiros Dantas, J. M., & Lavoie, J. M. (2021). CO2 capture and inorganic carbon assimilation of gaseous fermentation effluents using Parachlorella kessleri microalgae. *Journal of CO2 Utilization*, *50*, 101581.

Bharti, R. K., Srivastava, S., & Thakur, I. S. (2014). Proteomic analysis of carbon concentrating chemolithotrophic bacteria Serratia sp. for sequestration of carbon dioxide. *PloS One*, *9*(3), e91300.

Bhatia, S. K., Bhatia, R. K., Jeon, J. M., Kumar, G., & Yang, Y. H. (2019). Carbon dioxide capture and bioenergy production using biological system–A review. *Renewable and Sustainable Energy Reviews*, *110*, 143–158.

Chai, K. F., Ng, K. R., Samarasiri, M., & Chen, W. N. (2022). Precision fermentation to advance fungal food fermentations. *Current Opinion in Food Science*, *47*, 100881.

Chung, I. K., Beardall, J., Mehta, S., Sahoo, D., & Stojkovic, S. (2011). Using marine macroalgae for carbon sequestration: A critical appraisal. *Journal of Applied Phycology*, *23*(5), 877–886.

Das, S., Diels, L., Pant, D., Patil, S. A., & Ghangrekar, M. M. (2020). Microbial electrosynthesis: A way towards the production of electro-commodities through carbon sequestration with microbes as biocatalysts. *Journal of the Electrochemical Society*, *167*(15), 155510.

Gougoulias, C., Clark, J. M., & Shaw, L. J. (2014). The role of soil microbes in the global carbon cycle: Tracking the below-ground microbial processing of plant-derived carbon for manipulating carbon dynamics in agricultural systems. *Journal of the Science of Food and Agriculture*, *94*(12), 2362–2371.

Guhanraj, R., & Dhanasekaran, D. (2024). Probiotic functional gene explorations in the genome of *Limosilactobacillus fermentum* GD5MG. *Microbial Pathogenesis*, *192*, 106686.

Hayden, H. L., Mele, P. M., Bougoure, D. S., Allan, C. Y., Norng, S., Piceno, Y. M., … & Hovenden, M. J. (2012). Changes in the microbial community structure of bacteria, archaea and fungi in response to elevated CO_2 and warming in an Australian native grassland soil. *Environmental Microbiology*, *14*(12), 3081–3096.

Hu, G., Li, Y., Ye, C., Liu, L., & Chen, X. (2019). Engineering microorganisms for enhanced CO2 sequestration. *Trends in Biotechnology*, *37*(5), 532–547.

Hu, Y., & Holden, J. F. (2006). Citric acid cycle in the hyperthermophilic archaeon Pyrobaculum islandicum grown autotrophically, heterotrophically, and mixotrophically with acetate. *Journal of Bacteriology*, *188*(12), 4350–4355.

Hügler, M., Huber, H., Molyneaux, S. J., Vetriani, C., & Sievert, S. M. (2007). Autotrophic CO2 fixation via the reductive tricarboxylic acid cycle in different lineages within the phylum Aquificae: Evidence for two ways of citrate cleavage. *Environmental Microbiology*, *9*(1), 81–92.

Hügler, M., Huber, H., Stetter, K. O., & Fuchs, G. (2003). Autotrophic CO2 fixation pathways in archaea (Crenarchaeota). *Archives of Microbiology*, *179*(3), 160–173.

Hügler, M., Wirsen, C. O., Fuchs, G., Taylor, C. D., & Sievert, S. M. (2005). Evidence for autotrophic CO2 fixation via the reductive tricarboxylic acid cycle by members of the ε subdivision of proteobacteria. *Journal of Bacteriology*, *187*(9), 3020–3027.

Jiang, Y., Zhang, W., Wang, J., Chen, Y., Shen, S., & Liu, T. (2013). Utilization of simulated flue gas for cultivation of Scenedesmus dimorphus. *Bioresource Technology*, *128*, 359–364.

Jonas, L., Lee, Y. Y., Mroz, R., Hill, R. T., & Li, Y. (2025). *Nannochloropsis oceanica* IMET1 and its bacterial symbionts for carbon capture, utilization, and storage: Biomass and calcium carbonate production under high pH and high alkalinity. *Applied and Environmental Microbiology*, *91*(5), e00133–25.

Kaladharan, P., Veena, S., & Vivekanandan, E. (2009). Carbon sequestration by a few marine algae: Observation and projection. *Journal of the Marine Biological Association of India*, *51*(1), 107–110.

Khadem, A. F., Pol, A., Wieczorek, A., Mohammadi, S. S., Francoijs, K. J., Stunnenberg, H. G., … & Op den Camp, H. J. (2011). Autotrophic methanotrophy in Verrucomicrobia: Methylacidiphilum fumariolicum SolV uses the Calvin-Benson-Bassham cycle for carbon dioxide fixation. *Journal of Bacteriology*, *193*(17), 4438–4446.

Khan, M. K., Teng, J. Z., Khan, M. I., & Khan, M. O. (2019). Impact of globalization, economic factors and energy consumption on CO2 emissions in Pakistan. *Science of the Total Environment*, *688*, 424–436.

Kumar, M., Morya, R., Gnansounou, E., Larroche, C., Thakur, I. S. (2017) Characterization of carbon dioxide concentrating chemolithotrophic bacterium *Serratia sp.* ISTD04 for production of biodiesel. *Bioresource Technology*, *243*, 893–897.

Kumar, M., Sundaram, S., Gnansounou, E., Larroche, C., & Thakur, I. S. (2018). Carbon dioxide capture, storage and production of biofuel and biomaterials by bacteria: A review. *Bioresource Technology*, *247*, 1059–1068.

Liu, E. J., Tseng, I. T., Chen, Y. L., Pang, J. J., Shen, Z. X., & Li, S. Y. (2020). The physiological responses of Escherichia coli triggered by phosphoribulokinase (PrkA) and Ribulose-1, 5-Bisphosphate carboxylase/oxygenase (Rubisco). *Microorganisms*, *8*(8), 1187.

Liu, X., Luo, H., Yu, D., Tan, J., Yuan, J., & Li, H. (2022). Synthetic biology promotes the capture of CO2 to produce fatty acid derivatives in microbial cell factories. *Bioresources and Bioprocessing*, *9*(1), 124.

López, C. G., Fernández, F. A., Sevilla, J. F., Fernández, J. S., García, M. C., & Grima, E. M. (2009). Utilization of the cyanobacteria Anabaena sp. ATCC 33047 in CO2 removal processes. *Bioresource Technology*, *100*(23), 5904–5910.

Maheshwari N, Kumar M, Thakur IS, Srivastava S (2018) Carbon dioxide biofixation by free air CO2 enriched (FACE) bacterium for biodiesel production. *Journal of CO2 Utilization*, *27*, 423–432.

Mitchell, J. H., Freedman, A. H., Delaney, J. A., & Girguis, P. R. (2024). Co-expression analysis reveals distinct alliances around two carbon fixation pathways in hydrothermal vent symbionts. *Nature Microbiology*, *9*(6), 1526–1539.

Nogalska, A., Zukowska, A., & Garcia-Valls, R. (2018). Atmospheric CO2 capture for the artificial photosynthetic system. *Science of the Total Environment*, *621*, 186–192.

Nowicka, B., & Kruk, J. (2016). Powered by light: Phototrophy and photosynthesis in prokaryotes and its evolution. *Microbiological Research*, *186*, 99–118.

Oliver, A., Camarena-Bernard, C., Lagirarde, J., & Pozzobon, V. (2023). Assessment of photosynthetic carbon capture versus carbon footprint of an industrial microalgal process. *Applied Sciences*, *13*(8), 5193.

Onyeaka, H., & Ekwebelem, O. C. (2023). A review of recent advances in engineering bacteria for enhanced CO2 capture and utilization. *International Journal of Environmental Science and Technology*, *20*(4), 4635–4648.

Onyeaka, H., Miri, T., Obileke, K., Hart, A., Anumudu, C., & Al-Sharify, Z. T. (2021). Minimizing carbon footprint via microalgae as a biological capture. *Carbon Capture Science & Technology*, *1*, 100007.

Padhy, S. R., Bhattacharyya, P., Dash, P. K., Roy, K. S., Neogi, S., Baig, M. J., … & Mohapatra, T. (2020). Enhanced labile carbon flow in soil-microbes-plant-atmospheric continuum in rice under elevated CO_2 and temperature leads to positive climate change feed-back. *Applied Soil Ecology*, *155*, 103657.

Pekkoh, J., Ruangrit, K., Aurepatipan, N., Duangjana, K., Sensupa, S., Pumas, C., … & Srinuanpan, S. (2024). CO_2 to green fuel converter: Photoautotrophic-cultivation of microalgae and its lipids conversion to biodiesel. *Renewable Energy*, *222*, 119919.

Quraishi, M., Wani, K., Pandit, S., Gupta, P. K., Rai, A. K., Lahiri, D., … & Prasad, R. (2021). Valorisation of CO2 into value-added products via microbial electrosynthesis (MES) and electro-fermentation technology. *Fermentation*, *7*(4), 291.

Saini, R., Kapoor, R., Kumar, R., Siddiqi, T. O., & Kumar, A. (2011). CO2 utilizing microbes—a comprehensive review. *Biotechnology Advances*, *29*(6), 949–960.

Sha, Z., Bai, Y., Lan, H., Liu, X., Li, R., & Xie, Y. (2020). Can more carbon be captured by grasslands? A case study of Inner Mongolia, China. *Science of the Total Environment*, *723*, 138085.

Shu, W. S., & Huang, L. N. (2022). Microbial diversity in extreme environments. *Nature Reviews Microbiology*, *20*(4), 219–235.

Singh, S. P., & Singh, P. (2014). Effect of CO2 concentration on algal growth: A review. *Renewable and Sustainable Energy Reviews*, *38*, 172–179.

Steffens, L., Pettinato, E., Steiner, T. M., Mall, A., König, S., Eisenreich, W., & Berg, I. A. (2021). High CO2 levels drive the TCA cycle backwards towards autotrophy. *Nature*, *592*(7856), 784–788.

Tabita, F. R. (2009). The hydroxypropionate pathway of CO2 fixation: Fait accompli. *Proceedings of the National Academy of Sciences*, *106*(50), 21015–21016.

Treece, T. R. (2023). *Engineering Microorganisms for Carbon Capture and Chemical Production* (Doctoral dissertation, University of California, Davis).

Ummalyma, S. B., Maharajan, T., Sreelekshmy, B. R., & Eswaran, K. (2025). Carbon sequestration by microalgae and engineering of microalgal cells by CRISPR technology for efficient CO2 capture for carbon neutral bioproducts. *ACS ES&T Water*, 2(12), 1–15

Ungerer, J., Tao, L., Davis, M., Ghirardi, M., Maness, P. C., & Yu, J. (2012). Sustained photosynthetic conversion of CO_2 to ethylene in recombinant *cyanobacterium Synechocystis* 6803. *Energy & Environmental Science*, *5*(10), 8998–9006.

Valenzuela, J., Mazurie, A., Carlson, R. P., Gerlach, R., Cooksey, K. E., Peyton, B. M., & Fields, M. W. (2012). Potential role of multiple carbon fixation pathways during lipid accumulation in Phaeodactylum tricornutum. *Biotechnology for Biofuels*, *5*(1), 40.

Vega, L. P., Garcia, D. F., Sierra-Parada, R., & Pirazan, I. (2024). CO_2 capture capacity measurement using multitemporal analysis and biophysical variables in a tropical humid forest in the Colombian Andes. *Sustainability*, *16*(11), 4809.

Wang, X., Feng, Y., Liu, J., Lee, H., Li, C., Li, N., & Ren, N. (2010). Sequestration of CO2 discharged from anode by algal cathode in microbial carbon capture cells (MCCs). *Biosensors and Bioelectronics*, *25*(12), 2639–2643.

Wu, C., Lo, J., Urban, C., Gao, X., Humphreys, J., Shinde, S., … & Xiong, W. (2020). A synthetic acetyl-CoA bi-cycle synergizes the Wood–Ljungdahl pathway for efficient carbon conversion in syngas fermentation. *Journal of CO_2 Utilization*, *37*, 200–210.

Xiberras J, Klein M, De Hulster E, Mans R, Nevoigt E (2020) Engineering *Saccharomyces* cerevisiae for succinic acid production from glycerol and carbon dioxide. *Frontiers in Bioengineering and Biotechnology*, *8*, 566.

Xie, T., & Wu, Y. (2014). The role of microalgae and their carbonic anhydrase on the biological dissolution of limestone. *Environmental Earth Sciences*, *71*(12), 5231–5239.

Xue, S., Yang, X., Liu, G., Gai, L., Zhang, C., Ritsema, C. J., & Geissen, V. (2017). Effects of elevated CO_2 and drought on the microbial biomass and enzymatic activities in the rhizospheres of two grass species in Chinese loess soil. *Geoderma*, *286*, 25–34.

Yaashikaa, P. R., Saravanan, A., Kumar, P. S., Thamarai, P., & Rangasamy, G. (2024). Role of microbial carbon capture cells in carbon sequestration and energy generation during wastewater treatment: A sustainable solution for cleaner environment. *International Journal of Hydrogen Energy*, *52*, 799–820.

Yu, T., Wang, S., Yang, H., Sun, Y., Chen, Z., Xu, G., & Zhang, C. (2024). Carbon capture and resource utilization by algal–bacterial consortium in wastewater treatment: A mini-review. *Water*, *16*(15), 2197.

Zhang, Q., Cheng, C. L., Nagarajan, D., Chang, J. S., Hu, J., & Lee, D. J. (2017). Carbon capture and utilization of fermentation CO2: Integrated ethanol fermentation and succinic acid production as an efficient platform. *Applied Energy*, *206*, 364–371.

15 Life Cycle Assessment of AI-Powered Fermentation Systems

Environmental Trade-offs between Digitalization and Sustainability

A. Kosiha, G. Kalaiarasi, P. Gowthami, D. Yuvaraj, R. Suresh Babu, and Panneerselvam Theivendren

15.1 INTRODUCTION TO AI AND BIOTECHNOLOGY FOR SUSTAINABLE INDUSTRY

The need to have sustainable industrial processes is a global issue that has triggered unprecedented innovation at the confluence of biotechnology and artificial intelligence. One of the oldest biotechnological processes in the history of mankind, fermentation is currently at the center of this change. The conventional fermentation systems, though established, have serious challenges such as batch-to-batch variability, poor resource utilization, and low adaptability on-the-fly. These restrictions cause high environmental footprints marked by high consumption of energy, high consumption of water, and high production of wastes [1].

AI has turned out as a radical technology that can deal with these problems by having advanced algorithms that can optimize processes in real time and control them and modify parameters. Complex and multidimensional fermentation data can be analyzed using machine learning models to determine the best operating conditions, anticipate equipment failures, as well as move process parameters in real time with accuracy that is unattainable using traditional control methods. It is hoped that the introduction of AI technologies deep learning models, neural networks, and reinforcement learning will revolutionize the biomanufacturing industry because it will increase productivity, product consistency, and decrease environmental effects [2].

However, there are environmental concerns with the use of systems made by AI. The AI models, training, and inference are computationally expensive demanding huge energy as well as greenhouse emissions. Data centers, cloud infrastructure,

DOI: 10.1201/9781003684374-15

and hardware accelerators consume large amount of electricity and water and are significant contributors to the growing problem of e-waste. The life cycle thinking is in need of intensive scrutiny of this dilemma—a circumstance in which the technologies created to help increase sustainability will tend to put a strain on the environment of their own.

15.1.1 Objectives and Scope

It details a comprehensive LCA of computer-based systems of fermentation, including an environmental impact of the entire product life cycle, ranging from system design, production, through end-of-life disposal [3]. Its primary focuses are to:

- Characterize and contrast the environmental effects of the AI-controlled and conventional systems of fermentation on various impact categories
- Identify environmental hotspots and major trade-offs between advances of operational efficiency and infrastructure burdens of computation
- Evaluate the net environmental benefit of AI integration considering both direct operational improvements and indirect infrastructure impacts
- Provide data-informed recommendations for the responsible use of AI technologies to biomanufacturing
- Create methodological templates for the LCA practitioners of digital-biological hybrid systems

System boundaries include industrial-scale fermentation processes for the generation of bio-based chemicals, alternative protein production, and pharmaceuticals. Boundaries include all the significant material and energy flows throughout the product life cycle of raw material production, through manufacturing, use, and end-of-life disposal, as outlined by ISO 14044 and ISO 14040.

15.2 LIFE CYCLE ASSESSMENT METHODOLOGY

15.2.1 LCA Framework and Standards Life

LCA provides us with a procedural method of measuring the impact of the environment of every life stage of a product or process. This study adheres to the international standards ISO 14040:2006 (principles and framework) and ISO 14044:2006 (requirements and guidelines), in which four stages are required, namely goal and scope definition, the life cycle inventory analysis, the life cycle impact assessment, and the interpretation [4].

The LCA model is developed so that it is thoroughly evaluated by setting clear boundaries, determining all the inputs and outputs, and using scientifically tested models to characterize inventory data in the form of an environmental impact. In the case of fermentation systems that utilize AI, this model will need to be able to support both traditional bioprocessing elements and new elements of computational infrastructure; it will be necessary to take the allocation processes and system boundaries into account.

15.2.2 Definition of Goal and Scope

The objective of this LCA is to systemically evaluate and compare the environmental integrity of AI-based fermentation systems and classical fermentation systems, and reveal environmental trade-offs between improvements of operational efficiency and demands of computational infrastructure. This chapter aims to be useful to researchers, industry practitioners, technology developers, and policymakers responsible for sustainable biomanufacturing.

Functional Unit: Functional unit has been referred to as the system production of 1,000 kg of fermented product (bio-based chemical or alternative protein) of pre-established purity and quality (≥95% purity) [5]. Functional unit allows comparing different system layouts on the same basis while considering the yield, quality, and operable efficiency.

System Boundaries: System boundary adopts the cradle-to-grave approach and covers four dissimilar life cycle stages:

1. System Design Phase: Lab-based research, strain construction, CRISPR engineering, prototype development of the bioreactor, and process modeling
2. AI Model Development Stage: Data gathering and preparation, model architecture design, computer-based training, validation, and tuning
3. Phase of Operational Deployment: Fermentation processes, real-time inference and control through AI, downstream processing, waste treatment, and auxiliary services
4. End-of-Life Stage: Hardware decommissioning, e-waste disposal for recycling, facility decontamination

The time horizon accounts for a ten-year system operating life for bioreactor plants and three to five years for computer hardware for AI computation, corresponding to practical replacement intervals. Location scale takes industrial plants situated in locations of typical electricity grid compositions, and sensitivity analysis considers deviations of different energy sources.

15.2.3 Life Cycle Inventory Analysis

Life cycle inventory (LCI) analysis consists of the systematic gathering and quantification of all the respective inputs (materials, energy, water) and outputs (products, emissions, wastes) of the system boundaries established. For the systems of fermentation powered by AI, the LCI data are gathered through the following sources [6]:

- Data from piloted and industrially scaled-down plants' measurements, i.e., records of energy utilization, records of substrate consumption, product yield, and waste production data
- Literature, industry studies, and databases (ecoinvent v3.8, GaBi databases) for the background processes such as power generation, material production, and transportation

- Energy profiles of the training and inference of the AI models, operating parameters of the data centers, hardware production impact, and cooling demands
- Process simulation and digital twin outputs for situations where it is impractical or impossible to have direct measurement

Data quality checking conforms to predefined rules assessing temporal, geographic, and technological representativeness, completeness, precision, and consistency. Uncertainty analysis applies a Monte Carlo simulation with 10,000 iterations to propagate parameter uncertainties through effect estimates [7].

15.3 FERMENTATION SYSTEM TECHNOLOGIES BASED ON AI

15.3.1 Overview of Applications of AI

The integration of AI has given a paradigm shift to bioprocess engineering, so as to have functionalities beyond the constraints of the classical control systems. Application of AI to fermentation entails numerous functional applications: real-time process optimization, predictive maintenance, quality prediction, strain design, and metabolic pathway engineering. These applications make use of the machine learning algorithms' property of revealing complex, nonlinear patterns of the high-dimensional data of the fermentation and extracting patterns and optimization strategies, where classical methods would be impractical [8].

The primary advantage of AI-based systems lies in their ability to perform adaptive learning and dynamic optimization. Unlike conventional rule-based control systems that operate using predefined algorithms, AI-driven systems continuously learn from accumulated operational data and refine their decision-making processes over time. This adaptive capability enables continuous improvement in process performance, allowing the system to respond effectively to changing operating conditions, optimize resource utilization, and maintain stable and efficient production. As a result, AI-based systems enhance process robustness, productivity, and overall operational efficiency. It is this adaptability where the most benefit occurs within the fermentation processes, where they have inherent biological variability, substrate quality variability, and changing operational conditions.

15.3.2 Qualities and Forms of AI Algorithms

The reinforcement learning (RL) was especially convenient as far as optimization of real-time parameters was concerned, since algorithms based on reinforcement learning find the most appropriate control policy by trial-and-error interacting with the fermentation system. Embedded appliances with edge computing are used to implement RL algorithms that are used to adjust the parameters of the bioreactor to achieve the desired conditions in real time (temperature: −0.5°C, pH: −0.2 units, and agitation rate): −0.5°C, −0.2 units, and 5 milliseconds, respectively. This is a fine level of precision control that reduces failure rates on batches by 60% and inconsistency of yield by 30 [9].

Process foresight and metabolic pathway understanding Artificial Neural Network (ANNs) and specifically Deep Learning Models can capture complex nonlinear interactions and, therefore, have the capacity to do so. These models have been successfully used to make predictions of fermentation products, establish the optimal feeding program and identify the optimal cultivation conditions. They, however, have giant computations of 150–600 kWh per model to which environmental trade-offs are critical [10].

Strain selection and regression problems can be solved using the computationally tractable alternatives, which include Random Forests (RFs) and Support Vector Machines (SVMs). They are also desirable in environments where the computer resources are limited because of their relatively low training energy requirements of 5–50 kWh/mode and capability to support the data of any dimensionality [11].

The new generation of applications of enzyme design and CRISPR optimization are Generative Adversarial Networks (GANs) and Convolutional Neural Networks (CNNs). GANs are used to create new versions of enzymes and cellular metabolic pathways, and CNNs are used to make predictions of off-target CRISPR edits, cutting the design cycle months in half 70% reduction of design cycle [10]. These subsequent-generation applications can be described as computation-intensive (100–800 kWh of training energy) but can provide transformative capability of synthetic biology and metabolic engineering [12].

15.3.3 System Components and Architecture

Fermentation systems are AI-based systems that incorporate numerous senses, computations, and actuation units into the closed-loop framework of system control. System structuring incorporates the hierarchical structure which incorporates:

Sensor Layer: Sensors that are clever are constantly checking the vital parameters such as temperature, pH, dissolved oxygen (DO), pressure, optical density, nutrient, and gas composition. More advanced sensor systems are described to include microprocessor signal conditioning, self-diagnosis, and wireless so as to minimize wire-inning and maximize reliability [13].

Data Preprocessing and Input: Signals are sent to programmable logic controllers and field computers which digitize them, condition and filter the signals and preliminary data transformation. This low-level software package will bind the physical sensors and the higher-level AI algorithms and maintain the integrity and consistency of the data.

AI Inference Engine: Machine learning models are trained on edge computers, or cloud computers, to generate real-time predictions and control decisions. The deployment architecture provides a compromise between the computational through-put, latency requirements, and the energy efficiency. Edge deployment provides a trade-off between communication delays and communication energy whereas cloud deployment makes available greater computational resources to large models.

Control Actuation: The control decisions are carried out by intelligent actuators that respond to the control decisions by controlling the setting of the process parameters with pumps, valves, heating, cooling, and agitation drives. The more advanced actuators have internalized microprocessors, distributed control logic, and status feedback, as well as diagnostics.

Updating and Model Training. Special-purpose computing hardware typically cloud data centers or special AI servers do a model retraining periodically based on the historical data. This process of learning is also continuous, and the model is tuned or re-tuned to the new conditions of operation, providing maximum performance over prolonged operating periods.

15.3.4 Computational Needs and Infrastructure

Computational requirements of computer-based fermentation systems involve both development and use stages, both of which have unique environmental signatures. Table 15.1 assigns quantitative values to energy usage, greenhouse emissions, water use, and lifetimes of the different elements of computer infrastructure [14].

Train the Models: Then train and construct the models. This is the most computationally intensive process, costing 500–1,500 kWh/model of electricity, depending on complexity of the algorithms, size of the datasets, and architectural demands. Full-scale deep learning networks to predict metabolic networks or sophisticated process optimization would require days of computation on the GPU, with 250–750 kg CO2-equivalent emissions/training run. Models, though, once trained, can be recycled hundreds of times over plants and campaigns, paying back this front-end cost.

TABLE 15.1
Computational Infrastructure Environmental Impact

S. No.	Infrastructure Component	Energy Consumption	GHG Emissions (kg CO_2-eq)	Water Consumption	Lifespan/Cycle
1	AI Model Training (per model)	500–1500 kWh	250–750	1,000–3,000 L	One-time
2	Data Center Operations (annual)	150–450 MWh	75,000–225,000	150–450 ML	Continuous
3	Edge Computing Devices (per unit)	300–800 kWh	150–400	500–1,500 L	3–5 years
4	Cloud Computing (per TB data)	50–200 kWh	25–100	100–400 L	Per use
5	GPU Accelerators (per unit/year)	2000–5000 kWh	1,000–2,500	3,000–8,000 L	3–5 years
6	Network Infrastructure (per site)	10–30 MWh	5,000–15,000	20–60 ML	10–15 years
7	Cooling Systems (annual)	60–180 MWh	30,000–90,000	80–240 ML	Continuous
8	Backup Power Systems (annual)	20–60 MWh	10,000–30,000	15–45 ML	Continuous

Inference and Control: Real-time inference control requires relatively little energy per operation, approximately 2.9 Wh per cycle of inference, but its continuous 24/7 operation results in a substantial cumulative energy demand. For an industrial plant which has hundreds of fermentation tanks and conducts the control update every five minutes, the inference energy consumption throughout the year ranges from 300 to 800 kWh.

Data Infrastructure: Data storage, transmission, and management infrastructure adds 10–30 MWh annually to each industrial facility. Data centers of cloud-deployed configurations use 150–450 MWh annually, associated cooling loads of 60–180 MWh, and are responsible for 40% of the total energy needs of the data center.

Hardware Life Cycle: Custom-designed AI accelerators (GPUs, TPUs) run at all times, consuming 2,000–5,000 kWh per year per unit with comparably short lifespans of 3–5 years because of speedy technological progress. This faster obsolescence leads to increasing electronic waste issues, compelling systematic end-of-life management practices.

15.4 LIFE CYCLE INVENTORY

15.4.1 System Design Phase

Lab experiment, microbial strain building, bioreactor engineering, and process modeling are part of the system design stage. For conventional fermentation systems, this phase comprises, primarily, strain selection by conventional mutagenesis and screening, bioreactor definition, and experimental design-based process parameter optimization.

AI-powered systems introduce additional intricacy through the means of machine learning-guided prediction and sophisticated modeling of the computational operations by CRISPR-associated strain design. Deep learning networks that are trained with transcriptome data identify the most effective promoter-gene pairing, and convolutional networks predict off-target CRISPR edit effects, reducing the requirements for experiment trials but increasing the computational complexity.

Energy Use: Conventional system design is estimated to use around 8,000–12,000 kWh through 12–18-month development cycles mostly for laboratory instruments, incubators, analytical tools, and pilot-scale fermentation tests. AI-based design adds 20–25% (10,000–15,000 kWh total) for computational modeling, in silico screening, and AI-aided strain optimization but decreases laboratory experimental needs by 30–40%. Material Input: Both processes utilize pilot-scale equipment construction material, laboratory chemicals (buffers, media components, reagents), and consumables (pipette tips, filters, culture vessels). Reagent usage is reduced by computer pre-screening of processes via AI-based techniques, but requires additional electronic hardware for sensor integration and edge computing device. Emissions: System emissions are limited and are largely indirect energy-related emissions by power generation. Wastes from laboratory chemicals must be disposed of and treated, burdening the environment equally.

15.4.2 AI Model Development Phase

The phase of AI model development is a unique aspect of intelligent fermentation systems without a direct analogy in conventional strategies. The phase includes data gathering and curation, model design architecture, hyperparameter tuning, training, validation, and deployment readiness.

Data Preprocessing and Collection: Historical fermentation data needs to be gathered, cleaned, normalized, and formatted for machine learning usage. Preprocessing for extensive datasets covering multiple operational campaigns requires 50–150 kWh based on dataset size and complexity.

Model Training: Computational training of machine learning models forms the major energy contribution of this stage. Energy costs for training vary significantly on algorithm type [15]:

- Basic models (Random Forest, SVM): 5–50 kWh per model
- Neural networks: 20–150 kWh per model
- Deep learning architectures: 150–600 kWh per model
- Generative models (GANs): 200–800 kWh per model

For an average industrial application with an ensemble of three to five models spanned across various control objectives (parameter optimization, quality prediction, and failure detection), total training energy varies from 500 to 1,500 kWh.

Infrastructure Needs: Model training has traditionally used cloud compute platforms or specialized GPU servers, adding related energy for cooling, data transfer, and infrastructure overhead. For cloud-based training, indirect components add an additional 40–60% over direct computational energy.

Greenhouse Gas Emissions: The carbon footprint of model training is highly dependent upon electricity grid mix. Training a full suite of models in areas with fossil fuel-based grids (carbon intensity ~0.5 kg CO_2/kWh) has an emissions impact of around 250–750 kg CO_2-equivalent. Using renewable energy-fueled data centers can lower that by 90–95% [16].

15.4.3 Operational Deployment Phase

The operational deployment stage is the dominant phase in overall life cycle environmental effects for both conventional and AI-based fermentation systems, covering all operations related to routine production during the facility operating life.

15.4.3.1 Fermentation Operations

Substrate Consumption: Raw material feedstock is a key environmental hotspot [17]. Conventional systems generally have substrate conversion efficiencies of 65–75%, whereas AI-optimized systems enhance this to 85–95% through dynamic feed optimization and metabolic state monitoring. For the production of 1,000 kg product, substrate needs are lowered from 1,330–1,540 kg

TABLE 15.2
Comparative Environmental Performance—Traditional versus AI-Powered Fermentation

S. No.	Performance Metric	Traditional Fermentation	AI-Powered Fermentation	Improvement (%)	References
1	Energy consumption (MJ/kg product)	145–180	72–95	47–53%	[18]
2	Water consumption (L/kg product)	35–50	15–22	56–64%	[19]
3	GHG emissions (kg CO_2-eq/kg product)	12.5–16.2	4.2–6.8	58–66%	[20]
4	Substrate yield efficiency (%)	65–75	85–95	15–27%	[21]
5	Batch failure rate (%)	8–15	3–6	60–67%	[22]
6	Process cycle time (hours)	96–120	72–84	25–30%	[23]
7	Waste generation (kg/kg product)	0.35–0.45	0.12–0.18	60–74%	[24]
8	Operational labor (person-hours/ batch)	24–32	8–12	62–67%	[22]
9	Product purity (%)	92–95	96–99	3–4%	[25]
10	Batch-to-batch consistency (% variation)	±8–12	±2–4	50–75%	[26]

(conventional) to 1,050–1,175 kg (AI-driven), lowering both direct material impacts as well as upstream agricultural/extraction burdens.

Energy Use: Fermentation processes use energy for heating, mixing, aeration, and sterile conditions. Table 15.2 lists holistic comparative performance data indicating AI-based systems lower energy use from 145–180 MJ/kg product (conventional) to 72–95 MJ/kg product (AI-based) with 47–53% improvement through:

- Optimized temperature profiles, reducing heating/cooling cycles
- Adaptive aeration rates balancing oxygen requirement with metabolic condition
- Lower batch failures, avoiding wasted production cycles
- Optimized growth time reduction

Water Use: Process water, cooling water, and cleaning procedures form significant water requirements. Conventional systems utilize 35–50 L/kg product, whereas AI systems attain 15–22 L/kg product due to optimized cooling processes, decreased frequency of cleaning from enhanced process control, and reduced water utilization in a failed batch.

Greenhouse Gas Emissions: Direct CO_2 emissions due to aerobic fermentation are generally biogenic (carbon-neutral), but energy-related indirect emissions are predominant in the carbon footprint. The classical systems emit 12.5–16.2 kg CO_2-eq/kg product, brought down to 4.2–6.8 kg CO_2-eq/kg product in AI-based systems 58–66% reduction [27].

15.4.3.2 AI Inference and Control

Real-time AI inference in process control brings constant computational energy needs during the operational period. Edge computers running control algorithms spend 300–800 kWh per year per fermentation tank. In industrial plants with five to ten parallel fermentation trains, total AI inference energy amounts to 1,500–8,000 kWh per year.

Data transfer between sensors, edge devices, and cloud infrastructure for model refresh adds a further 5–10% to computational energy demand. Cloud-enabled systems providing remote monitoring and advanced analytics consume higher levels of energy, but allow for centralized optimization across many facilities [28].

15.4.3.3 Downstream Processing

Product recovery, purification, and formulation processes are energy-intensive and produce waste streams. AI optimization cascades from fermentation to downstream operations through predictive scheduling, routing decisions based on quality, and coupled process control. These efficiencies usually decrease downstream energy use by 15–25% and waste production by 20–35% compared to conventional sequential batch operation.

15.4.4 End-of-Life Phase

End-of-life stage includes decommissioning, material recovery, waste management, and site remediation processes. This stage shows significant differences between legacy and AI-based systems owing to electronic hardware elements.

Hardware Decommissioning: AI computation hardware (edge devices, GPUs, servers) needs special treatment because of toxic materials such as lead, mercury, rare earth elements, and flame retardants. Collection, transport, and dismantling operations use 50–150 kWh per ton of electronic devices [29].

Material Recovery: High-tech recycling processes, such as hydrometallurgical and pyrometallurgical, can reclaim precious materials (rare earths, precious metals, copper) from electronic waste, covering virgin material extraction environmental impacts. Nevertheless, present worldwide e-waste recycling is low (~17% officially recycled), and the majority of material flows into informal recycling channels or landfills.

Disposal of Bioreactor: Stainless steel bioreactor tanks and related piping are generally recyclable with better than 90% recovery for steel. AI-based systems add other sensors, actuators, and control equipment to manage electronic waste treatment procedures.

Decontamination of Site: Decontamination and cleaning processes to eliminate microbial contamination and leftover chemicals use water, energy, and cleaning chemicals. Environmental footprints are similar for conventional and AI-based facilities since fermentation processes vary little in risk of contamination.

15.5 ENVIRONMENTAL IMPACT ASSESSMENT RESULTS

15.5.1 Global Warming Potential

Global warming potential (GWP) in terms of kg CO_2-eq emissions is the most well-known environmental impact category and acts as a key measure for climate change reduction programs [30]. Figure 15.1 shows the complete GHG emissions profile by life cycle phase for conventional and AI-based fermentation systems for a representative three-year system operating time.

Traditional fermentation systems produce total cumulative GHG emissions of 4,730 kg CO_2-eq in three years, broken down into:

- System Design: 150 kg CO_2-eq (3.2%)
- Operations: 4,500 kg CO_2-eq (95.1%)
- End-of-Life: 80 kg CO_2-eq (1.7%)

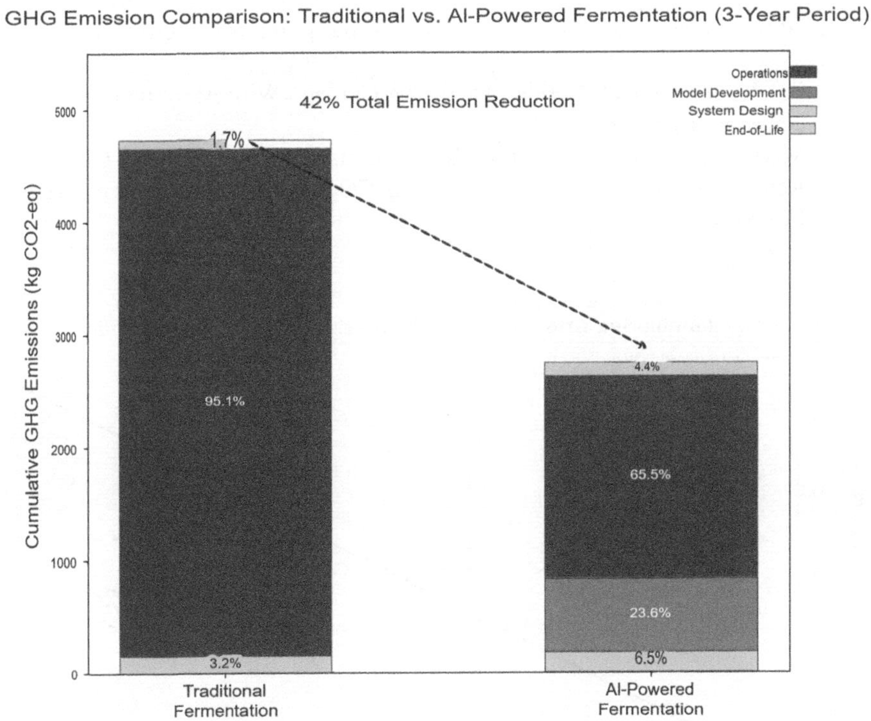

FIGURE 15.1 Life cycle greenhouse gas emissions distribution across different stages for traditional and AI-powered fermentation systems.

The operational stage by far accounts for overall emissions, echoing the energy requirements of constant fermentation operations like heating, aeration, mixing, and downstream processing.

AI-powered fermentation systems produce 2,750 kg CO_2-eq total 42% savings with significantly different breakdown:

- System Design: 180 kg CO_2-eq (6.5%)
- Model Development: CO_2-eq (23.6%)
- Operations: 1,800 kg CO_2-eq (65.5%)
- End-of-Life: 120 kg CO_2-eq (4.4%)

In spite of greater initial emissions from computational hardware (830 kg CO_2-eq versus 150 kg CO_2-eq), net GHG savings of 58% over the course of three years are realized from the operational efficiency gains enabled by AI optimization. The point at which cumulative emissions of AI-based systems fall below that of conventional systems is about 1.2 years of operation (Figure 15.2).

Sensitivity to Energy Sources: The carbon footprint of electricity generation plays a crucial role in these outcomes. Analysis by various grid mixes shows:

- Fossil fuel-dominated grids (0.5–0.7 kg CO_2-eq/kWh): AI systems reduce emissions by 52–58%
- Mixed grids (0.3–0.4 kg CO_2-eq /kWh): Emissions reduced by 58–64%
- Renewable-dominated grids (0.05–0.1 kg CO_2-eq/kWh): Emissions reduced by 65–72%

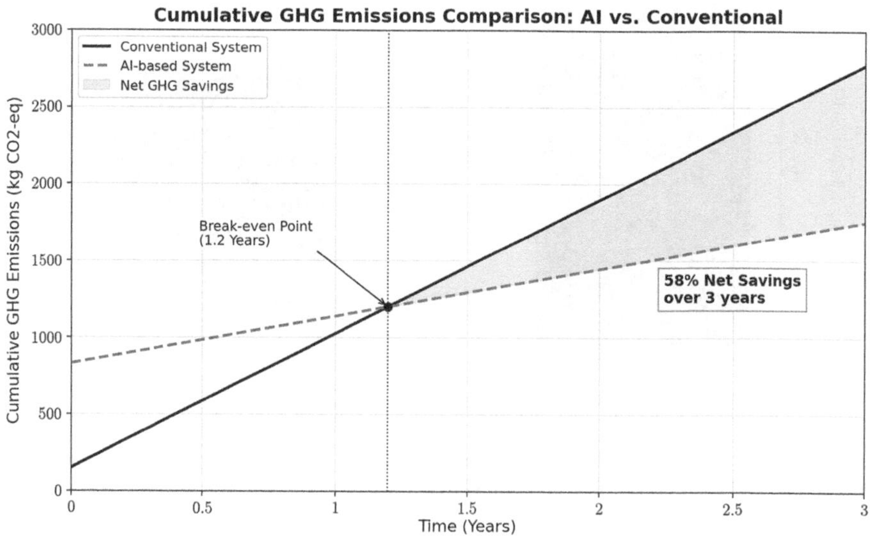

FIGURE 15.2 Cumulative greenhouse gas emissions over ten-year operational lifetime comparing traditional and AI-powered fermentation systems.

Supplying computational infrastructure using renewable power maximizes climate dividends while keeping the carbon footprint of digitalization low [31].

15.5.2 Acidification Potential

Acidification potential (AP), in units of kg SO_2 equivalents, expresses the contribution of acidifying emissions such as sulfur dioxide, nitrogen oxides, and ammonia to acidification of soils and waters. They mainly arise from the combustion of fossil fuels in generating electricity and transportation.

Conventional fermentation processes produce 0.085–0.110 kg SO_2 eq per kg product, whereas AI-based processes yield 0.038–0.052 kg SO_2-eq per kg product, a 54–60% decrease. The reduction comes primarily from diminished operational energy use, as acidifying emissions have a high association with fossil fuel-derived electricity generation. Substrate transportation and growth make secondary acidification contributions through agricultural emissions and vehicle exhaust [32].

15.5.3 Eutrophication Potential

Potential for eutrophication (EP), in terms of kg PO_4^{3-}-equivalents, represents the excess supply of nutrients (phosphorus and nitrogen) to aquatic and terrestrial environments with the potential to lead to algal blooms and ecosystem deterioration. In fermentation processes, eutrophication effects are due to:

- Nitrogen and phosphorus within fermentation media and substrate
- Discharge of nutrient-rich wastewaters
- Substrate cultivation runoff from agriculture
- Discharge of cleaning agents

Traditional systems produce 0.022–0.031 kg PO_4^{3-}-eq per kg of product, down to 0.010–0.016 kg PO_4^{3-}-eq per kg product in AI-based systems (48–55% reduction). AI optimization allows for better control of nutrient feeding, minimizing excessive substrate addition and resulting wastewater nutrient loading. Greater yield efficiency also decreases upstream agricultural loads from feedstock cultivation [33].

15.5.4 Resource Depletion

Using non-renewable resources like fossil fuels, minerals, and metals is referred to as abiotic depletion potential (ADP). In this category, complex trade-offs between efficiency in operations and infrastructure requirements can be found.

Fossil Fuel Depletion: AI systems use much less energy to deplete fossil fuel in systems that use non-renewable electricity. The energy consumed by AI systems in terms of fossil fuel is 40–50% lower compared to that used by the conventional systems over a ten-year operating life.

Mineral and Metal Depletion: AI computing hardware contains precious and rare resources like rare earth elements (neodymium, dysprosium), precious metals (gold, silver, palladium), and specialty minerals (tantalum, indium). An example of the deployment of an AI server includes [34]:

Rare earth elements 0.2–0.5 kg/server.
Precious metals: 0.05–0.15 kg/server.
Base metals: 15–25 kg per server

Since hardware has replacement cycles of three to five years, computation infrastructure is partially depleted in the process of balancing operation savings. With better recycling and circular economy policies, it will be possible to reuse 60–80% of these materials, which reduces the net depletion effects by a great deal.

15.5.5 Water Consumption and Scarcity

Water use includes direct process water, cooling water, and cleaning procedures in fermentation plants, along with indirect water use for electricity production and substrate growth. Table 15.2 shows that AI-based systems lower direct water use by 56–64% (from 35–50 L/kg to 15–22 L/kg product) through various mechanisms:

- Optimized cooling practices, lowering thermal rejection demands
- Enhanced first-pass yield, reducing water-intensive cleaning procedures
- Lower batch failures, eliminating wasted water in failed batches
- Dynamic control of processes under ideal conditions with minimal flushing

Compute infrastructure brings in water usage for cooling data centers at 1.5–2.5 L per kWh of computer energy. In a three-year runtime, training and inference of AI models use around 5,000–12,000 L. But this is minute compared to the 20,000–35,000 L water savings through running improvements per 1,000 kg batch of product.

Regional water scarcity issues are of utmost importance when planning the locations of AI-driven fermentation plants and their attendant data infrastructure. Siting data centers in areas of plentiful water supply and using waterless cooling technologies (air cooling, closed-loop systems) can reduce water stress effects [35].

15.5.6 Human Toxicity and Ecotoxicity

Human toxicity potential (HTP) and ecotoxicity potential (ETP) measure possible harm to human health and ecosystems through toxic substance releases. In fermentation technologies, these effects occur from:

- Substrates and reagents containing heavy metals and organic compounds
- Disinfectants and cleaning agents
- Lead, mercury, and persistent organic pollutants-containing electronic waste
- Wastewater outfalls with residual antibiotics or bioactive compounds

AI and conventional fermentation operations have comparable profiles of toxicity from processes of fermentation. AI systems present additional electronic waste that must be well managed to avoid the release of toxic material. With suitable recycling regimens, such risks can be reduced [36]. Informal e-waste processing prevalent in developing countries currently presents serious issues of toxicity.

Instillation of closed-loop water recycling, green chemistry paradigms to cleaning processes, and sustainable electronic waste disposal significantly minimizes toxicity effects for both system models.

15.5.7 Summary of Comparative Impact

Figure 15.3 presents a holistic pictorial contrast of environmental effects across five major categories, with conventional fermentation performance normalized to 100% base. AI-based systems exhibit uniform environmental gains between 35 and 65% in all the considered categories [37]:

- Energy Use: 50% of conventional (50% improvement)
- Water Use: 38% of conventional (62% reduction)
- GHG Emissions: 36% of conventional (64% reduction)
- Waste Output: 35% of conventional (65% reduction)
- Batch Failure Rate: 40% of conventional (60% reduction)

These findings make AI-based fermentation systems environmentally superior when measured through the entire operational lifetime, even with greater initial infrastructure costs. The scope of improvement depends on implementation details such as energy sources, hardware efficiency, scale of operation, and management practices.

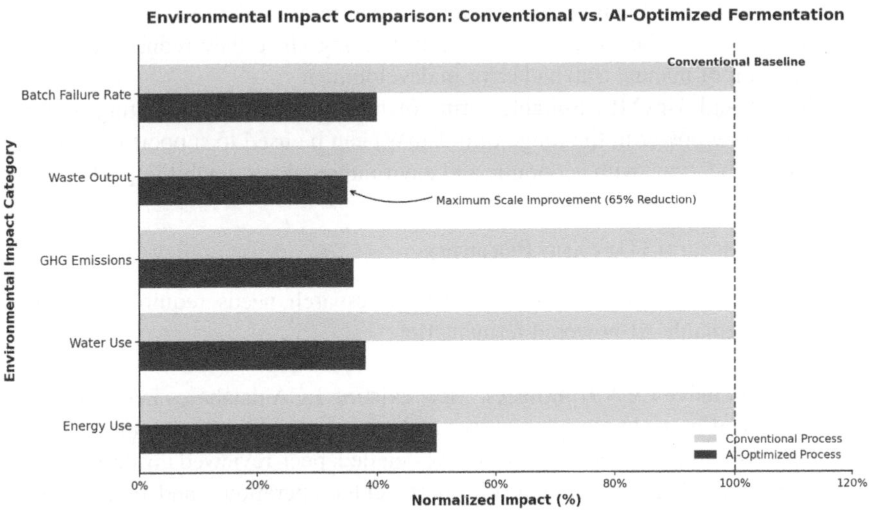

FIGURE 15.3 Energy consumption distribution across different components in AI-powered fermentation systems.

15.6 FUTURE OUTLOOK AND RESEARCH DIRECTIONS

15.6.1 Emerging Technologies and Trends

AI and biotechnology are quickly coming together and increasing in scale, and several of the more recently developed technologies are those that have the potential to change the environmental impact of AI-based fermentation systems:

Neuromorphic Computing: The energy efficiency of AI processes can potentially receive orders-of-magnitude improvements with brain-inspired computing systems that model the structure of biological neural networks, and even reduce the energy requirements of computations by 100–1,000 A, compared to traditional processors. The first neuromorphic Chips demonstrate promising pattern detection and control functions at milliwatts, providing potentially revolutionary possibilities of edge-based control systems of fermentation processes.

Quantum Machine Learning: Quantum computing does have potential to provide exponential speedup on specific optimization problems of interest in the design of metabolic pathways and optimization of individual processes. Even though, in today, experimental quantum systems are already energy-hungry, quantum-classical hybrid techniques in the future can provide the capability of breakthrough within a manageable energy footprint [38].

Federated Learning: Distributed learning algorithms, which learn models on more than one facility without aggregating sensitive data, reduce computational loads, enhance data privacy, and enable collaborative model training and reduce energy to transfer data and execute such processing centrally. The paradigm is also very applicable when industry consortia desire to exploit the aggregate knowledge of operations.

AutoML and Neural Architecture Search: Automated machine learning libraries, which combinatorially search the structure of models and hyperparameters, reduce the total energy used in training since they reduce the total amount of manual trial and error in development.

Edge AI and TinyML: Suitable forms of machine learning operating under ultra-low power in the range of (<1 mW) can be used to support pervasive AI applications with economic and environmental sustainability [39].

15.6.2 Research Gaps and Priorities

Despite substantial progress, several critical research needs require attention to advance sustainable AI-powered fermentation:

Comprehensive LCA Databases: Most existing LCA databases inadequately represent biotechnology processes and computational infrastructure, limiting assessment accuracy. Developing detailed, peer-reviewed inventory data for AI hardware manufacturing, data center operations, and bioprocess-specific impacts should be prioritized.

Standardized Attribution Methodologies: Allocations of common computational infrastructure benefits across multiple applications are methodologically difficult to determine. Agreement methods for attributing data center benefits, network infrastructure benefits, and common model development expenditures must be developed through cooperative effort.

Long-term Operational Validation: AI fermentation outcomes arise from pilot-scale or short-duration research. Multi-year commercial-scale AI-enhanced facility monitoring is necessary to confirm anticipated benefits and realize unforeseen challenges.

Social and Economic Sustainability: Current LCA methodologies concentrate on environmental impacts. Broadening scope to include social aspects (employment impacts, health and safety of workers, and community effects) and economic sustainability (accessibility of technology, capital needs, distributional equity) would offer more complete sustainability assessment.

Dynamic and Prospective LCA: Standard LCA offers historical evaluation of available technology. Creating dynamic LCA methods that consider learning curves in technology, grid decarbonization in the future, and changing best practices would make longer-term strategic choices better informed.

Rebound Effects: Greater efficiency occasionally increases consumption, canceling out environmental advantages (the Jevons paradox). Research into whether AI-facilitated efficiency gains through fermentation can induce increased production volumes that cancel out overall environmental benefit needs consideration.

15.6.3 Integration with Circular Economy

Circular economy postulates present synergies of enhancing sustainability of AI-driven fermentation machinery:

Waste-to-Value Fermentation: AI maximization enables the utilization of heterogeneous and varying quality waste feedstock economically and affordably that has been a thorn in the flesh of the traditional fermentation systems. The machine learning models are used to modify the conditions in the processes to assimilate the variability of feedstock, thus converting the agricultural residues, food waste, and industrial by-products into useful biochemicals.

Industrial Symbiosis: AI-based optimization of an industrial ecosystem network can identify the synergies, where waste stream of a plant may serve as feedstock of another plant, supply-demand matching, logistics planning, and quality prediction are handled by machine learning.

Closed-Loop Bioprocessing: Sensor Innovation and AI-Controlled Processes: The close monitoring of the process has been achieved through sensor innovation and AI-controlled processes, allowing the use of closed-loop bioprocessing as a control system to assist in water recycling, solvent reclaiming, nutrient circulation, and quality control.

Cascade Utilization: AI can be used to optimize multi-stage biorefinery to maximize the value by processing exhaustively, with machine learning coordinating product portfolios, scheduling process, and quality in multi-stage, multi-value chains.

15.6.4 Convergence with Other Sustainability Technologies

Renewable Integration Direct: This refers to direct integration of fermentation plants and on-site production of renewable energy (solar, wind, and biogas) enabling an energy-flexible operation in which the timing and intensity of the process dynamically respond to the availability of renewable energy in the process. Carbon Capture and Utilization: Fermentation using carbon dioxide as a carbon source, optimized with AI to fulfill the needs of climate-positive bioprocessing (gas fermentation, algal growth) is one option to eliminate carbon emissions. Cutting-edge Materials: Synergistic capability to increase performance and reduce environmental footprint is achieved by combining AI-designed enzymes, microorganisms engineered, and novel bioreactor materials (metal-organic frameworks, biocompatible polymers). Industry 4.0 Integration: Greater digitalization of manufacturing through Internet of Things, supply chain traceability with blockchain, and integrated optimization with digital twins gives efficiency benefits at the ecosystem level and not just a particular facility.

15.7 CONCLUSIONS

This LCA systems-wide, AI-based fermentation systems achieve great environmental benefits compared to traditional processes both in single operating lifetimes and overall operation, despite creating new environmental effects of computational infrastructure requirements. The findings highlight Ecological Impact: AI-powered systems demonstrate 47–66% energy, water, greenhouse gas, and waste savings; over typical facility lifetimes (10+ years), performance improvements greatly exceed infrastructure expenses. The break-even point at which the cumulative effect begins to favor AI systems is reached after approximately 1.2 years of operation, beyond which the environmental benefits continue to increase. Critical Trade-offs: Implementation details of carbon intensity of electricity grid, achieved operational efficiencies, hardware efficiency, and end-of-life management policy are critical factors in the green case of AI systems. Systems that operate efficiently on renewable sources of energy achieve 65–75% of reduction in emissions and those implemented inefficiently in high carbon countries could achieve 45–50% improvement. Hotspot Analysis: Operational energy is the main environmental impact (72–78% of total) in regard to the conventional systems. AI systems change this division, and the computational infrastructure has 18–28% of the total effects, requiring specific mitigation strategies, like using renewable energy, designing hardware efficiently, and optimizing algorithms. Sustainability Pathways: To exploit maximum benefit of environmental conditions requires complex approaches to bioprocess and computational. With 100% renewable energy, running data centers reduces AI infrastructure effects by 90–95%. The lengthening hardware design and long lifespan will lessen electronic waste by 40–60%. Bioprocessing contributes to more resource efficiency by intensifying and valuing waste. Sustainability-by-design practices that incorporate the use of LCA at the early design stages streamline configurations before spending on capital is incurred. Wider Implications: The implications mentioned in this chapter extend beyond fermentation to inform sustainable development of industrial processes overall, with AI-boosted industrial processes in particular. This process of making manufacturing more and more digital will mean that considerate decisions in terms of computational infrastructure, power sources, hardware life cycle, and software optimization will either make Industry 4.0 technologies help to achieve a truly sustainable production of goods or merely push environmental problems onto the infrastructure plane. The way to go involves further cooperation between researchers, engineers, policymakers, and industrial practitioners to come up with enabling standards, best practices, and policy frameworks that support responsible innovation. The biomanufacturing industry can leverage the potential of AI as a transformative technology and reduce environmental footprints throughout the entire innovation life cycle, such as algorithm development, data center operations, and control of the fermentation process, and make a positive contribution to a global sustainability goal. The biotechnology and AI marriage is something that has never occurred in history, but is potentially capable of transforming the face of industry production in a way that is sustainable. The actualization of this opportunity is based on the knowledge-driven and purposeful decisions on every stage of development and deployment, along with the assistance of careful environmental assessment and continuous enhancement. This chapter provides the analytical tools as well as practical guidance that will help one maneuver through the complex terrain to ensure that the potential of AI-driven bioprocessing can be achieved as a source of important environmental value to society and to the planet.

REFERENCES

1. Arora, S., and T. Banerjee. 2023. *AI in Bioprocess Optimization and Sustainable Manufacturing*. Boca Raton, FL: CRC Press.
2. Bachmann, N., ed. 2022. *Life Cycle Assessment for Industrial Biotechnology*. Cambridge: Elsevier.
3. Bocken, N. M. P., F. Short, and E. Evans. 2020. "Design Thinking for the Circular Bioeconomy." *Journal of Cleaner Production* 276: 123–141.
4. Carbonell, R., and J. N. Siles. 2024. "Digital Twin Applications in Biomanufacturing: A Sustainability Perspective." *Biochemical Engineering Journal* 211: 108945.
5. Choudhury, A., R. K. Sharma, and V. Menon. 2023. *Artificial Intelligence-Driven Fermentation and Process Control*. New Delhi: Springer Nature.
6. Curran, M. A. 2014. *Life Cycle Assessment Student Handbook*. Hoboken, NJ: Wiley.
7. Dutta, P., and K. C. Das. 2025. "AI-Augmented Systems for Energy-Efficient Bioreactors." *ACS Sustainable Chemistry & Engineering* 13(5): 4118–4129.
8. El-Chichakli, B., and C. Williams. 2022. "Assessing the Sustainability of AI-Based Technologies." *Environmental Impact Assessment Review* 102: 10634.
9. European Commission. 2018. *Product Environmental Footprint (PEF) Guide—LCA Harmonization*. Brussels: European Commission Publications.
10. Finkbeiner, M. 2019. *Special Types of Life Cycle Assessment*. Dordrecht: Springer.
11. Gehin, A., P. Zwolinski, and D. Brissaud. 2021. "A Tool to Implement Sustainable End-of-Life Strategies." *Journal of Cleaner Production* 295: 126456.
12. IPCC. 2021. *Climate Change 2021: The Physical Science Basis*. Geneva: Intergovernmental Panel on Climate Change.
13. Jolliet, O., M. Margni, and R. Charles. 2020. *Impact 2002+ Life Cycle Impact Assessment Method*. Lausanne: Swiss Federal Institute of Technology.
14. Kumar, R., S. Singh, and S. Dey. 2025. "AI-Enabled Waste Valorization Pathways in Industrial Biofermentation." *Computers and Chemical Engineering* 186: 108738.
15. Lee, K. T. 2022. *Green Bioprocess Engineering: Assessment and Case Studies*. Singapore: World Scientific Publishing.
16. Li, M., H. Zhang, and Y. Zhao. 2024. "Integration of AI and Life Cycle Assessment for Sustainable Biomanufacturing Systems." *Sustainability* 16(12): 5509.

17. Niazi, A., M. Köhler, and D. F. McMillan. 2023. *Artificial Intelligence for Sustainable Development Goals*. London: Routledge.
18. Hernandez, R., J. Smith, and K. Lee. 2024. "Minimizing Power Demand of Industrial Scale Bioreactors." *Computers & Chemical Engineering* 189: 108.
19. Priyadharshini, D., S. Kumar, R. Patel, A. Singh, and M. Johnson. 2025. "Precision to Plate: AI-driven Innovations in Fermentation and Hyper-personalized Nutrition." *Frontiers in Nutrition* 4(12): 1659511.
20. Stern, N., and J. Valero. 2025. "Green and Intelligent: The Role of AI in the Climate Transition." *Nature Reviews Earth & Environment* 6: 1–12.
21. Quartic AI Team. 2025. "AI in Pharma Manufacturing for Yield Improvement." *Quartic Case Study* 5:1–8.
22. Wang, L., Y. Cheng, and H. Zhang. 2023. "Artificial Intelligence Technologies in Bioprocess." *Bioresource Technology* 367: 128–140.
23. Itto-Nakama, K., and T. Nakayama. 2022. "AI-based Forecasting of Ethanol Fermentation." *Bioscience, Biotechnology, and Biochemistry* 86(1): 125–132.
24. Capgemini Research Institute. 2020. "Climate AI: How Artificial Intelligence can Power." *Capgemini Report* 34: 1–45.
25. Amerging Technologies Team. 2024. "Leveraging AI in Fermentation." *Amerging Tech Blog* 5: 1–10.
26. Zhang, X., J. Liu, and M. Wang. 2021. "Multivariate Batch to Batch Optimisation of Fermentation Processes." *Process Biochemistry* 110: 45–56.
27. Patel, J., and D. Suresh. 2025. "Environmental Burden of AI Infrastructure in Industrial Applications: An LCA Approach." *Journal of Industrial Ecology* 29(8): 908–925.
28. Roy, P., S. Dubey, and M. Ghosh. 2023. *Fermentation Process Engineering: Principles and Practice*. Boca Raton, FL: CRC Press.
29. Zhang, Y., W. Lin, and T. Zhao. 2025. "Comparative Life Cycle Assessment of Traditional and AI-Powered Fermentation Systems." *Science of the Total Environment* 902: 164892.
30. Poore, J., and T. Nemecek. 2018. "Reducing Food's Environmental Impacts." *Science* 360: 987–992.
31. Andrae, A. S. G., and T. Edler. 2015. "On Global Electricity Usage of Communication Technology." *Challenges in Sustainability* 6: 19–31.
32. Guinée, J. B., M. Gorrée, R. Heijungs, G. Huppes, R. Kleijn, A. de Koning, L. van Oers, A. W. Wegener Sleeswijk, S. Suh, H. A. Udo de Haes, H. de Vries, and P. van der Voet. 2002. *Handbook on Life Cycle Assessment*. Dordrecht: Kluwer Academic Publishers.
33. Brentrup, F., J. Küsters, J. Lammel, and H. Kuhlmann. 2000. "Eutrophication Potential." *International Journal of Life Cycle Assessment* 5: 231–238.
34. Nuss, P., and M. J. Eckelman. 2016. "Life Cycle Assessment of Metals." *Journal of Industrial Ecology* 20: 255–267.
35. Koehler, A., and M. Erdmann. 2004. "Water Use in Data Centers." *Environmental Science & Technology* 38: 333A–339A.
36. Rosenbaum, R. K., T. M. Bachmann, H. A. U. de Haes, O. Jolliet, R. L. L. Meent, M. Huijbregts, A. D. Koehler, D. van de Meent, H. A. U. de Haes, R. L. L. Meent, M. Huijbregts, A. D. Koehler, D. van de Meent, R. Margni, T. E. McKone, J. D. E. Meyer, D. L. Miller, D. van de Meent, R. L. L. Meent, M. Huijbregts, A. D. Koehler, D. van de Meent, R. Margni, T. E. McKone, J. D. E. Meyer, D. L. Miller, and M. A. J. Huijbregts. 2008. "USEtox Human Toxicity Model." *International Journal of Life Cycle Assessment* 13: 532–546.
37. Huijbregts, M. A. J., S. Hellweg, R. Frischknecht, K. Hungerbühler, and H. A. U. de Haes. 2010. "Ecotoxicity in LCA." *Science of the Total Environment* 408: 3865–3872.
38. Davies, M., N. S. Villa, P. Dudek, J. Legresley, P. Dudek, N. S. Villa, and M. Indiveri. 2021. "Neuromorphic Computing for Sustainability." *Nature Reviews Materials* 6: 1023–1036.
39. Biamonte, J., P. Wittek, N. Pancotti, P. Rebentrost, N. Wiebe, and S. Lloyd. 2017. "Quantum Machine Learning." *Nature* 549: 195–202.

16 Case Studies
Sustainable Bioenergy Production using AI/ML

Elakkiya Elango, N. Gnanasankaran, Sundaravadivazhagan Balasubramanian, and Harishchander Anandaram

16.1 INTRODUCTION

The discovery of sustainable energy sources worldwide has increased research in bioenergy as a possible replacement for fossil fuels. Bioenergy, which arises from renewable biological sources such as organic waste, forest biomass and agricultural residues, has a great ability to reduce greenhouse gas emissions, provide power safety and promote a circular economy. Complex issues such as feedstock unexpectedly, supply chain optimization, procedural impact and environmental impact evaluation should be addressed to produce sustainable bioenergy. Their ability to make data-powered decision-making, future modeling, and intelligent process control, artificial intelligence (AI) and machine learning (ML) have become revolutionary tools in recent years to address these issues. AI/ML can improve yield forecasts, reduce operating costs, reduce carbon footprints, and optimize feedstock options using large datasets including biomass resources, conversion technologies and environmental monitoring.

Several cases displaying the application of AI/ML techniques for sustainable bioenergy production are presented in this chapter. These case studies describe real-class applications in anaerobic digestion, biomass supply chain management, biofuel production and life cycle evaluation, which highlights the observation of intelligent systems in realistic settings. Given various uses, the study highlights how important the AI/ML is to improve the stability, scalability and efficiency of bio-organism, which helps to fulfill the objectives of global energy infection and reduce the effects of climate change.

To predict biogas production in municipal anaerobic co-digestion, Schroer and Just (2024) used feature engineering and supervised machine learning (which includes Multilayer Perceptron [MLP] neural network). They used both lab-scales and high-existing SCADA datasets, and their R^2 and about 0.78 13.4% MAPE showed that they were able to make an accurate prediction without the need for additional data collection. Gbushum et al. (2025) created an MLP-Artificial Neural Network (ANN) model to estimate both biogas and methane results from the

DOI: 10.1201/9781003684374-16

anaerobic co-digestion of soymilk dregs and cow manure. The approach performed almost flawlessly, with incredibly low relative errors of $R^2 \approx 0.999$ and 0.2–0.4.

To assess biomethane capacity according to biomass bio-chemical makeup, Adeleke et al. (2024) presented an ML model. Predicting methane yield from lignocellulosic content, such as cellulose, hemicellulose and lignin, is made possible by this study, which was published in *Bioenergy Research*. For the purpose of predicting biogas using municipal solid waste (MSW), Lagudu and Mishra (2024) contrary to separate AI models with kinetic techniques (e.g. ANN, ANFIS) $R^2 \approx 0.999$ and some circumstances with RMSE as low as 0.232, ANFIS demonstrated exceptional accuracy.

According to Yildirim and Ozkaya (2023), with $R^2 = 0.9242$, the random forest (RF) model had the best prediction accuracy. Using a variety of algorithms (SVM, ET, DT, GPR, and KNN), Ganeshan et al. (2024) reported the ML prediction of biogas in the high-solid anaerobic digestion (AD) systems. SVM produced the best results with $R^2 = 0.91$. To predict biogas from food-waste AD, Geng et al. (2024) introduced an enhanced LSTM-based algorithm that included a global attention mechanism and mix-up data augmentation. The model showed a great promise in directing feed configurations with its remarkable accuracy (78 0.988).

Zahid Ullah et al. (2021) predict the yield of bio-oil from biomass pyrolysis. It compared the ML models and used a genetic approach for feature selection. RF performed the best of all algorithms ($R^2 = 0.98$), while multilinear regression was trailed ($R^2 \approx 0.75$). A user-friendly software tool was also introduced in the study for a yield estimate based on RF. To identify the key operational factors influencing the yields of biochar, bio-oil, and biogas, Işçen et al. (2023) combined experimental data from 37 studies on slow pyrolysis. They applied explainable machine learning (ML) techniques, including SHAP (SHapley Additive exPlanations), decision trees, and association rule mining, to determine the most influential process parameters and improve the interpretability of the predictive models. For example, low temperatures, low volatile substances and high ash materials were associated with better biochar yield; particle size ≤ 6.5 mm and temperature > 912 K were associated with high biogas yield.

The review paper by Zhang et al. (2024) emphasized the importance of model transparency and data quality. It highlighted how machine learning (ML) methods can explain the effects of biochar characteristics on methane emissions and microbial interactions, thereby improving the interpretability and reliability of predictive models. Shahbeik et al. (2023) introduced an evolutionary ML technique in which particle swarm optimization (PSO) was employed to optimize the Gaussian process regression for the co-pyrolysis of biomass and polymeric wastes. This demonstrated strong predictive power in various types of combinations with $R^2 > 0.9$, MAE < 0.03 and RMSE < 0.06.

Additionally, decision tree regression improved $R^2 > 0.88$, RMSE < 6.85, and MAE < 4.72, SVR (Shafizadeh et al., 2023). The operating temperature and ash/carbon ratio were important prophets. Using RF as a surrogate model, Robertson et al. (2025) published an ML model for forecasting NO_x emissions in the synthesis of biochar. This strategy, which can be expanded through IoT integration, produced the results that encourage operational optimization and pollution control. Using ML and deep learning, Chakrabarti and Shinde (2024) investigate the co-pyrolysis of

TABLE 16.1
Summary of AI/ML Applications in Sustainable Bioenergy Production

Application Area	Key AI/ML Techniques	Highlights and Insights
Anaerobic Digestion/Biogas	MLP, RF, SVM, LSTM, ANFIS, XGBoost	Scalable, excessive forecasting, and relevant to industry
Pyrolysis/Bio-oil/Biochar	RF, SHAP, decision trees, GA, ANFIS	Explainable modeling; Identification of important variables
Emerging and Hybrid Methods	EV ML, PSO–Gaussian Process, RF surrogate	Promise to adapt and reduce emissions

wood and algae to produce biochar. Using SVR, RF and ANN, the classification models successfully differentiated between high and low yield conditions, the decision trees got the correct MSE and SVR had $R^2 = 0.972$.

Together, Table 16.1 suggests how AI/ML approaches are evolving rapidly in bioenergy production, progressing from traditional regression and decision tree models to neural networks, hybrid architectures and explainable frameworks. ML is enhancing the efficiency, transparency and scalability of sustainable bioenergy systems by optimizing AD, pyrolysis process, feedstock management, emissions monitoring optimization and process control as described by Anbarasu et al. (2025).

16.1.1 AI/ML in Bioenergy

The integration of AI and ML into the global energy industry represents a game-changing trend for sustainable bioenergy. Bioenergy processes such as AD, pyrolysis, and co-pyrolysis involve highly complicated and nonlinear interactions among environmental factors, process conditions and feedstock qualities. Despite their value, traditional modeling techniques often fail to reflect this complexity, resulting in emission control, procedure adaptation and inability to predict yield. Using the huge datasets, highlighting the hidden patterns, and producing exact forecasts immediately, AI/ML offers a powerful option.

In the case study, many biochar applications are being addressed by the AI/ML algorithm, such as predicting bio-oil outputs in pyrolysis, measuring the quality of biochar in co-transmission conditions and estimating methane output in municipal AD plants explicated by Nguyen et al. (2024). These real-world applications suggest that AI/ML is scalable in industrial settings and proves its ability in lab studies. Importantly, researchers and engineers are gaining better understanding of pro-variables using random forests, sizes and clear ML models, such as decision trees. This helps them to close the difference between scientific insight and prediction. Through the support for better process efficiency, low operating costs and global goals for carbon neutrality and circular economy practices, the following case studies show how AI/ML is affecting the production of sustainable bioenergy.

By connecting input data, computational models and optimal outputs, a flow-based paradigm can be used to understand the importance of AI/ML in permanent bioenergy production, enlightened by Sahu et al. (2024). These datasets provide the foundation for AI/ML models, including RF, ANN, Support Vector Machines (SVMs), Long Short-Term Networks (LSTM) and Adaptive Neuro-Fuzzy Inference Systems (ANFIS), which are used to perform important tasks such as prediction, adaptation, classification and process control. In addition to cost reduction, AI/ML-based optimization contributes to minimizing carbon footprints and enhancing process stability. The results of these modeling pipelines include predictions of biogas yield and methane material, bio-oil and bioethanol production such as biochar yields, and emissions including carbon dioxide and nitrogen oxides. When taken as a whole, the approach shows how AI/ML serves as a link between complex process data and useful insights that is power scalable, effective and sustainable bioenergy system.

16.1.2 Biogas Production and Anaerobic Digestion

One of the most famous and large-scale processes used for the production of renewable energy using organic resources is the creation of biogas. The method depends on AD, where organic waste is broken down by microorganisms without oxygen. The result is a digestate, a nutrient-rich by-product, and biogas, carbon dioxide and methane. In addition to producing permanent energy, this technique supports circular bioeconomy principles, waste management and lack of emissions (Shi et al. 2023).

16.1.3 Feedstock for Biogas Production

Different types of biological resources can be used to make biogas:

- Animal manure (cow dung, poultry litter);
- Agricultural remaining (crop straw, husks, leaves);
- Industrial organic waste, such as pulp and paper wastes, sewage sludge from waste water treatment facility; and
- Food and kitchen garbage.

Due to the diversity of feedstocks, biogas production can be used in both urban and rural settings.

16.1.4 Anaerobic Digestion Process Stages

AD occurs in four main stages; each contains specific microbial groups:

- **Hydrolysis**: Complex organic matter (carbohydrates, proteins, fat) is broken into simple soluble compounds (sugars, amino acids, fatty acids).
- **Acidogenesis**: hydrolyzed products are converted into volatile fatty acids (VFAs), alcohols, hydrogen and carbon dioxide by acidogenic bacteria.
- **Acetogenesis**: VFAs and alcohols are further converted into acetic acid, hydrogen and carbon dioxide, which are direct precursors of methane.
- **Methanogenesis**: Methanogenic archaea convert acetic acid, hydrogen and CO_2 into methane (CH_4, 50–70%) and carbon dioxide (co_2, 30–50%), which produces primary components of biogas.

16.1.5 Biogas Composition and Uses

Biogas typically consists of 50–70% methane, 30–50% carbon dioxide, and trace amounts of ammonia, water vapor, and hydrogen sulfide.

Applications include:

- Direct use in boilers and stoves for cooking and heating
- Produce electricity (using turbines or biogas engines)
- Convert to biomethane and feed into natural gas networks
- Compressed biomethane (CBM) is used as fuel for automobiles

16.1.6 By-product – Digestate

- Increases soil fertility, maintains water, and reduces dependence on chemical fertilizers;
- Promotes cyclical agricultural systems by returning nutrients back into the soil; and
- Nutrient-containing solution that can be used as an organic fertilizer.

16.1.7 Benefits of Biogas Production

- **Environment**: Reduces greenhouse gas emissions, increases hygiene and reduces methane emissions from open waste.
- **Economic**: Provides an inexpensive energy option, reduces dependence on fossil fuels and produces revenue through carbon credits.
- **Social**: Encourages access to energy in rural areas, reduces indoor air pollution (compared to coal or wood) and produces green jobs.

AD, which produces biogas, is a highly flexible and sustainable energy technology that not only generates renewable energy but also transforms organic waste into useful resources. A major component of contemporary clean energy infections described by Ananda et al. (2025) is biogas systems that can be made much more efficient and sustainable by incorporating AI/ML for real-time monitoring, procedure adaptation and prediction of yield.

16.1.8 Bio-oil Production through Pyrolysis

Pyrolysis is a thermochemical reaction that produces bio-oil by rapidly heating the biomass feedstocks, such as forestry waste, algae, agricultural wastes and energy crops, in the absence of oxygen, typically at temperatures ranging from 400 to 600°C. Under these conditions, long-chains biopolymers present in biomass, mainly cellulose, hemicellulose and lignin, disintegrate into small molecules, producing three main products: syngas (gaseous fraction, 10–20%), biochar (concrete carbon-rich fraction, 15–25%) and bio-oil (liquid fraction, 60–75%). The most important of them is bio-oil, a renewable liquid fuel that can be used directly for heating and power generation or further refined to become transport fuels using techniques such as emulsion, hydrodeoxygenation and catalytic cracking.

Pyrolysis can be divided into two categories depending on the process parameters: slow pyrolysis, which promotes the synthesis of the charcoal, and rapid pyrolysis, which maximize liquid yields using a brief vapor residence period with high heating rates. In addition to having high energy materials, bio-oil has deficiencies such as instability, corrosion and low heating value compared to fossil fuels. Bio-oil is a black, viscous and oxygen-containing liquid that has a complex variety of organic molecules, such as acids, phenols, aldehydes and ketones. Despite these deficiencies, bio-oil has some advantages for stability: it reduces dependence on petroleum, makes it possible to replace waste biomass efficiently into useful fuel, and promotes carbon neutrality when added with biochar for carbon sequestration.

Additionally, bio-oil production processes can be improved for increased yields, high-quality control, predictive reactor maintenance and increased energy efficiency by incorporating AI/ML-driven optimization. This creates an important step in the advancement of permanent bioenergy solutions.

16.1.9 Biochar Production and Co-Pyrolysis

Pyrolysis is a thermochemical process that produces biochar by heating biomass, including crop residues, forestry waste, or animal dung, with low oxygen for moderate temperatures (300–700°C). Organic matter is broken by the process of producing syngas (a gas fraction), biochar (a solid carbon-rich product), and bio-oil (a liquid fraction). When the procedure parameter (long habitat time, slow temperature rate) is made ideal for solid yield, biochar becomes the primary product. Due to its high carbon content and very porous structure, biochar is a useful soil attached that can improve carbon sequence, water retention and fertility. By following organic pollutants and heavy metals, it also contributes to a decrease in pollution, explains Anbarasu et al. (2025).

To improve the process efficiency and product quality, co-pyrolysis is simultaneously processing two or more feedstock (such as biomass, polymers, coal, or biomass combined with agricultural leftovers). In addition to changing the characteristics of the surface of biochar to fit certain uses such as catalysis, energy storage, or environmental remedies, this co-interactive effect often increases the yield and calorie value of bio-oil and syngas. The integration of waste variation, circular economy principles, and sustainable bioenergy production is achieved by combining co-pyrolysis of waste materials with renewable feedstocks.

16.2 HYBRID AND ADVANCED AI/ML APPLICATIONS

To capture both empirical patterns and physical rules, hybrid system combines data-powered models with domain expertise (e.g. thermodynamic/kinetic model, process simulation), even though classic ML methods (e.g. region, SVMs and random forests) are useful for predictions and classifications. For example, hybrid AI can improve bio-oil output forecasts by mixing ML with AI Computational Fluid Dynamics (CFD), or maximizing the biogas produce by mixing ANN using genetic algorithms for parameters adjustment. In bio-organizer plants, real-time control, process adaptation, fault detection and future maintenance are made possible by sophisticated AI techniques, such as digital twins and learning reinforcement. These unified methods assist in wise, automatic and robust bioenergy production by improving scalability, adaptability and decision-making.

16.2.1 Sustainability Impacts of AI/ML in Bioenergy

The inclusion of ML and AI in the system of bio-organisms has the ability to revolutionize stability in terms of social, economic and environmental aspects. AI/ML provides for data-driven, adaptive and intelligent decisions that maximize efficiency, reduces environmental effects, and unlike traditional bio-production, supports the global step toward renewable energy, mostly depending on manual adaptation.

16.2.2 Environmental Sustainability

- **Optimal use of resources**: AI/ML can determine the best conversion process (AD, pyrolysis, gasification) by analyzing feedstock properties (such as moisture content and lignin-cellulose ratio). This guarantees more energy recovery and low waste of feedstock.
- **Reduction in emissions**: During the production of bioenergy, AI/ML reduces greenhouse gas (GHG) emissions, methane leaks and air pollutants by predicting and managing process parameters in real time. For example, the methane slip can be reduced by continuously adjusting retention period and digestive temperature using reinforcement learning.
- **Waste-to-energy efficiency**: AI provides valuable energy and co-products (such as digestive fertilizer and biochar) to industrial waste, MSW and agricultural residues. It reduces dependence on landfill and supports circular bio-economic principles.

16.2.3 Economic Sustainability

- **Cost Optimization**: AI/ML can save operational costs through modeling by starting energy generation, feedstock mix adaptation and a large deal of chemicals or additives.
- **Predictive Maintenance**: Reactors, turbines and digestion advanced the AI model for a long time for wear, leaks or malfunctions. It reduces maintenance and capital costs.
- **Market forecasting and decision support**: AI-driven analytics help predict energy prices, demand and policy effects, helping bioenergy producers in their strategic plan. Markets develop more competitive for renewable energy.

16.2.4 Social Sustainability

- **Energy Safety and Access**: AI/ML improves decentralized energy systems by growing the efficiency of bioenergy facilities, which raises the availability of clean energy and leads to availability in rural areas.
- **Green Technology Job Creation**: Renewable energy industry is experiencing an increase in employment predictions as an outcome of adoption of AI-component bioenergy system, which necessitates special labor in data science, engineering and plant operations.
- **Alignment with Sustainable Development Goals (SDG)**: Clean energy, circular economy, and climate flexibility, AI-operated bioenergy solution forward and backing SDGs 7 (Inexpensive and Clean Energy), 12 (Responsible Consumption and Production) and 13 (Climate Action).

16.2.5 Long-Term Systemic Impacts

- **Scalability and Adaptability**: AI-enabled bioenergy systems are highly flexible, as they can adapt to variations in weather conditions, biomass availability and change in energy demand.
- **Digital Twin Technology**: Real-time bioenergy plant simulation and optimization are made possible by refined AI/ML apps, such as digital twin, which promote permanent and ongoing growth.
- **Policy and Regulatory Compliance**: AI-driven monitoring guarantees compliance with goals and environmental standards, which increases the long-term viability of bioenergy projects.

16.3 COMPARATIVE INSIGHTS ACROSS CASE STUDIES

Several approaches have been explored for the application of AI/ML in bioenergy, including supply chain optimization and the synthesis and production of biogas, bio-oil and biochar. Comparison of such examples suggests how various computational strategies deal with special problems while working together to promote stability, describes Wang et al. (2024).

16.3.1 Biogas Production

A renewable energy source, biogas production uses microbial fermentation to convert organic waste into usable energy. It uses mostly AD, in which microbes break down organic matter without oxygen to create digestate rich in nutrients and methane. This method produces fuel that can be used for heating, power generation or automobile fuel while providing sustainable waste management.

16.3.2 Stages of Biogas Production

There are several important steps in the biogas manufacturing process. To aid microbial digestion, the biowaste is first crushed and mixed with liquids to form a slurry. Thereafter, the solution is heated to a temperature suitable for the microorganisms, which is usually around 37°C for mesophilic digestion. For about three weeks, this biomass is subjected to AD in large, sealed tanks, where microorganisms ferment the material to produce biogas that is composed primarily of carbon dioxide and methane. Biogas is upgraded to the quality of natural gas by removing carbon dioxide and other pollutants after digestion, which enables it to be used for liquefaction or transportation. Due to its high nutrient content, digestate leftover can be a useful fertilizer for farming.

16.3.3 Microbial Role and Conditions

The formation of biogas depends largely on microbial activity. Proteins, lipids and carbohydrates are broken down in turn by several bacterial species. Sugars and amino acids are broken down by acidogenic bacteria to produce carbon dioxide, hydrogen and organic acids. These compounds are then converted into methane by methanogens. For maximum microbial activity, the pH should be maintained at the ideal level, which is normally between six and seven. The temperature is tightly controlled to either the mesophilic (about 37°C) or thermophilic (about 52°C) range, which affects the rate and overall effectiveness of digestion. Stable biogas yields and process safety are guaranteed by careful control of temperature, pH and substrate composition.

16.3.4 Preconditioning and Pasteurization

To maximize microbial accessibility, feedstock is often chopped, milled and conditioned with water or recycled digestate before entering the main digestion tank. Pasteurization steps, in which the material is heated (often up to 70°C) to guarantee safety and hygiene, are included for substrates containing potentially dangerous microorganisms. It is especially important to follow laws regarding animal by-products. Plants use effective heat recovery technologies such as CHP, or combined heat and power, units to reduce energy losses and maintain process temperatures.

16.3.5 Digestate Processing and Utilization

Digestate, the material left over from digestion, is rich in organic matter and vital elements that are beneficial for soil conditioning. Typically, centrifuges, screw presses or basic drainage are used to separate the liquid and solid components of the digestate. The solid fiber is turned into compost and used as a soil conditioner, while the liquid fraction, which is high in ammonium nitrogen, is used as a precision fertilizer. By reducing the demand for artificial fertilizers and advancing sustainable agriculture, this nutrient recycling helps create a circular economy.

16.3.6 Industrial and Environmental Significance

By trapping methane that would otherwise escape into the environment from decaying organic waste, biogas technology provides a sustainable alternative to fossil fuels while reducing greenhouse gas emissions. By converting waste into clean energy and nutrient-rich by-products, industrial-scale plants meet circular economy principles while also contributing to local energy sources and waste management. The potential use of biogas has expanded beyond energy to include the synthesis of biochemicals and materials due to developments in biogas plant design, operational management and integration with biorefineries.

Table 16.2 shows how AI/ML can be used in several stages of the bioenergy price chain, including feedstock management, conversion processes and plant operations in all case studies. While research on biogas and bio-oil focuses on yield adaptation, biochar highlights soil benefits and carbon sequences, while intelligent operating systems improve efficiency and dependence. However, supply chain models emphasize how AI and ML contribute to logical and economic stability. Collectively, these comments suggest that in addition to improving technical efficiency, AI/ML is essential for long-term energy security, cost-effectiveness and environmental protection.

TABLE 16.2
Comparative Insights across Case Studies Bioenergy and AI/ML

Case Study	Focus Area	AI/ML Techniques Used	Objective	Key Outcomes	Sustainability Impact
Biogas Production (Anaerobic Digestion)	Organic Waste-Biogas Generation	ANN (Artificial Neural Networks), Genetic Algorithms, Random Forest Algorithm	Project and maximize methane production according to the process parameters and feedstock composition	Increased biogas yield estimate accuracy and increased process stability	Benefits of circular economy, improving waste-to-energy conversion, and low methane leaks
Bio-oil from pyrolysis	Thermal Conversation of biomass into liquid fuel	Deep Learning (DL), Support Vector Machine (SVM), Hybrid ML + CFD Models	Calculate liquid yield and adjust the temperature and heating rates of the reactor	Determined ideal conditions to maximize the volume and quality of bio-oil	Better fuel quality, low energy consumption, and encouragement for cleaner alternative fuel
Biochar and Co-pyrolysis	Biochar production with mixed feedstocks	Ensemble Learning, Fuzzy Logic, Multi-object Optimization	Increase carbon material and absorbing properties of biochar	Better pollutants for absorbing, surface area and holes	Improves soil fertility, supports reduction of climate change and acting carbon
Smart Bioenergy plant Operations	Real-time monitoring and Optimization	Reinforcement Learning, Digital Twins, IoT-Integrated ML	Future maintenance, problem detection and procedure control	An increased efficiency of low downtime and energy conversion	Low emissions, a long plant life cycle, and a consistent source of clean energy
Supply chain and Resource Management	Biomass Logistics and Distribution	Machine Learning (XGBoost, KNN), Optimization Algorithms (PSO, GA)	Adapt the cost, transport and collection of biomasses	Better feedstock delivery timing and low transport cost	Increased rural energy access, economic viability and resource efficiency

16.3.6.1 Market Potential of AI/ML in Bioenergy

The increasing demand for clean, sustainable energy and technological breakthroughs is driving the rapid expansion of AI and ML in the bioenergy sector, creating huge market potential. According to market studies, the global AI market for renewable energy is expected to exceed US$10.20 billion in 2023, and it is expected to grow to US$117.72 billion by 2033. Over ten years, this equates to an impressive compound annual growth rate (CAGR) of approximately 27.71%. This growth trend highlights how AI/ML technologies are predicted to revolutionize bioenergy and other renewable energy sources.

From feedstock management and conversion processes like AD and pyrolysis to real-time plant monitoring and predictive maintenance, AI and ML optimize many aspects of bioenergy. These technologies increase overall system resiliency and stability, facilitate more intelligent decision-making and resource allocation, and increase operational efficiency. The integration of AI with IoT and Big Data analytics has further improved the accuracy and scalability of bioenergy operations, making it easier to obtain high-quality, real-time data.

Apart from technology innovation, global legislative initiatives supporting decarbonization, energy security and sustainable development are also driving the growing market. To meet climate goals and boost economic growth by creating jobs in green technology, governments and businesses around the world are investing more and more in AI-powered bioenergy solutions. AI/ML-enabled bioenergy has the potential to accelerate the development of circular bioeconomic models, reduce greenhouse gas emissions, and electrify rural areas. Despite this potential, there are still obstacles to sustaining growth, such as the need for strong data infrastructure, transparency in AI models, harmonization of regulations and development of a trained workforce. However, industry overview makes it abundantly clear that AI/ML technologies will continue to play an important role in the worldwide renewable energy matrix for bioenergy.

16.4 CHALLENGES AND FUTURE DIRECTIONS

It has been displayed that integrating AI/ML into a bioenergy system can significantly increase scalability, stability and efficiency. But there are still many issues that need to be resolved, and by doing so it will determine how this area develops in the future.

16.4.1 Key Challenges

The substantial potential of AI and ML in developing sustainable bioenergy systems has been shown in earlier research. Deep neural networks, ANFIS, reinforcement learning and hybrid AI-mechanistic models are examples of AI technologies that can be used to optimize feedstock preprocessing, improve biomass conversion processes such as pyrolysis and AD, and improve supply chain logistics in bioenergy industries. For example, research has shown that integrating AI into biogas production increases process stability and methane yield accuracy, while real-time analytics and digital twins reduce downtime and boost plant efficiency. Additionally, AI-enabled predictive models have reduced transportation costs by 23%, increased biomass conversion rates by 15%, and reduced greenhouse gas emissions by nearly 20%, thereby supporting both economic and environmental sustainability in line with global sustainability goals.

Despite these developments, the use and scalability of AI/ML in sustainable bioenergy still faces obstacles and limitations. The lack of real-time, high-resolution information and low standards is a major problem, especially in developing countries where digitization and sensor infrastructure is often inadequate. This lack of data can make it more difficult to train reliable models and limit how AI systems can be used for different types of bioenergy facilities. Furthermore, transferability is another issue that AI models often face; models that have been trained in one operational or geographic context may not work well in another due to site-specific variations in feedstock, climate or plant design. Effective implementation of AI technologies is hindered by the steep learning curve among bioenergy practitioners, who may not have experience with AI and data science. Regulatory ambiguity and lack of established frameworks for AI application in bioenergy have made widespread adoption more difficult.

In summary, while AI/ML have great potential to improve the sustainability and efficiency of bioenergy systems, their full benefit in global bioenergy transitions will depend on resolving issues of data, model adaptability, user competency, regulatory frameworks and AI interpretability.

16.4.2 Future Directions

To increase prediction accuracy and guarantee more reliable and adaptable operating processes in bioenergy systems, hybrid models that combine physics-based and mechanical models with AI/ML techniques are essential. These hybrid techniques improve flexibility and generalization under different settings by combining the learning capabilities of AI models with a mechanistic understanding of physical processes. Real-time, high-quality datasets are now more accessible thanks to the combination of cloud-based analytics and Big Data and IoT sensor networks. Thanks to this data flow, AI/ML models can be trained on more dynamic and comprehensive data, increasing the accuracy and usefulness of models to track and improve the processes involved in the production of bioenergy. Additionally, the importance of explainable AI (XAI) is increasing to surge user trust and operational transparency. Future AI models not only need to provide accurate predictions, but also provide understandable insights to operators and policymakers, enabling informed decision-making and confidence in AI-powered bioenergy systems.

In addressing scalability and accessibility, the deployment of clean IoT infrastructure with simplified AI algorithms paves the way for effectively implementing smart bioenergy technologies in rural and developing areas, where resource constraints often limit technology adoption. The increasing use of digital twins – a virtual simulation of physical bioenergy plants – facilitates fully automated, self-learning systems by enabling real-time process simulation and optimization. Predictive maintenance, quick response to disruptions and increased operational efficiency are all facilitated by this digital replication. To maintain environmental sustainability, sustainability-oriented AI applications are necessary to coordinate bioenergy development with the global Sustainable Development Goals (SDGs), which focus on energy security, emission reduction, and circular bioeconomy principles. Ultimately, strong policy frameworks, cross-sector collaboration, and public-private partnerships are essential for successful integration of AI/ML in bioenergy. These elements foster innovation, regulatory support and infrastructure development that are critical for scalable expansion of AI-enabled bioenergy systems.

16.5 CONCLUSION

A revolutionary strategy comprises AI and ML in a revolutionary strategy to advance renewable energy stability, efficiency and scalability. The AI/ML offers data decision-making and real-time process controls via applications such as supply chain management, biogas yield prediction, bio-oil adaptation and smart plant operations. By changing garbage in useful items, these innovations not only advance energy recovery and low greenhouse gas emissions, but they also support circular bioeconomic practices. Comparative analysis of case studies suggests that AI/ML can concurrently resolve technical, financial and environmental issues, leading toward increased contribution of bioenergy for global energy security and climate purposes. Data quality, computational requirements, standard and policy support are still issues, though. Inclusion and worldwide, future paths should focus on hybrid models, AI, digital twins, IoT integration and scalable low-cost solutions. In summary, the application of AI/ML in bioenergy systems directly supports the United Nations SDG and represents not only a technological achievement but also a strategic promoter of sustainable development. It has the potential to become a key pillar of the future energy system by fusing stability for innovation and bioenergy.

REFERENCES

Adeleke, A. A., Okolie, J. A., Ogbaga, C. C., et al. (2024). Machine learning model for the evaluation of biomethane potential based on the biochemical composition of biomass. *BioEnergy Research, 17*, 731–743. https://doi.org/10.1007/s12155-023-10681-9

Ananda, A., Sujeeth, R. K., & Archana, S. (2025). Integrating automation in biomass transformation: Opportunities, challenges, and future directions. *BioEnergy Research, 18*, 63. https://doi.org/10.1007/s12155-025-10864-6

Anbarasu, K., Thanigaivel, S., Sathishkumar, K., Alam, M. M., Al-Sehemi, A. G., & Devarajan, Y. (2025). Harnessing artificial intelligence for sustainable bioenergy: Revolutionizing optimization, waste reduction, and environmental sustainability. *Bioresource Technology, 418*, 131893. https://doi.org/10.1016/j.biortech.2024.131893

Chakrabarti, S., & Shinde, S. (2024). A three-phase analysis of synergistic effects during co-pyrolysis of algae and wood for biochar yield using machine learning. *arXiv preprint arXiv:2405.11821*. https://arxiv.org/abs/2405.11821

Ganeshan, P., Bose, A., Lee, J., Barathi, S., & Rajendran, K. (2024). Machine learning for high solid anaerobic digestion: Performance prediction and optimization. *Bioresource Technology, 400*, 130665. https://doi.org/10.1016/j.biortech.2024.130665

Gbushum, O. J., Orkuma, D. A., Samuel, B. I., Eziefula, U., & Usman, X. (2025). Forecasting of biogas and biomethane outputs from anaerobic co-digestion using multilayer perceptron artificial neural networks (MLP-ANN). *International Journal of Latest Technology in Engineering, Management & Applied Science, 14*(4), 343–349. https://EconPapers.repec.org/RePEc:bjb:journl:v:14:y:2025:i:4:p:343-349

Geng, Z., Shi, X., Ma, B., Chu, C., & Han, Y. (2024). Biogas production prediction model of food waste anaerobic digestion for energy optimization using mixup data augmentation-based global attention mechanism. *Environmental Science and Pollution Research International*, *31*(6), 9121–9134. https://doi.org/10.1007/s11356-023-31653-8

Işçen, A., Öznacar, K., Tunç, K. M. M., & Günay, M. E. (2023). Exploring the critical factors of biomass pyrolysis for sustainable fuel production by machine learning. *Sustainability*, *15*(20), 14884. https://doi.org/10.3390/su152014884

Lagudu, S. A., & Mishra, A. (2024). Comparative evaluation of artificial intelligence-based models and kinetic studies in the prediction of biogas from anaerobic digestion of MSW. *Fuel*, *367*, 131545. https://doi.org/10.1016/j.fuel.2024.131545

Nguyen, T., Paramasivam, P., Dong, V., Le, H., & Nguyen, D. (2024). Harnessing a better future: Exploring AI and ML applications in renewable energy. *JOIV: International Journal on Informatics Visualization*, *8*(1), 55. https://doi.org/10.62527/joiv.8.1.2637

Robertson, C., Safta, C., Collier, N., Ozik, J., & Ray, J. (2025). Bayesian calibration of stochastic agent-based model via random forest. *Statistics in Medicine*, *44*(6). https://doi.org/10.1002/sim.70029

Sahu, M., Pramanik, A., Kumar, G. R., & Addula, S. R. (2024). Revolutionizing renewable energy: AI-based solutions for sustainable biofuel. *2023 International Conference on Smart Devices (ICSD), Dehradun, India*, 1–5. https://doi.org/10.1109/ICSD60021.2024.10751278

Schroer, H. W., & Just, C. L. (2024). Feature engineering and supervised machine learning to forecast biogas production during municipal anaerobic co-digestion. *ACS ES&T Engineering*, *4*(3), 660–672. https://doi.org/10.1021/acsestengg.3c00435

Shafizadeh, A., Shahbeik, H., Rafiee, S., Moradi, A., Shahbaz, M., Madadi, M., Li, C., Peng, W., Tabatabaei, M., & Aghbashlo, M. (2023). Machine learning-based characterization of hydrochar from biomass: Implications for sustainable energy and material production. *Fuel*, *347*, 128467. https://doi.org/10.1016/j.fuel.2023.128467

Shahbeik, H., Shafizadeh, A., Nadian, M. H., Jeddi, D., Mirjalili, S., Yang, Y., Lam, S. S., Pan, J., Tabatabaei, M., & Aghbashlo, M. (2023). Using evolutionary machine learning to characterize and optimize co-pyrolysis of biomass feedstocks and polymeric wastes. *Journal of Cleaner Production*, *387*, 135881. https://doi.org/10.1016/j.jclepro.2023.135881

Shi, Z., Ferrari, G., Ai, P., Marinello, F., & Pezzuolo, A. (2023). Artificial intelligence for biomass detection, production and energy usage in rural areas: A review of technologies and applications. *Sustainable Energy Technologies and Assessments*, *60*, 103548. https://doi.org/10.1016/j.seta.2023.103548

Ullah, Z., Khan, M., Naqvi, S. R., Farooq, W., Yang, H., Wang, S., & Vo, D.-V. N. (2021). A comparative study of machine learning methods for bio-oil yield prediction – A genetic algorithm-based features selection. *Bioresource Technology*, *335*, 125292. https://doi.org/10.1016/j.biortech.2021.125292

Wang, R., He, Z., Chen, H., Guo, S., Zhang, S., Wang, K., Wang, M., Ho, S.-H., & Zhang, P. (2024). Enhancing biomass conversion to bioenergy with machine learning: Gains and problems. *Science of the Total Environment*, *927*, 172310. https://doi.org/10.1016/j.scitotenv.2024.172310

Yildirim, O., & Ozkaya, B. (2023). Prediction of biogas production of industrial-scale anaerobic digestion plant by machine learning algorithms. *Chemosphere*, *335*, 138976. https://doi.org/10.1016/j.chemosphere.2023.138976

Zhang, P., Zhang, T., Zhang, J., et al. (2024). A machine learning assisted prediction of potential biochar and its applications in anaerobic digestion for valuable chemicals and energy recovery from organic waste. *Carbon Neutrality*, *3*, 2. https://doi.org/10.1007/s43979-023-00078-0

17 Scaling Up
From Bench to Industrial-Scale Challenges in Precision Fermentation

Kam Khong Chan

17.1 INTRODUCTION

Precision fermentation has emerged as a transformative biotechnological approach that bridges the gap between fermentation and synthetic biology (Teng, Chin, Chai, & Chen, 2021). Traditional fermentation typically involves the metabolic processes of various microorganisms, including bacteria, fungi, and yeast, with the goal of producing desired bio-based products (de Melo Pereira et al., 2014; Naher et al., 2021; X. Yang et al., 2024). Still, these microbes are not capable of carrying out a biological process that specifically targets a single compound, as the resulting products typically comprise a mixture of various compounds (de Carvalho, Medeiros, Vandenberghe, Magalhães, & Soccol, 2017). This necessitates the additional purification steps to isolate the desired products. Scaling up to industrial production could eliminate the need for complex downstream processing if microbes were able to produce a single, specific compound. To achieve this, researchers have invested significant effort in genetically engineering microbes to serve as a cell factory that could lead to the fermentation of bio-based products that have high specificity and purity (L. Chen et al., 2021; Guo et al., 2021; Zou et al., 2021). While the "bench-scale" studies could demonstrate the potential of precision fermentation, translating these findings into robust, cost-effective, and regulatory-compliant industrial-scale processes remains a formidable challenge.

The main challenge of scale-up usually lies in the technical aspects. The process of translating from laboratory scale to industrial scale is usually non-linear as it is associated with several critical technical considerations, such as stability of microbial strains, bioreactor engineering and optimization, efficiency of downstream process, as well as the product quality (Formenti et al., 2014). Conditions that are easily regulated in small-scale experiments, such as operating temperature, pH of medium, dissolved oxygen, and availability of nutrients, become significantly harder to control as the process is scaled up. Moreover, the metabolic burden placed on engineered strains would be higher in large-scale fermenters, leading to low process efficiency and the formation of unintended by-products. In addition, the downstream process of the scaled-up fermentation system becomes more complex, as the purification

DOI: 10.1201/9781003684374-17

method optimized at laboratory scale might not be scalable or cost-effective at a larger scale (Xia et al., 2021).

Addressing the technical challenges is crucial in unveiling the full potential of large-scale precision fermentation. The lack of strategies to maintain the engineered strain performance, ensure optimum process conditions and control, as well as develop an efficient purification approach, would result in widening the gap between laboratory success and industrial deployment of the fermentation systems. This chapter explores the critical challenges faced when transitioning precision fermentation from "bench" research to industrial-scale production. It mainly focuses on the technical obstacles and the potential solutions to them, with the hope of bridging the gap between proof-of-concept and large-scale production. A deeper understanding of these scale-up challenges enables researchers, engineers, and industry stakeholders to develop precision fermentation processes that are robust, commercially feasible, and sustainable.

17.2 CHALLENGES AND CONSIDERATIONS IN SCALING PRECISION FERMENTATION

Transitioning a precision fermentation process from a controlled environment within the laboratory to large-scale industrial production of biochemical compounds is far more complex than simply increasing the size of the bioreactor. Each stage of the scale-up process, spanning from bench experiments to pilot-scale trial runs, could have the potential of encountering the issues associated with the biological behavior of microbes, as well as the physical and operational conditions of the bioprocess; these occurrences may impact the performance of the precision fermentation. At the laboratory scale, fermentation works are carried out in small and well-mixed reactors, where conditions of the fermentation are strictly regulated to ensure the cell viability and productivity (Amanullah, Buckland, & Nienow, 2003). Parameters such as temperature, pH, oxygen availability, as well as nutrient feed can be adjusted easily and precisely, enabling ease of control of the operational aspects of fermentation experiments. With this, the engineered microorganisms are able to perform at their maximum capability, resulting in high product yield and purity.

As the fermentation process is scaled up to the point of pilot scale or beyond, new challenges start to emerge. For instance, the oxygen availability, being one of the most crucial components in cell respiration, would be greatly affected. This is due to the low uniformity of the mixing process within the large bioreactor, leading to oxygen gradients at different parts within the bioreactor (Formenti et al., 2014). In addition, the operational dynamics become more complex once the bioreactor is being scaled up, as the large fermentation systems require complex control systems that can cater to the long mixing period, higher energy input, and the additives needed to maintain the environment for the microbes to survive. These operational parameters are normally regulated by the sensors; however, these sensors work reliably in small bioreactors, but may be less accurate when employed in large bioreactors, leading to the complexity in terms of process control and real-time monitoring (Mitra & Murthy, 2022).

These critical issues associated with the scale-up process of fermentation imply that there is no guarantee of a smooth transition from bench to industrial-scale precision fermentation. In this regard, it is crucial to understand the key technical challenges involved in the scaling up of precision fermentation. This chapter is centered around two key aspects, namely:

- Genetic engineering
- Bioprocess control

Addressing these challenges is critical to ensuring that the scaled-up precision fermentation processes remain technically robust, practical, and economically viable at an industrial scale.

17.3 GENETIC ENGINEERING OF MICROBES

Microbial hosts are normally subjected to microbial engineering to produce a targeted, specific biomolecule as a product, with minimal by-product formation in precision fermentation. Although proof-of-concept production at the laboratory scale is established, transitioning to commercial-scale manufacturing introduces significant technical hurdles. Synthetic biology and metabolic engineering have broadened the options for strain optimization; however, the performance of genetically engineered microbes at large scale remains limited by multiple factors, with one of the factors lying in their genetic stability. The preservation of genetic information is critical for achieving predictable and reliable performance of the microbes as fermentation cell factories. This genetic information is normally stored in the microbe's genomic DNA and is often supplemented by introducing additional sequences into the genome (Darmon & Leach, 2014).

17.3.1 Case Studies of Genetic Engineering for Precision Fermentation

There are several studies focusing on genetically engineering the microbes with the aim of targeting specific fermented products (Table 17.1). Hildebrand et al. have employed a gene knockout strategy to eliminate the pyruvate formate-lyase (PFL) and lactate dehydrogenase (LDH) metabolic pathways of *Escherichia coli*, resulting in no formation of lactate together with low acetate yield at the end of fermentation. This strategy has maximized the ethanol production up to 97.5% (Hildebrand, Schlacta, Warmack, Kasuga, & Fan, 2013). Also, Wang et al. have utilized the same engineering strategy to delete the gene encoding the phosphotransacetylase (PTA) pathway of the *Bacillus subtilis*, potentially reducing the acetate production. In addition, they have also overexpressed the acetolactate synthase (ALS) encoded with the acetolactate synthase S (*alsS*) gene, resulting in the achievement of a 130% increase in diacetyl yield (Y. Wang, Sun, Zheng, Zhang, & Bao, 2019). In addition, the gene silencing technique has been used by Gheinani et al., where they attempted to silence the lactate dehydrogenase A (*ldhA*) gene of *Rhizopus oryzae* with the use of synthetic RNA, in which they have successfully reduced the formation of lactic acid (85.7% reduction), while enhancing the ethanol yield by 15.4% (Gheinani, Jahromi,

TABLE 17.1
Summary of the Studies on Genetic Engineering of Microbes in Fermentation

Microbial Strain	Engineering Strategy	Aim	Targeted Product	References
Escherichia coli	Gene knockout, metabolic pathway engineering	Eliminate lactate production and reduce the acetate formation	Ethanol	Hildebrand et al. (2013)
Bacillussubtilis	Gene knockout, gene overexpression	Induce overexpression of α-acetolactate synthase	Diacetyl	Y. Wang et al. (2019)
Rhizopus oryzae	Gene silencing	Reduce the formation of lactic acid	Ethanol	Gheinani et al. (2011)
Clostridium saccharoperbutyl-acetonicum	Genome editing	Suppress the formation of acetate and butyrate	Butanol	S. Wang et al. (2017)
Lactobacillus brevis	Directed evolution, site-specific mutagenesis	Modification of the protein amino acid sequence	γ-aminobutyricacid	Shi et al. (2014)
Saccharomyces cerevisiae	Metabolic pathway engineering	Overexpress the PPP-related enzymes to enhance the rate of non-oxidative PPP	Ethanol	Hasunuma et al. (2011)

Feuk-Lagerstedt, & Taherzadeh, 2011). Wang et al. (2017) approached the modification of the gene using a developed genome editing system, namely the clustered regularly interspaced short palindromic repeats with CRISPR-associated (CRISPR-Cas9) system. They utilize the CRISPR-Cas9-based genome editing system to delete the PTA and butyrate kinase (BUK) genes in *Clostridium saccharoperbutylacetonicum*; however, the acetate and butyrate production are not completely suppressed. Still, their findings have shown that the butanol yield after fermentation has been enhanced by 13% when using the genome-edited microbial strain as compared to that of the wild-type strain (S. Wang, Dong, Wang, Tao, & Wang, 2017). Shi et al. (2014) have employed the directed evolution and site-specific mutagenesis techniques to enhance the enzymatic activity of glutamate decarboxylase (GAD) from *Lactobacillus brevis*. With this approach, the catalytic activity of mutated GAD was remarkably improved, recording a 43.2-fold increase in γ-aminobutyricacid production (Shi et al., 2014). Also, Hasunuma et al. (2011) have attempted to engineer the metabolic pathway of the xylose-fermenting yeast, namely *Saccharomyces cerevisiae*. They have over-expressed the pentose phosphate pathway (PPP) enzymes (e.g., transaldolase or transketolase), resulting in the suppression of the inhibiting effects of acetic acid and formic acid. This results in higher ethanol yield from the xylose substrate (Hasunuma et al., 2011). These studies have shown success in terms of elevating the yield of the targeted fermented products, as well as improving the tolerance of the microbial strain toward external stress.

17.3.2 Challenges of Genetic Engineering

17.3.2.1 Metabolic Pathway Engineering

Metabolic engineering has been utilized to modify native strains, with the aim of increasing the production of bioproducts (L. Chen et al., 2021; Jeffries & Jin, 2004; Soma, Yamaji, & Hanai, 2022). For strain metabolic engineering to be considered successful, the overall process must remain cost-effective at large-scale production. The engineering of metabolic pathways usually begins with the creation or overexpression of the genes from the target pathway, followed by the deletion of competing pathways that consume precursors, products, or cofactors (Montaño López, Duran, & Avalos, 2022). Production yields can then be further enhanced through metabolic strategies such as eliminating toxic intermediates, strengthening the driving force for pathway flux, and extending the cellular production phase (Yan & Fong, 2017). Even so, metabolic engineering still faces challenges that could hinder its application in large-scale precision fermentation. The engineered metabolic pathway could compete with the microbial host's native cellular resources (e.g., cofactors, ATP, polymerase, and ribosomes). This occurrence leads to the metabolic burden on the microbes due to the uptake of cellular resources by the engineered cellular pathways, disrupting the cellular functions and the balance of cofactors (Volk et al., 2022). Past studies have shown that the challenges related to the metabolic burden as well as the genetic instability can be overcome by employing the chromosomal integration approach, where the engineered genes were introduced into the chromosome instead of plasmid of the microbes (N. Chen, Wang, Zhao, & Deng, 2018; Escalante et al., 2010; Siddiqui, Choksi, & Smolke, 2014; Wu et al., 2018). This approach could be beneficial for the upscaling of precision fermentation, utilizing the engineered microbes for producing bioproducts.

17.3.2.2 Genome Editing

In the realm of gene editing, CRISPR-Cas9 has become a transformative tool for engineering the genetic material of microbes, owing to its high precision and efficiency (P. Yang, Condrich, et al., 2024). This particular tool enables the guide RNA to direct the Cas9 nuclease to a targeted genomic site, generating a double-strand break. The cell's natural repair pathways then resolve this break, resulting in the genetic material being additionally inserted, deleted, or modified (Rabaan et al., 2023). Despite its unprecedented accuracy in modifying the DNA sequences, the gene-edited microbes would still be associated with some challenges when put into application in precision fermentation. A major limitation of applying CRISPR-Cas9 in gene editing is bearing the risks of off-target effects, in which the genomic regions resembling the intended target sequence are cleaved by the Cas9. Such unintended activity can introduce unwanted mutations and cause low stability of genetic material, especially in filamentous bacteria. Nonetheless, this can be easily solved by optimization of the CRISPR-Cas9 protocol (Ropero-Pérez, Marcos, Manzanares, & Garrigues, 2024).

TABLE 17.2
Challenges and Solutions Associated with Genetic Engineering in Scaling Up Precision Fermentation

Strategy	Scale-Up Challenges	Potential Solutions
Metabolic pathway engineering	• Genetic instability • Metabolic burden	• Chromosomal integration of biochemical pathways
Genome editing	• Genetic instability • Low efficiency and accuracy when dealing with non-model microbes	• Optimization of CRISPR-Cas9 protocol • Application of SAGE

In addition, the CRISPR-Cas9 is normally optimized for a few model microorganisms, such as E. coli, but the accuracy and efficiency of this technique are greatly reduced when dealing with other types of non-model microorganisms. This challenge is further exacerbated by the limited understanding of genetic and metabolic pathways in non-model microorganisms, making it more difficult to engineer these organisms for a better precision fermentation process (Yan & Fong, 2017; P. Yang et al., 2024). This challenge can be addressed through serine recombinase-assisted genome engineering (SAGE), which enables the stable incorporation of multiple genetic sequences into the chromosomes of bacteria, thus expanding the toolkit for engineering a wide range of bacterial hosts (Elmore et al., 2023). With this method, the gene-edited non-model microorganisms would perform optimally and efficiently during the fermentation process (Table 17.2).

17.4 BIOPROCESS CONTROL

Scaling up a fermentation process is typically intended to produce larger product quantities while maintaining, or ideally improving, specific yields and product quality. However, scaling up the precision fermentation from bench to industrial scale is not a straightforward transition, as this process would behave unpredictably in a large-scale system, even though it performs optimally in a small and well-optimized laboratory setting. At larger scales, changes in geometric and physical conditions can lead to less uniform mixing and altered physiological environments. These factors may affect process stability and reproducibility, potentially lowering specific yields, increasing the formation of side products, as well as influencing batch-to-batch consistency (Schmidt, 2005). In this regard, addressing the key bioprocess control challenges in large-scale precision fermentation is essential to enable a smoother transition from laboratory innovations to economically viable and practical industrial processes. Bioprocess control refers to the regulation of conditions that create a nearly optimal environment for the growth of microorganisms, allowing them to healthily reproduce and synthesize the desired product (Alford, 2006). This section encompasses the key bioprocess challenges that could hinder the scaling up of precision fermentation, as well as some strategies for tackling the process limitations (Figure 17.1).

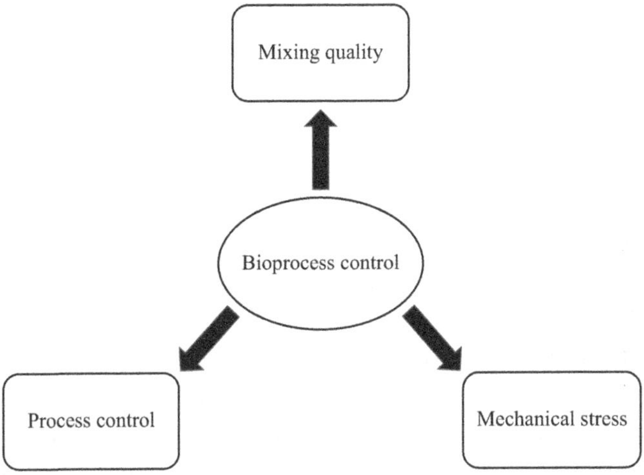

FIGURE 17.1 Key aspects of the bioprocess control in precision fermentation.

17.4.1 Mixing Quality of the Bioreactor

Mixing is considered one of the critical parameters in large-scale fermentation, essential in ensuring uniform distribution of nutrients (e.g., glucose), dissolved oxygen, pH, and temperature within the fermentation culture. Yet, as the process is scaled from bench to industrial bioreactors, achieving consistent and efficient mixing becomes progressively more challenging (Formenti et al., 2014). As the scale of the fermentation system is magnified, the flow pattern of the medium might vary, and the physical difference of the system could have a significant effect on the homogeneity of the medium. While bench-scale systems often achieve near-perfect mixing, large bioreactors require substantially more time to reach optimum mixing conditions. Insufficient mixing in such systems may impede the effective distribution of oxygen and nutrients, ultimately causing localized cell starvation within the culture (Gabelle, Augier, Carvalho, Rousset, & Morchain, 2011; Beth Helene Junker, 2004). As a result, the cell growth would be negatively impacted, leading to low process efficiency. Still, there are ways to tackle the mixing limitations for a better fermentation process.

A study encompasses the development of an aerobic fermentation system equipped with multistage propellers and internal deflectors. This design significantly improved the oxygen mass transfer coefficient (kLa) within the bioreactor, resulting in an 11% increase in L-lysine production (183.5 g/L) compared to a conventional setup (J. Wang et al., 2023). Another approach to enhance fermentation process efficiency is coupling stirred-tank bioreactors with other types of bioreactors. For instance, a study has shown the integration of a stirred-tank bioreactor with multi-column bioreactors, achieving higher overall ethanol yields and productivity, with reported values of 120.96 g/L and 2.52 g/L/h, respectively. This bioreactor configuration was chosen to leverage the uniform mixing characteristics of stirred-tank bioreactors alongside the low power and water consumption advantages of multi-column bioreactors

(Sriputorn et al., 2021). Also, an approach named very high gravity fermentation has been developed by the researchers to improve the fermentation process. This particular approach requires a high degree of sugar concentration (> 250 g/L) to increase the ethanol yield, with the associated benefits of low energy consumption, low waste production, and lower risk of bacterial contamination (Liu, Wang, Lin, & Bai, 2012; Puligundla, Smogrovicova, Mok, & Obulam, 2019). Such approaches can effectively address scale-up challenges associated with bioreactor mixing, enhancing the distribution of nutrients and oxygen and thereby improving the bioprocess performance of large-scale precision fermentation systems.

17.4.2 Measurement of the Process Control

Process control is an essential element in ensuring that the bioreactor operation is kept at desired operating process conditions, as well as meeting the requirements of being environmentally safe, economically feasible, and of high product quality. Bioreactor control for the fermentation process typically involves managing the complex dynamic behavior of microorganisms, as well as the model mismatches (Formenti et al., 2014). In most cases, the industrial fermenters consist of control systems that are aimed at regulating the fermentation process by keeping the key control variables at a predefined set of points or trajectories to tackle unpredictable disturbances. Disturbances that drive the bioreactor system away from its optimum condition are typically managed by maneuvering key control parameters such as the flow rate of nutrients, operating temperature, agitation, pH of the medium, and concentration of dissolved oxygen (Murthy, Johnston, Rausch, Tumbleson, & Singh, 2012). The proportional integral derivative (PID) controller is commonly utilized in many industrial plants for regulatory level process control, and this particular device is regarded as an effective device for single-input single-output linear systems (Mitra & Murthy, 2022). However, conventional PID controllers are often inadequate for ensuring optimal operation in precision fermentation, especially when the process involves genetically engineered microorganisms.

As the dissolved oxygen and pH are the important parameters to keep the performance of the microbes during fermentation, other often ignored key variables, such as substrate and product changes, as well as the by-product production, are not captured by the sensors (Formenti et al., 2014; Mitra & Murthy, 2022). Moreover, monitoring cell density in large bioreactors is challenging, which limits further improvements in fermentation product yield. In addition, the models used to design PID controllers are often insufficiently accurate for the highly dynamic nature of bioprocesses, making process control even more difficult at large scale (Alford, 2006; Smets, Claes, November, Bastin, & Van Impe, 2004). This may result from the lack of available process variables needed to construct the equations, or could be due to the model mismatches during the upscaling process. In this context, it is crucial to optimize the feed rate of the sugar substrate necessary for the cells to perform biological activities. In fed-batch bioreactor systems, this control helps prevent product accumulation as the process progresses (Sarkar & Modak, 2003). Also, the optimization of the fed-batch fermentation systems is normally done through the optimization of substrate feed rate, ensuring an optimum substrate-to-product conversion, which will not result in the product inhibition of the cells (Chai, Teo, Tan, & Tham, 2022). Oftentimes, the genetically engineered microbes used in precision fermentation would reduce the production of side-products, and hence, regulating the substrate feed rate might be sufficient for a better fermentation process (Pereira et al., 2025).

The better monitoring of the product and cell formation rate can be done by utilizing the online model-based monitoring. This particular approach utilizes the mechanistic model in an online platform, leading to the improvement of model prediction using the online parameter estimation (Mears, Stocks, Albaek, Sin, & Gernaey, 2017b). This approach was successfully demonstrated by Mears et al. at a pilot scale in a fermentation process utilizing the filamentous fungus. In their study, the model structure incorporated online parameter estimation, allowing the use of real-time measurement data to update time-varying parameters such as the growth rate of microbes, as well as the formation rate of the fermented products (Mears, Stocks, Albaek, Sin, & Gernaey, 2017a). Still, with these strategies, more understanding is required on the complex dynamic systems of bioprocess, and more effort should be poured into developing a robust model that could result in a proficient control system for large-scale precision fermentation systems.

17.4.3 Mechanical Stress

Mechanical stress within the bioreactor is often attributed to the consistent mechanical agitation during the fermentation process. Within the lab-scale process, fermentation studies were normally performed under shaking conditions to ensure a homogenous medium, as well as to foster the substrate-to-product conversion. However, the stirring action caused by the mechanical stirrer and the aeration introduced into the bioreactor would result in a shear stress that might damage the cells, leading to poor fermentation performance (Hewitt & Nienow, 2007). Mammalian cells are generally more vulnerable to mechanical stress arising from agitation-induced turbulence and bubble bursting during aeration (Hu, Berdugo, & Chalmers, 2011). Nonetheless, most fermentation studies avoid mammalian cells and instead employ smaller microorganisms such as yeasts and bacteria (Andrade Barreto et al., 2023; Cassimiro et al., 2022; De Montijo-Prieto et al., 2023; Li et al., 2021; Mendoza Salazar, Martínez Álvarez, Ardila Castañeda, & Lizarazo Medina, 2022). According to Thomas, cells with sizes smaller than the Kolmogorov microscale of turbulence, such as yeasts and bacteria, exhibit greater resistance to mechanical stress within the bioreactor environment (Thomas, 1990). Still, filamentous microbes are highly prone to mechanical stress due to their fragile and hyphal structure. The morphology of the filamentous microbes could be altered into pellets by the shear stress caused by stirring and aeration, potentially reducing the bioprocess efficiency (B. H. Junker et al., 2004). Moreover, the development of the microbial pellets increases the viscosity of the fermentation medium and impedes the effective oxygen transfer and nutrients within the fermentation medium (Iram, Özcan, Yatmaz, Turhan, & Demirci, 2022).

Fortunately, the approach of adding microparticles into the fermentation medium could enhance the mass transfer rate of nutrients and oxygen within the system, as well as control the growth of the microbes. Driouch et al. (2012) have utilized titanium silicone oxide as the microparticle for the bioefficiency of *Aspergillus niger* using a stirred tank bioreactor setup. Their findings have shown that the microparticle-enhanced *Aspergillus niger* is capable of achieving an enzyme titre of 1080 U/mL, which is found to be seven-fold higher than the conventional bioprocess (Driouch, Hänsch, Wucherpfennig, Krull, & Wittmann, 2012). In addition, Zhang et al. (2021) have attempted to add zinc oxide nanoparticles into the anaerobic fermentation system, and this approach has led to an enhancement of hydrogen production, recording a hydrogen yield of 74.0 mL-H_2/g-VS (2.5 times higher than control). Their findings have shown that the nanoparticles did not pose cytotoxic effects on the microbial community within the waste-activated sludge (Zhang, Wei, & Ni, 2021). Still, the type of microparticles cannot be generalized across all fermentation processes, as each microparticle possesses distinct properties such as size of particle, elemental composition (e.g., Al, Zn, Si, or Ti), and density. Moreover, the role of a given microparticle may vary depending on the growth characteristics of the filamentous organisms. Hence, despite the vulnerability of filamentous microorganisms toward the mechanical stress within the large bioreactor, one can consider selecting the right microparticle to enhance the productivity of the precision fermentation, compensating for the effects of mechanical stress on the fermentation system.

17.5 CONCLUSION AND FUTURE OUTLOOK

The scaling up of the precision fermentation system is considered to be challenging in translating the bench success into an industrial application. The key issues, such as inefficiency of bioreactor mixing, limitations of conventional process control strategies, presence of mechanical stress within the bioreactor, as well as the metabolic burden and genetic instability of the engineered microbes, underscore the complexity of operating the precision fermentation process at large scale. Even so, the scaling-up process of precision fermentation presents an avenue for further innovation, allowing for process enhancement with significant research efforts. That said, effective strategies are essential to tackle the process scale-up related challenges.

Looking forward, the next phase of progress in precision fermentation will likely be shaped by the integration of advanced process control with both digital and biological innovations. Strain engineering and synthetic biology will intersect with process engineering, enabling the engineering of microbes that are not only aimed for higher yields and product titres but also for resilience against shear stress, oxygen fluctuations, as well as possessing greater endurance against the burden in a large-scale fermentation system. Moreover, precision fermentation must align with the goals of sustainability and the circular bioeconomy. Future scale-up strategies will not emphasize only technical feasibility but also reducing energy consumption, efficient substrate utilization, and minimal waste generation. The future of precision fermentation scale-up lies in bridging biology, engineering, and digital technologies. By combining resilient microbial systems, smart control frameworks, and sustainable design principles, precision fermentation can transition from niche production to a cornerstone of next-generation biomanufacturing.

REFERENCES

Alford, J. S. (2006). Bioprocess control: Advances and challenges. *Computers & Chemical Engineering*, *30*(10), 1464–1475. https://doi.org/10.1016/j.compchemeng.2006.05.039

Amanullah, A., Buckland, B. C., & Nienow, A. W. (2003). Mixing in the fermentation and cell culture industries. *Handbook of Industrial Mixing: Science and Practice*, 1071–1170. https://doi.org/10.1002/0471451452.ch18

Andrade Barreto, S. M., Martins da Silva, A. B., Prudêncio Dutra, M. d. C., Costa Bastos, D., de Brito Araújo Carvalho, A. J., Cardoso Viana, A., … dos Santos Lima, M. (2023). Effect of commercial yeasts (Saccharomyces cerevisiae) on fermentation metabolites, phenolic compounds, and bioaccessibility of Brazilian fermented oranges. *Food Chemistry*, *408*, 135121. https://doi.org/10.1016/j.foodchem.2022.135121

Cassimiro, D. M. d. J., Batista, N. N., Fonseca, H. C., Naves, J. A. O., Dias, D. R., & Schwan, R. F. (2022). Coinoculation of lactic acid bacteria and yeasts increases the quality of wet fermented Arabica coffee. *International Journal of Food Microbiology*, *369*, 109627. https://doi.org/10.1016/j.ijfoodmicro.2022.109627

Chai, W. Y., Teo, K. T. K., Tan, M. K., & Tham, H. J. (2022). Fermentation process control and optimization. *Chemical Engineering & Technology*, *45*(10), 1731–1747. https://doi.org/10.1002/ceat.202200029

Chen, L., Yan, W., Qian, X., Chen, M., Zhang, X., Xin, F., Ochsenreither, K. (2021). Increased lipid production in Yarrowialipolytica from acetate through metabolic engineering and cosubstrate fermentation. *ACS Synthetic Biology*, *10*(11), 3129–3138. https://doi.org/10.1021/acssynbio.1c00405

Chen, N., Wang, J., Zhao, Y., & Deng, Y. (2018). Metabolic engineering of Saccharomyces cerevisiae for efficient production of glucaric acid at high titer. *Microbial Cell Factories*, *17*(1), 67. https://doi.org/10.1186/s12934-018-0914-y

Darmon, E., & Leach, D. R. F. (2014). Bacterial genome instability. *Microbiology and Molecular Biology Reviews*, *78*(1), 1–39. https://doi.org/10.1128/mmbr.00035-13

de Carvalho, J. C., Medeiros, A. B. P., Vandenberghe, L. P. S., Magalhães, A. I., & Soccol, C. R. (2017). 34 - Approaches for the isolation and purification of fermentation products. In A. Pandey, S. Negi, & C. R. Soccol (Eds.), *Current developments in biotechnology and bioengineering* (pp. 783–805). Elsevier.

de Melo Pereira, G. V., Soccol, V. T., Pandey, A., Medeiros, A. B. P., Andrade Lara, J. M. R., Gollo, A. L., & Soccol, C. R. (2014). Isolation, selection, and evaluation of yeasts for use in the fermentation of coffee beans by the wet process. *International Journal of Food Microbiology*, *188*, 60–66. https://doi.org/10.1016/j.ijfoodmicro.2014.07.008

De Montijo-Prieto, S., Razola-Díaz, M. d. C., Barbieri, F., Tabanelli, G., Gardini, F., Jiménez-Valera, M., … Gómez-Caravaca, A. M. (2023). Impact of lactic acid bacteria fermentation on phenolic compounds and antioxidant activity of avocado leaf extracts. *Antioxidants*, *12*(2), 298. https://doi.org/10.3390/antiox12020298

Driouch, H., Hänsch, R., Wucherpfennig, T., Krull, R., & Wittmann, C. (2012). Improved enzyme production by bio-pellets of Aspergillus niger: Targeted morphology engineering using titanate microparticles. *Biotechnology and Bioengineering*, *109*(2), 462–471. https://doi.org/10.1002/bit.23313

Elmore, J. R., Dexter, G. N., Baldino, H., Huenemann, J. D., Francis, R., Peabody, G. L., … Coleman-Derr, D. (2023). High-throughput genetic engineering of nonmodel and undomesticated bacteria via iterative site-specific genome integration. *Science Advances*, *9*(10), eade1285.

Escalante, A., Calderón, R., Valdivia, A., de Anda, R., Hernández, G., Ramírez, O. T., … Bolívar, F. (2010). Metabolic engineering for the production of shikimic acid in an evolved Escherichia coli strain lacking the phosphoenolpyruvate: Carbohydrate phosphotransferase system. *Microbial Cell Factories*, *9*(1), 21. https://doi.org/10.1186/1475-2859-9-21

Formenti, L. R., Nørregaard, A., Bolic, A., Hernandez, D. Q., Hagemann, T., Heins, A. L., … Krühne, U. (2014). Challenges in industrial fermentation technology research. *Biotechnology Journal*, *9*(6), 727–738. https://doi.org/10.1002/biot.201300236

Gabelle, J. C., Augier, F., Carvalho, A., Rousset, R., & Morchain, J. (2011). Effect of tank size on kLa and mixing time in aerated stirred reactors with non-Newtonian fluids. *The Canadian Journal of Chemical Engineering*, *89*(5), 1139–1153. https://doi.org/10.1002/cjce.20571

Gheinani, A. H., Jahromi, N. H., Feuk-Lagerstedt, E., & Taherzadeh, M. J. (2011). RNA silencing of the lactate dehydrogenase gene in Rhizopus oryzae. *Journal of RNAi and Gene Silencing: An International Journal of RNA and Gene Targeting Research*, *7*, 443. https://pubmed.ncbi.nlm.nih.gov/21769297/

Guo, Q., Zheng, L.-J., Luo, X., Gao, X.-Q., Liu, C.-Y., Deng, L., … Zheng, H.-D. (2021). Engineering Escherichia coli for D-allulose production from D-fructose by fermentation. *Journal of Agricultural and Food Chemistry*, *69*(45), 13578–13585. https://doi.org/10.1021/acs.jafc.1c05200

Hasunuma, T., Sanda, T., Yamada, R., Yoshimura, K., Ishii, J., & Kondo, A. (2011). Metabolic pathway engineering based on metabolomics confers acetic and formic acid tolerance to a recombinant xylose-fermenting strain of Saccharomyces cerevisiae. *Microbial Cell Factories*, *10*(1), 2. https://doi.org/10.1186/1475-2859-10-2

Hewitt, C. J., & Nienow, A. W. (2007). The Scale-Up of Microbial Batch and Fed-Batch Fermentation Processes. In S. Sariaslani & G. M. Gadd (Eds.), *Advances in applied microbiology* (Vol. 62, pp. 105–135). Academic Press.

Hildebrand, A., Schlacta, T., Warmack, R., Kasuga, T., & Fan, Z. (2013). Engineering Escherichia coli for improved ethanol production from gluconate. *Journal of Biotechnology*, *168*(1), 101–106. https://doi.org/10.1016/j.jbiotec.2013.07.033

Hu, W., Berdugo, C., & Chalmers, J. J. (2011). The potential of hydrodynamic damage to animal cells of industrial relevance: Current understanding. *Cytotechnology*, *63*(5), 445–460. https://doi.org/10.1007/s10616-011-9368-3

Iram, A., Özcan, A., Yatmaz, E., Turhan, İ., & Demirci, A. (2022). Effect of microparticles on fungal fermentation for fermentation-based product productions. *Processes*, *10*(12), 2681. https://doi.org/10.3390/pr10122681

Jeffries, T. W., & Jin, Y. S. (2004). Metabolic engineering for improved fermentation of pentoses by yeasts. *Applied Microbiology and Biotechnology*, *63*(5), 495–509. https://doi.org/10.1007/s00253-003-1450-0

Junker, B. H. (2004). Scale-up methodologies for Escherichia coli and yeast fermentation processes. *Journal of Bioscience and Bioengineering*, *97*(6), 347–364. https://doi.org/10.1016/S1389-1723(04)70218-2

Junker, B. H., Hesse, M., Burgess, B., Masurekar, P., Connors, N., & Seeley, A. (2004). Early phase process scale-up challenges for fungal and filamentous bacterial cultures. *Applied Biochemistry and Biotechnology*, *119*(3), 241–277. https://doi.org/10.1007/s12010-004-0005-x

Li, S., Tao, Y., Li, D., Wen, G., Zhou, J., Manickam, S., … Chai, W. S. (2021). Fermentation of blueberry juices using autochthonous lactic acid bacteria isolated from fruit environment: Fermentation characteristics and evolution of phenolic profiles. *Chemosphere*, *276*, 130090. https://doi.org/10.1016/j.chemosphere.2021.130090

Liu, C.-G., Wang, N., Lin, Y.-H., & Bai, F.-W. (2012). Very high gravity ethanol fermentation by flocculating yeast under redox potential-controlled conditions. *Biotechnology for Biofuels*, *5*(1), 61. https://doi.org/10.1186/1754-6834-5-61

Mears, L., Stocks, S. M., Albaek, M. O., Sin, G., & Gernaey, K. V. (2017a). Application of a mechanistic model as a tool for online monitoring of pilot-scale filamentous fungal fermentation processes—The importance of evaporation effects. *Biotechnology and Bioengineering*, *114*(3), 589–599. https://doi.org/10.1002/bit.26187

Mears, L., Stocks, S. M., Albaek, M. O., Sin, G., & Gernaey, K. V. (2017b). Mechanistic fermentation models for process design, monitoring, and control. *Trends in Biotechnology, 35*(10), 914–924. http://doi.org/10.1016/j.tibtech.2017.07.002

Mendoza Salazar, M. M., Martínez Álvarez, O. L., Ardila Castañeda, M. P., & Lizarazo Medina, P. X. (2022). Bioprospecting of indigenous yeasts involved in cocoa fermentation using sensory and chemical strategies for selecting a starter inoculum. *Food Microbiology, 101*, 103896. https://doi.org/10.1016/j.fm.2021.103896

Mitra, S., & Murthy, G. S. (2022). Bioreactor control systems in the biopharmaceutical industry: A critical perspective. *Systems Microbiology and Biomanufacturing*, *2*(1), 91–112. https://doi.org/10.1007/s43393-021-00048-6

Montaño López, J., Duran, L., & Avalos, J. L. (2022). Physiological limitations and opportunities in microbial metabolic engineering. *Nature Reviews Microbiology*, *20*(1), 35–48. https://doi.org/10.1038/s41579-021-00600-0

Murthy, G. S., Johnston, D. B., Rausch, K. D., Tumbleson, M. E., & Singh, V. (2012). Design and evaluation of an optimal controller for the simultaneous saccharification and fermentation process. *Applied Biochemistry and Biotechnology*, *166*(1), 87–111. https://doi.org/10.1007/s12010-011-9406-9

Naher, L., Fatin, S. N., Sheikh, M. A. H., Azeez, L. A., Siddiquee, S., Zain, N. M., & Karim, S. M. R. (2021). Cellulase enzyme production from filamentous fungi Trichoderma reesei and Aspergillus awamori in submerged fermentation with rice straw. *Journal of Fungi*, *7*(10), 868. https://doi.org/10.3390/jof7100868

Pereira, A. A., Yaverino-Gutierrez, M. A., Monteiro, M. C., Souza, B. A., Bachheti, R. K., & Chandel, A. K. (2025). Precision fermentation in the realm of microbial protein production: State-of-the-art and future insights. *Food Research International*, *200*, 115527. https://doi.org/10.1016/j.foodres.2024.115527

Puligundla, P., Smogrovicova, D., Mok, C., & Obulam, V. S. R. (2019). A review of recent advances in high-gravity ethanol fermentation. *Renewable Energy*, *133*, 1366–1379. https://doi.org/10.1016/j.renene.2018.06.062

Rabaan, A. A., Al Fares, M. A., Almaghaslah, M., Alpakistany, T., Al Kaabi, N. A., Alshamrani, S. A., … Hakami, A. R. (2023). Application of the CRISPR-Cas system to mitigate superbug infections. *Microorganisms*, *11*(10), 2404. https://doi.org/10.3390/microorganisms11102404

Ropero-Pérez, C., Marcos, J. F., Manzanares, P., & Garrigues, S. (2024). Increasing the efficiency of CRISPR/Cas9-mediated genome editing in the citrus postharvest pathogen Penicillium digitatum. *Fungal Biology and Biotechnology*, *11*(1), 8. https://doi.org/10.1186/s40694-024-00179-0

Sarkar, D., & Modak, J. M. (2003). Optimisation of fed-batch bioreactors using genetic algorithms. *Chemical Engineering Science*, *58*(11), 2283–2296. https://doi.org/10.1016/S0009-2509(03)00095-2

Schmidt, F. R. (2005). Optimization and scale-up of industrial fermentation processes. *Applied Microbiology and Biotechnology*, *68*(4), 425–435. https://doi.org/10.1007/s00253-005-0003-0

Shi, F., Xie, Y., Jiang, J., Wang, N., Li, Y., & Wang, X. (2014). Directed evolution and mutagenesis of glutamate decarboxylase from Lactobacillus brevis Lb85 to broaden the range of its activity toward a near-neutral pH. *Enzyme and Microbial Technology*, *61*, 35–43. http://doi.org/10.1016/j.enzmictec.2014.04.012

Siddiqui, M. S., Choksi, A., & Smolke, C. D. (2014). A system for multilocus chromosomal integration and transformation-free selection marker rescue. *FEMS Yeast Research*, *14*(8), 1171–1185. https://doi.org/10.1111/1567-1364.12210

Smets, I. Y., Claes, J. E., November, E. J., Bastin, G. P., & Van Impe, J. F. (2004). Optimal adaptive control of (bio)chemical reactors: Past, present, and future. *Journal of Process Control*, *14*(7), 795–805. https://doi.org/10.1016/j.jprocont.2003.12.005

Soma, Y., Yamaji, T., & Hanai, T. (2022). Dynamic metabolic engineering of Escherichia coli improves fermentation for the production of pyruvate and its derivatives. *Journal of Bioscience and Bioengineering*, *133*(1), 56–63. https://doi.org/10.1016/j.jbiosc.2021.09.015

Sriputorn, B., Laopaiboon, P., Phukoetphim, N., Uppatcha, N., Phuphalai, W., & Laopaiboon, L. (2021). Very high gravity ethanol fermentation from sweet sorghum stem juice using a stirred tank bioreactor coupled with a column bioreactor. *Journal of Biotechnology*, *332*, 1–10. https://doi.org/10.1016/j.jbiotec.2021.03.012

Teng, T. S., Chin, Y. L., Chai, K. F., & Chen, W. N. (2021). Fermentation for future food systems: Precision fermentation can complement the scope and applications of traditional fermentation. *EMBO Reports*, *22*(5), e52680. https://doi.org/10.15252/embr.202152680

Thomas, C. R. (1990). Problems of shear in biotechnology. In A. E. Humphrey & N. R. Bott (Eds.), *Chemical engineering problems in biotechnology* (pp. 23–94). Oxford: Elsevier Applied Sciences.

Volk, M. J., Tran, V. G., Tan, S.-I., Mishra, S., Fatma, Z., Boob, A., Zhao, H. (2022). Metabolic engineering: Methodologies and applications. *Chemical Reviews*, *123*(9), 5521–5570. https://doi.org/10.1021/acs.chemrev.2c00403

Wang, J., Wang, S., Zhao, S., Sun, P., Zhang, Z., & Xu, Q. (2023). Productivity enhancement in L-lysine fermentation using an oxygen-enhanced bioreactor and an oxygen vector. *Frontiers in Bioengineering and Biotechnology*, *11*. https://doi.org/10.3389/fbioe.2023.1181963

Wang, S., Dong, S., Wang, P., Tao, Y., & Wang, Y. (2017). Genome editing in Clostridium saccharoperbutylacetonicum N1-4 with the CRISPR-Cas9 system. *Applied and Environmental Microbiology*, *83*(10), e00233-00217. https://doi.org/10.1128/AEM.00233-17

Wang, Y., Sun, W., Zheng, S., Zhang, Y., & Bao, Y. (2019). Genetic engineering of Bacillus sp. and fermentation process optimization for diacetyl production. *Journal of Biotechnology*, *301*, 2–10. https://doi.org/10.1016/j.jbiotec.2019.05.308

Wu, H., Li, Y., Ma, Q., Li, Q., Jia, Z., Yang, B., … Chen, N. (2018). Metabolic engineering of Escherichia coli for high-yield uridine production. *Metabolic Engineering*, *49*, 248–256. https://doi.org/10.1016/j.ymben.2018.09.001

Xia, J., Wang, G., Fan, M., Chen, M., Wang, Z., & Zhuang, Y. (2021). Understanding the scale-up of fermentation processes from the viewpoint of the flow field in bioreactors and the physiological response of strains. *Chinese Journal of Chemical Engineering*, *30*, 178–184. https://doi.org/10.1016/j.cjche.2020.12.004

Yan, Q., & Fong, S. S. (2017). Challenges and advances for genetic engineering of non-model bacteria and uses in consolidated bioprocessing. *Frontiers in Microbiology*, *8*, 2060.

Yang, P., Condrich, A., Lu, L., Scranton, S., Hebner, C., Sheykhhasan, M., & Ali, M. A. (2024). Genetic engineering in bacteria, fungi, and oomycetes, taking advantage of CRISPR. *DNA*, *4*(4), 427–454. https://doi.org/10.3390/dna4040030

Yang, X., Hong, J., Wang, L., Cai, C., Mo, H., Wang, J., & Liao, Z. (2024). Effect of lactic acid bacteria fermentation on plant-based products. *Fermentation*, *10*(1), 48. https://doi.org/10.3390/fermentation10010048

Zhang, Y.-T., Wei, W., & Ni, B.-J. (2021). Revealing the mechanism of zinc oxide nanoparticles facilitating hydrogen production in alkaline anaerobic fermentation of waste activated sludge. *Journal of Cleaner Production*, *328*, 129580. https://doi.org/10.1016/j.jclepro.2021.129580

Zou, S., Zhao, K., Tang, H., Zhang, Z., Zhang, B., Liu, Z., & Zheng, Y. (2021). Improved production of D-pantothenic acid in Escherichia coli by integrated strain engineering and fermentation strategies. *Journal of Biotechnology*, *339*, 65–72. https://doi.org/10.1016/j.jbiotec.2021.07.014

Index

Pages in *italics* refer to figures and pages in **bold** refer to tables.

F

G

H

I

L

M

O

P

Q

R

S

T

U

W